12th International Transportation Specialty Conference 2018

Held as Part of the Canadian Society for Civil Engineering Annual Conference 2018

Fredericton, Canada
13-16 June 2018

Editors:

Trevor Hanson

ISBN: 978-1-5108-8971-2

Printed from e-media with permission by:

Curran Associates, Inc.
57 Morehouse Lane
Red Hook, NY 12571

CURRAN ASSOCIATES INC. proceedings.com

Some format issues inherent in the e-media version may also appear in this print version.

Printed by Curran Associates, Inc. (2019)

For permission requests, please contact Canadian Society for Civil Engineering at the address below.

Canadian Society for Civil Engineering
521-300 rue St-Sacrement
Montréal, QC
H2Y 1x4

Phone: (514) 933-2634
Fax: (514) 933-3504

info@csce.ca

Additional copies of this publication are available from:

Curran Associates, Inc.
57 Morehouse Lane
Red Hook, NY 12571 USA
Phone: 845-758-0400
Fax: 845-758-2633
Email: curran@proceedings.com
Web: www.proceedings.com

TABLE OF CONTENTS

Building Tomorrow's Society
Bâtir la Société de Demain

Fredericton, Canada
June 13 to 16, 2018 / *13 au 16 juin, 2018*

A METHODOLOGY FOR FEASIBILITY ANALYSIS OF DEMAND-RESPONSIVE TRANSIT SERVICES

Yang, Yongzhe[1,3], and Mehran, Babak[2]
[1] PhD Student, Environmental Systems Engineering, Faculty of Engineering and Applied Science, University of Regina, 3737 Wascana Parkway, Regina SK S4S 0A2 Canada
[2] Assistant Professor, Environmental Systems Engineering, Faculty of Engineering and Applied Science, University of Regina, 3737 Wascana Parkway, Regina SK S4S 0A2 Canada
[3] yang207y@uregina.ca

Abstract: Under low demand conditions, bus transit operators have to pay a high cost per passenger to maintain desired level of service. Therefore, Demand-Responsive Transit (DRT) becomes an alternative option for transit agencies. Unlike regular transit services which run on fixed routes and schedules, DRT operates per user requests on variable routes or schedules to reduce operating costs while providing acceptable service level. DRT service is preferred when demand for public transportation is low. However, as the demand or bus ridership increases, a regular bus service is more efficient. Determining the critical demand levels for switching from a regular bus service to DRT is essential for transit operators. A methodology is proposed to evaluate and compare the oprerational costs of offering regular and demand responsive bus services. The proposed cost model for regular bus service has three independent parametrs i.e. fleet size, vehicle service hours, and vehicle traveled distance. Model parameters were estimted using the City of Regina's bus transit data. Two DRT service models were considered: 1) Contract-out Taxi Service (COTS) and, 2) In-house Paratransit Service (IHPS). The proposed cost models can be used to estimate expected operational costs of regular and demand responsive bus services as function of demand. A low demand bus route in Regina was used as a case study to demonstrate the application of the proposed methodology.

1 INTRODUCTION

Low ridership is a significant challenge for bus transit operators in small to medium municipalities in Canada, especially for newly developed areas, residential subdivisions, and communities with specific social-economic characteristics. Low ridership results in lower revenues for transit operators. Therefore, additional costs should be covered either by increasing the fares, or reducing the service frequency, which can affect the level of service and customer satisfaction. On the other hand, eliminating bus services in low demand areas is not an option because municipalities are required to provide public transportation services to their citizens as part of essential amenities to maintain acceptable living standards in urban areas. Thus, determining feasible and efficient public transit options for low demand areas is critical for bus transit agencies. Demand-Responsive Transit (DRT) is a viable option for transit agencies in low ridership situation. Unlike regular transit services which run on fixed routes and schedules, DRT operates per user requests on flexible routes or schedules to reduce operating costs and ideally maintain the same, or even better levels of service. It is anticipated that DRT service is preferred when demand for public transportation is low. However, as the demand increases, a regular bus service is more efficient. Determining the critical demand levels for switching from a regular bus service to DRT (and vice versa) is essential for transit operators. In this paper a methodology is proposed to evaluate and compare the oprerational costs of offering regular and demand responsive bus services. Contract-Out Taxi Service (COTS), and In-House Paratransit Service (IHPS) are considered as feasile DRT services to replace

Fixed-Route Bus Service (FRBS) in low demand conditions. Analytical models are developed to estimate the operating costs of regular and proposed demand responsive transit services. A low demand bus route in the City of Regina is used as an example to demonstrate model applications. The rest of the paper is organized as below. After a brief review of the existing literature, the cost models for FRBS, COTS, and IHPS services are described in Section 3. Section 4 describes a case study to demonstrate the aplications of the proposed analyese methods. Sensitivity analyses are carried out in Section 5 to evalute the cost models' responses to variations in input variables. The article ends with conclusions in Section 6 which provides a brief summary and recommendations for future research.

2 LITERATURE REVIEW

DRT is describes as a flexible route and/or schedule, shared ride, per user requests transportation services between public fix-route bus services and single-hired taxi services (KFH Group 2008). A study conducted by Rufolo (1995) in Poland concluded that DRT is suitable for low demand neighbourhoods compared to fixed route bus service, and it is preferable to use contract-out service rather than in-house management in terms of cost-effectiveness of DRT services. A survey conducted by Pamler et al. (2007) including 67 large U.S. transit agencies investigated the impact of implementation of new technologies for DRT across the country. Their study concluded that implementation of new technologies and practices can improve the productivity and reduce the costs of DRT services. Laws et al (2009) investigated publicly funded DRT schemes in England and Wales, and concluded that it is important to invest sufficient time in the design phase, and the system should be simple and meet the needs. Case studies from New Zealand and Australia indicated that the majority of efficacious DRT services are provided to commuters, schoolchildren, and shoppers. Moreover, many-to-one operation was found to be more successful than many-to-many services. It was also found that limited-stops services are more successful than door-to-door services (Scott, 2010).There are limited numbers of researches on the comparison of regular bus transit service and DRT. Diana et al. (2009) compared the performance of a competing fixed route transit and a DRT service in terms of traveled distance by these services while ensuring a comparable service to the same set of customers. They found that DRT provided higher level of service than fixed route transit under low demand scenarios. Li and Quadrifoglio (2010) proposed an analytical model to compare fixed route transit and DRT which could be used to determine when to switch from one to another during the day. Edwards and Watkins (2013) used a dataset including 10% of Metropolitan Atlanta Rapid Transit Authority passenger surveys, and developed a methodology to compare the performance of fixed route bus and DRT over a wide variety of streets and transit service layouts. Their methodology could be used by transit planners to determine whether DRT should be used instead of Fixed Route Transit to save costs and improve customer satisfaction. The review of the existing literature confirms that the existing methodologies for evaluating DRT are:

1. Mainly theoretical and based on simplified road networks, and passenger survey data,
2. Generally based on ideal scenarios, and
3. Often case specific and valid under specific circumstances.

This research aims to develop a general quantitative approach for comparing regular and alternative DRT services in terms of their operating costs. Furthermore, the proposed analysis methods in this research enable evaluation of the operating costs considering variations in future demand, OD flows, and other input parameters which can occur due to changes in socio-economic factors, land use patterns, and local policies. Therefore, the results of this study can be used by transit planners as a decision support tool to determine the conditions for switch between regular and demand responsive transit services in low demand conditions.

3 METHODOLOGY

In this study two DRT services are considered i.e. COTS and IHPS. In COTS DRT scheme, the transit agency contracts out the public transportation services for low demand neighbourhoods to a private taxi service provider who will be responsible for serving the demand for public transportation in designated zones based on a mutually agreed pricing scheme. IHPS based DRT relies on the existing paratransit service and fleet to serve low demand areas. In this scheme paratransit vehicles are deployed upon requests for ride to pick passengers from designated locations close to their actual trip origin. To compare and assess FRBS, COTS, and IHPS, the DRT services discussed in this paper are featured with i) fixed pick-up and drop-off locations (same as fixed bus stops), and ii) flexible schedules and routes. The number of passengers boarding at various bus stops is treated as demand or expected ridership for DRT as well. DRT services are assumed to operate per user requests. It is assumed that DRT users contact a call centre to arrange their trip by providing the information regarding their trip origin, destination, and preferred departure time. Pick-up and drop-off time and locations will be determined by the operator based on the user's preferred origin, destination, and departure time. A DRT service will be deployed as soon as adequate requests for service is received or the maximum wait time is reached. The users will be picked-up and dropped-off at predefined locations which are the bus stops nearest to their preferred trip origins and destinations, respectively. The DRT service will operate through the shortest paths between trip origins and destinations and may serve multiple pick up and drop off locations in a single trip.

3.1 Cost Model Assumptions

Operating cost models are developed for FRBS, COTS, and IHPS to describe the operating costs (dependent variable) as function of ridership and other service specific parameters (independent variables). Operating cost models enable detailed comparison of FRBS with COTS and IHPS to determine the ridership thresholds to switch from one service to another. FRBS cost model parameters may vary depending on land use activities and social economic conditions. Model parameters for COTS and IHPS may also vary depending on service provider, management scheme, city bylaws, etc. Hence, in this study the following assumptions have been made to enable development of general cost models for any specific transit route:

1. For FRBS, a linear relationship is assumed between the bus service design demand (P_{max}), and the annual average hourly boarding (R_h) along the route.

[1] $$R_h = \frac{R}{h_p + h_r}$$

Model parameters are defined in Table 1.

2. For COTS, the taxi companies follow the city's taxi bylaw (e.g. City of Regina) in pricing theirs services.
3. For IHPS, the unit cost per vehicle-hour (e.g. provided by the City of Regina), represents the unit costs for fleet maintenance and operations.
4. For FRBS and IHPS, only operating costs are considered. Capital and/or other costs are not considered in this analysis.
5. It is assumed that passenger demand (i.e. ridership) and origin-destination (OD) matrix are the same for FRBS, COTS, and IHPS services. To estimate the operating costs corresponding to lower or higher ridership values, the original OD matrix is rescaled, respectively.
6. For FRBS, a fixed route is considered with predefined stops. Similar predefined stops are considered for COTS and IHPS. However, it is assumed that the trips between stops are completed through the shortest paths.

3.2 Estimation of Design Demand and OD Flows

The operating costs for FRBS is directly related to bus fleet size. The design demand is one of the main input parameters for estimating the required fleet size for FRBS, which refers to the number of passengers onboard at Maximum Load Section (MLS). Passenger load diagram can be used to identify the MLS and the corresponding design demand if passenger boarding and alighting is known at each stop along the bus route. In this study, it is assumed that only passenger boarding data is available at each bus stop. Thus, a methodology is proposed to estimate passenger alighting at each bus stop using available passenger boarding data. The proposed method assumes that there is a symmetry between passenger boarding and alighting patterns at paired bus stops (i.e. closely located bus stops at opposite directions) during the morning and afternoon peak hours (Navick and Furth, 2002). For a bus route with stops distributed in each direction (e.g. north bound and southbound), the stops are grouped based on the direction of the bus route. Each stop in one direction (e.g. 1 for northbound) is paired with its most closely located stop on the opposite direction (e.g. 2 for southbound). Equations [2] is used to estimate passenger alighting at stop i based on passenger boarding at its paired stop j. In this equations it is assumed that the proportions of passenger boarding and alighting are equal for paired stops during AM and PM peak hours.

Table 1: Parameters, definitions, and units

Variables	Definition	Unit
α_1	unit cost of number of vehicles	$/veh/yr
α_2	unit cost of vehicle service hours	$/hr
α_3	unit cost of vehicle traveled distance	$/km
n	number of vehicles	veh
l_t	vehicle traveled distance	km/yr
h_v	number of vehicle service hours	hr/yr
v_p	number of operating vehicles in peak hours	veh
h_p	number of peak hours	hr/yr
v_r	number of vehicles in off-peak hours	veh
h_r	number of off-peak hours	hr/yr
γ	ratio of design demand to average hourly passenger boarding	-
L	cycle Length	km
R	annual ridership	prs/yr
s	average operating speed	km/hr
C	bus capacity	prs/veh
LF	bus loading factor	
β_0	initial cost (base fare)	$
β_1	unit cost for additional congestion charge	$
β_2	unit cost of additional distance related charge	$/hr
TC_r	travel time in congestion from location (i) to (j) for trip r	hr
γ_1	threshold of travel time in congestion without additional charge	hr
δ_1	unit travel time with additional charge	hr
L_r	distance from location (i) to (j) for trip r	km
γ_2	threshold of traveled distance without additional charge	km
δ_2	unit travelled distance with additional charge	km
τ	average travel distance per passenger	km
R	annual ridership	prs/yr
m	average passengers on board for each service	prs
ρ	cost per vehicle service hour	$/hr
T_r	total vehicle hours for trip r	hr
s_r	operating speed for trip r	km/hr
Ts_r	vehicle service hour for trip r	hr
l_r	trip distance from location (i) to (j) for trip r	km
Td_r	dead heading time trip r	hr
Td_{max}	average of maximum dead heading times for all trips	hr

[2] $$\frac{PA_i^{am/pm}}{PA_1^{am/pm}} = \frac{PB_j^{pm/am}}{PB_2^{pm/am}}$$

Where, $PA_i^{am/pm}$ = passenger alighting at stop (i) during AM or PM peak hours, $PA_1^{am/pm}$ = total passenger alighting for direction 1 (e.g. northbound) during AM or PM peak hours, $PB_j^{pm/am}$ = passenger boarding at paired stop (j), during PM or AM peak hours and $PB_2^{pm/am}$ = total passenger boarding for direction 2 (e.g. southbound) during PM or AM peak hours. In equation [2] the total passenger boarding and alighting are assumed to be equal in each direction during AM or PM peak hours ($PA_1^{am/pm} = PB_1^{am/pm}$; $PA_2^{am/pm} = PB_2^{am/pm}$). Considering day-to-day variations in peak hour passenger boarding and alighting patterns, the annual passenger boarding data should be analyzed to determine the day with maximum peak hour (e.g. AM or PM peak) passenger boarding. Consequently, equation [2] can be used to determine the MLS and its corresponding design demand. Another important input which is mainly required for estimating the operating costs of DRT services considered in this study is the annual OD matrix. The annual OD matrix represents annual passenger exchange rates between bus stops (i.e. trip origins and destinations). The annual OD matrix is needed for estimating important parameters such as demand for DRT and individual trip lengths. To estimate the annual OD matrix, first equation [3] is used to calculate annual passenger alighting rates at each bus stop based on passenger boarding data. Similar to equation [2], equation [3] is based on the assumption of symmetry between passenger boarding and alighting at paired stops. Consequently, the annual OD matrix is computed based on passenger boarding data and estimated passenger alighting rates using a methodology proposed by Tsygalnitsky (1977) which is based on fluid analogy.

[3] $\frac{PA_i^y}{PA_1^y} = \frac{PB_j^y}{PB_2^y}$

Where, PA_i^y = annual passenger alighting rate at stop (i), PA_1^y = annual passenger alighting rate for direction 1 (e.g. northbound), PB_j^y = annual passenger boarding at paired stop (j), and PB_2^y = annual passenger boarding for direction 2 (e.g. southbound).

In equation [3] it is assumed that the total annual passenger boarding and alighting are approximately equal in each direction (i.e. $PA_1^y = PB_1^y$ and $PA_2^y = PB_2^y$).

3.3 Operating Cost Models

3.3.1 FRBS

A variety of operating cost models have been proposed for FRBS in the literature (e.g. Cherwony and Mundle, 1980; Abbas and Abd-Allah, 1998). For a given bus route, there is a consensus that annual FRBS operating costs (C_R) are mainly function of the fleet size (n), total vehicle-hours (h_v), and the total vehicle-kilometres (l_t) of service, as shown in equation [4]. Model parameters can be calibrated using empirical data as demonstrated in section 4.

[4] $C_R = \alpha_1(n) + \alpha_2(h_v) + \alpha_3(l_t)$

Model parameters are defined in Table 1. It should be noted that independent parameters in equation [4] can be rewritten as function of the annual ridership. Thus, equation [4] can be used to evaluate the annual operating costs of RFBS as function of annual ridership. An example is provided in section 4.

3.3.2 COTS

COTS is a reasonable option for proving DRT services in low demand areas. A single taxi can serve multiple passengers depending on their departure times, and trip origins and destinations. The use of an efficient trip planning and routing algorithm is essential for improving the performance of COTS.

Based on the guidelines provided in taxi bylaws, the annual operating cost model for COTS could be formulated as in equation [5], which includes a base fare, time and distance related costs. In equation [5], for each trip (r), i and j represent the stop ID corresponding to the trip origin and destination, respectively. Model parameters are defined in Table 1.

[5] $C_R = \sum_{r=1}^{R} [\beta_0 + \beta_1 \left(\frac{TC_r - \gamma_1}{\delta_1}\right) + \beta_2 \left(\frac{L_r - \gamma_2}{\delta_2}\right)]$

3.3.3 IHPS

Paratransit service in the Canadian municipalities is intended for providing transportation to the people who are not able to use FRBS. It is a scheduled, ride shared, and door-to-door service. IHPS is operated by the City and its available paratransit fleets to deliver a scheduled, ride shared service as per customers' requests. The only difference is that IHPS transports clients from defined origins to defined destinations (similar to fixed bus route stops) instead of door-to-door service. To develop a cost model for IHPS, the unit costs of operation, such as the cost per vehicle service hour, and the cost per vehicle traveled distance are required. However, such detailed information may not be often available. Alternatively, equation [6] is proposed to estimate IHPS operating costs which is only based on the vehicle service hour (Ts_r) and deadheading time (Td_r). Deadheading time in this case refers to the periods when there are no passengers on board for each service. Canadian public transportation agencies generally estimate the unit cost per vehicle service hour for their paratransit services as part of their financial analysis process. Model parameters in equation [6] are defined in Table 1.

[6] $C_R=\rho\times\sum_{r=1}^{R}(T_r)=\rho\times\sum_{r=1}^{R}(Ts_r+Td_r)=\rho\times\sum_{r=1}^{R}(\frac{l_r}{s_r}+Td_r)$

4 MODEL APPLICATION

Bus route 14 in Regina is used as a case study to demonstrate the application of the proposed analysis methods in this paper. Route 14 is 9.84 km long with annual ridership of 9,954 passengers and 26 stops in both directions. Considering low ridership trends, provision of DRT service is a viable option to improve the economic efficiency of route 14. Bus route 14 is located in south east of Regina, and serves a mixed land use area as it passes through commercial, institutional, open space/recreational, and residential areas. Required data for model development and calibration included hourly, weekly, monthly passenger boarding at bus stops, operating costs data for bus routes and paratransit services, vehicle service hours for bus routes, and fleet size in peak and off-peak hours for bus routes, collected in 2015 by Regina Transit Services for all bus routes in Regina.

4.1 Calibration of FRBS Operating Cost Model

To calibrate the cost model indicated in Equation [4], Regina's bus route data in 2015 were analyzed. As shown in Table 2, significant correlations were observed between, fleet size (n), vehicle service hours (h_v), and vehicle traveled distance (l_t). Thus, independent parameters were examined one-by-one for model development.

Table 2: Correlation analysis for FRBS cost model parameters

	C_R	n	h_v	l_t
C_R	1			
n	0.8502	1		
h_v	0.9997	0.8506	1	
l_t	0.9774	0.7826	0.9773	1

It was found that FRBS operating cost can be modeled with high accuracy (adjusted R^2=0.95) as function of vehicle service hours (h_v). Equation [7] shows the regression model developed for FRBS operating costs. Model parameter was found statistically significant (p=0.05).

[7] $C_R=91.80\times h_v$

Vehicle service hours (h_v) in equation [7] can be rewritten as the sum of total service hours during peak and off-peak hours:

[8] $h_v=v_p\times h_p+v_r\times h_r$

In equation [8], v_p represents the required fleet size during peak hours which can be estimated using equation [9]:

[9] $$v_p = \frac{L \times P_{max}}{s \times C \times LF}$$

Assuming a linear relationship between design demand (P_{max}) and average hourly passenger boarding (R_h), and considering equation [1], equation [10] can be derived:

[10] $$\gamma = \frac{P_{max}}{R_h} = \frac{P_{max} \times (h_p + h_r)}{R}$$

By solving equation [10] for design demand (P_{max}), equation [11] can be derived:

[11] $$p_{max} = \frac{\gamma \times R}{h_p + h_r}$$

Consequently, equation [12] can be derived by combining equations [9] and [11]:

[12] $$v_p = \left[\frac{L \times \gamma \times R}{s \times C \times LF \times (h_p + h_t)}\right]^+$$

Finally equation [13] is derived by combining equations [7], [8], and [12]:

[13] $$C = 91.80 \times \left(\left[\frac{L \times \gamma \times R}{s \times C \times LF \times (h_p + h_t)}\right]^+ \times h_p + v_r \times h_r\right)$$

Definitions of the parameters are provided in Table 1. It should be noted that equation [13] relates annual bus ridership to total annual operating costs. The values of other dependent parameters in equation [13] were assumed or estimated by analyzing the data provided by the City of Regina, which are shown in Table 3. Bus design demand (P_{max}) at MLS and the ratio of design demand to average hourly passenger boarding (γ) for bus route 14 was estimated based on the methodology described in section 3.2 using 2015 bus ridership data.

Table 3: The values of the parameters used in Equation [13]

Parameter	L	γ	s	C	LF	h_p	v_r	h_r
Value	9.84 km	1.219	19.68 km/hr	*67	*0.9	251 hr/yr	1	1,096.83 hr/yr

* Assumed values

4.2 Calibration of COTS Operating Cost Model

Equation [5] was used as the basis for developing an operations cost model for COTS. The second term in equation [5] represents additional charges for the time spent in congestion. Bus route 14 serves a subdivision with very low traffic volumes and uncongested roads. Thus, the second term in equation [5] was excluded. The third term in equation [5] represents the distance based taxi fare component. As shown in equation [14] the average trip distance (τ) was estimated based on the annual OD matrix and the shortest distanced between all the stops. Bus ridership data in 2015 was analyzed based on the methodology described in section 3.2 to estimate the annual OD matrix.

[14] $$\tau = \frac{PD_S}{R}$$

PD_s represents the total passenger travelled distance which is calculated by summing the product of OD flows (F_{ij}) and the shortest distances (D_{ij}) from stop i to j. k represents the total number of stops.

[15] $$PD_S = \sum_{i=1}^{k} \sum_{j=1}^{k} (F_{ij} \times D_{ij})$$

Given m as the average number of passengers on board for each taxi trip, equation [5] can be written as:

$$[16]\ C_R \approx \frac{R}{m}[\beta_0+\beta_2\left(\frac{\tau-\gamma_2}{\delta_2}\right)]$$

Estimated and assumed model parameters are shown in Table 4.

Table 4: The Values of the Parameters Used in Equation [16]

Parameter	β_0	β_2	τ	γ_2	δ_2
Value	$4/prs	$0.25/m	1.425 km	0.120 km	0.138 km

* Assumed values

4.3 Calibration of IHPS Operating Cost Model

Similar to COTS operating cost model, τ and m were applied to simplify the original IHPS operating cost model formulated in equation [6]. Equation 17 shows the simplified IHPS operating cost model. All variables in equation [17] are defined in Table 1.

$$[17]\ C_R \approx \rho \times \frac{R}{m} \times (\frac{\tau}{s} + f \times Td_{max})$$

The values of ρ and s were set to $63.50/hr and 19.68 km/hr based on the data provided by the city of Regina. The deadheading time is defined as a percentage ($0 \leq f \leq 100$) of Td_{max} which is the average of maximum deadheading time for all trips. For each trip from stop i to j, given L_i and L_j as the distance from the dispatch center to stops i and j, and F_{ij} as the OD flow, the average maximum deadheading time Td_{max} can be estimated using equation [18]. k represents the total number of stops.

$$[18]\ Td_{max} = \frac{\sum_{i=1}^{k}\sum_{j=1}^{k}[F_{ij} \times (L_i + L_j)]}{s \times R}$$

4.4 Comparisons of FRBS, COTS, and IHPS Operating Costs as Function of Annual Ridership

Figure 6 shows variations of the operating cost models as function of annual ridership for FRBS, COTS, and IHPS services. Line 1 represents FRBS cost model (equation [13]). A single bus can serve both peak and off-peak hours on route 14 until the ridership reaches 132,573 passengers per year, at which an additional bus is required to serve peak hours. Line 2 and Line 3 are drawn based on Equation [17]. Line 2 represents the worst scenario for IHPS assuming m=2 and f=50%, which means that paratransit vehicles serve only two passenger on average per each service and the deadheading time is 50% of Td_{max}. Line 3 represents the best scenario for IHPS operation denoting that every paratransit vehicle carries 5 passengers on average while the deadheading time is 20% of Td_{max}. Equation [16] was used to produce Line 4 and Line 5 for COTS operating cost model. Line 4 represents the worst scenario for COTS operations and implies that every taxi delivers only one passenger at a time. Line 5 represents the best scenario (highest efficiency). In this scenario, each taxi is always fully loaded (m=3) between origin and destination stops. The analysis results indicate that the average number of passengers per vehicle (m) and deadheading time (for IHPS) are some of the most important parameters in determining the operating cost of DRT services. It should be noted that the best and worst scenarios are not realistic. However, they identify the feasible ranges for each DRT service cost model based on different values of m ($1 \leq m < 10$ for IHPS; $1 \leq m < 3$ for COTS). Based on Figure 6, the following results can be observed for route 14 (inequalities compare the cost performance of each service):

- COTS always performs better than IHPS considering the worst scenarios,
- IHPS always performs better than COTS considering the best scenarios,
- When $1 \leq R < 6{,}958$, COTS>IHPS>FRBS, considering the worst scenarios, and
- When $1 \leq R < 36{,}553$, COTS>IHPS>FRBS, considering the best scenarios.

Although some of this observations are obvious, the critical values of ridership for switching from one service to another can be determined based on some simple assumptions. As the annual ridership for route 14 is 9,954 a minimum required average vehicle occupancy rate can be calculated so that replacing FRBS with COTS or IHPS will result in cost savings and better economic performance. Figure 6 can be reproduced for any other bus route based on more realistic values for m and deadheading time. The results of the analysis can be used by transit planners to compare the efficiency of DRT and FRBS services in low demand areas.

5 SENSITIVITY ANALYSIS

The proposed analysis method enables evaluating the sensitivity of the operating costs due to variations in other input parameters. For example, Figure 7 shows the impact of changing the operating speed (s) of FRBS and IHPS by ±10% (indicated by 0.9s and 1.1s) on their operating costs. Change in the average operating speed affects FRBS cost model as the required fleet size will be affected. The addition of another bus is required in lower ridership if operating speed is reduced, and vice versa. On the other hand, lower operating speed will result in higher operating costs for IHPS, and vice versa. The new intersection points of operating cost models can be determined accordingly to compare the performance of each service based on the best and worst operating scenarios.

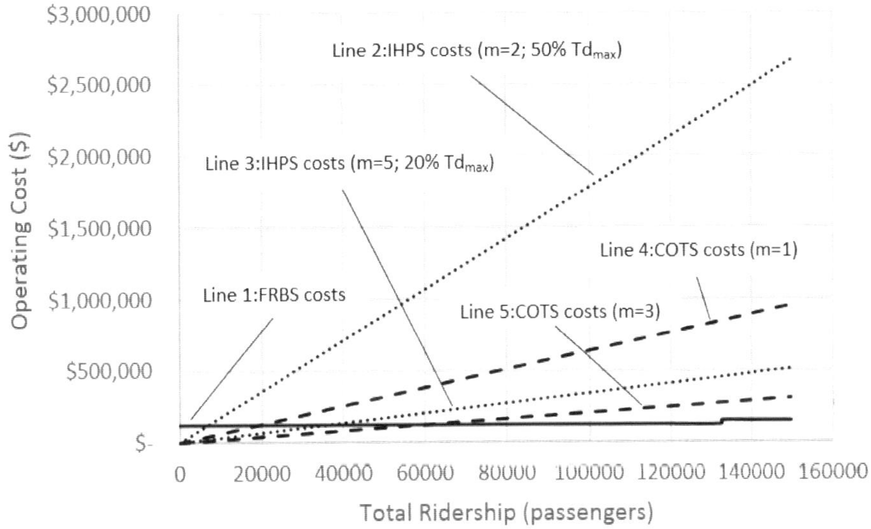

Figure 1: The relationship between operating costs and annual ridership for FRBS, COTS, and IHPS

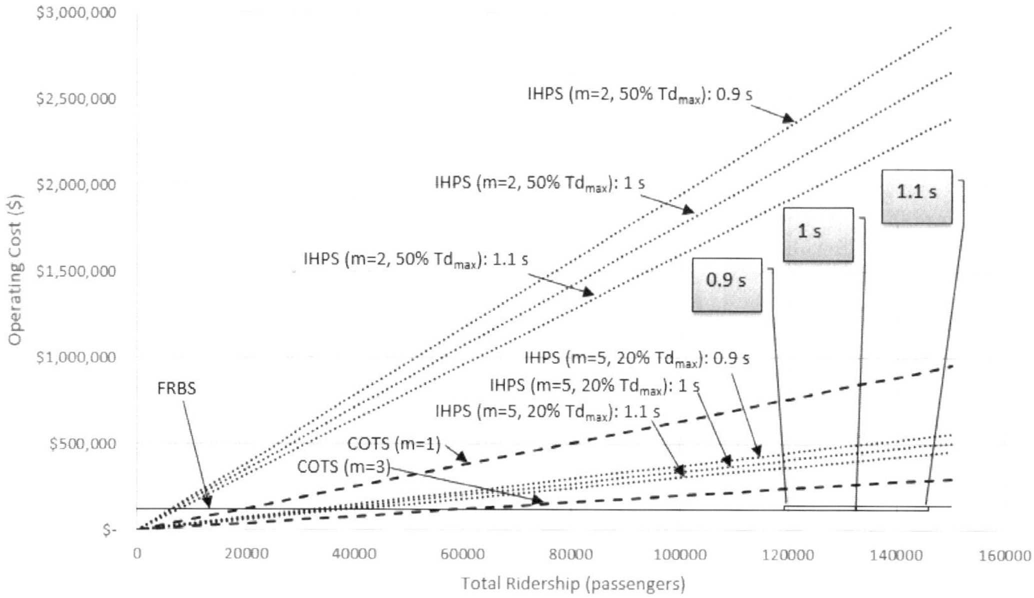

Figure 2: Sensitivity analysis for average driving speed

6 CONCLUSION

In this study feasibility of offering DRT services in low demand areas was investigated. Analytical models were proposed to evaluate and compare the operating costs of regular bus and demand responsive transit services. Two DRT services were considered i.e. COTS and IHPS. Appropriate cost models were developed for each service type to describe the operating costs as function of the annual ridership. The proposed models enable determining the critical values of annual ridership to switch from FRBS to COTS or IHPS and vice versa. Vehicle service hours was found to be the main independent parameter for modeling FRBS operating costs. The cost model for COTS included basic, time dependent, and distance based fare components. For IHPS the cost model was developed based on the unit cost for vehicle service hours. Applications of the proposed models were demonstrated using a low demand bus route in Regina. For DRT services it was shown that, operating costs are strongly related to average vehicle occupancy rate and deadheading time (for IHPS). The proposed methods in this research are based on several assumptions which should be refined and validated to generalize their application. For example, a linear relationship was assumed between the bus design demand and the average hourly ridership. This assumption should be evaluated using historical bus ridership data. In this study, passenger alighting was modeled using passenger boarding data based on the assumption of symmetry between passenger boarding and alighting ratios which should be validated using field surveys. It should be noted that the best and worst case scenarios for COTS and IHPS cost models were estimated based on extreme values for vehicle occupancy rate and deadheading time. However, the values of these parameters could be estimated based on more realistic assumptions. Finally, in this study only the operating costs were considered as the main decision criteria to switch from FRBS to DRT services. However, other important criteria such as level of service and user costs including wait time and in vehicle time should be included in the analysis, which will be considered in the future extensions of this research.

Acknowledgements

Authors would like to thank the City of Regina for providing the data needed for this study.

References

Abbas, K. A., and M. H. Abd-Allah. 1998. Estimation and Assessment of Cost Allocation Models for Main Transit Systems Operating in Cairo. *Transport Reviews.* 19(4): 353-375.

Cherwony, W., and S. Mundle. 1980. Transit Cost Allocation Model Development. *Journal of Transportation Engineering.* 106: 31-42.

Diana, M., L. Quadrifoglio., and C. Pronello. 2009. A Methodology for Comparing Distance Traveled by Performance-equivalent Fixed-Route and Demand Responsive Transit Services. *Transportation Planning and Technology.* 32(4): 377-399.

Edwards, D., and K. Watkins. 2013. Comparing Fixed Route and Demand-Responsive Feeder Transit System in Real-World Settings. *Transportation Research Record.* 3: 128-135. doi:10.3141/2352-15.

KFH Group. 2008. Guidebook for measuring, assessing, and improving performance of demand-response transportation. *TCRP report 124.* Washington DC: Transportation Research Board. 132.

Laws, R., M. Enoch., S. Ison., and S. Potter. 2009. Demand Responsive Transport: A review of Schemes in England and Wales. *Journal of Public Transportation.* 12(1): 19-37.

Li, X., and L. Quadrifoglio. 2010. Feeder Transit Services: Choosing Bewteen Fixed and Demand Responsive Policy. *Transportation Research Part C.* 18(5): 770-780.

Navick, D. S., and P. G. Furth. 2002. Estimating Passenger Miles, Origin-Destination Patterns, and Loads with Location-Stamped Farebox Data. *Transportation Research Record 1799*, TRB, National Research Council, Washington, D.C. 107-113.

Palmer, K., M. Dessouky., and Z. Zhou. 2008. Factors Influencing Productivity and Operating Cost of Demand Responsive Transit. *Transportation Research Part A.* 42(3): 503-523.

Rufolo, Anthony M., James G. Strathman, Eric Kuhner, and Zohngren Peng. 1995. Assessment of Demand Responsive Versus Fixed-Route Transit Service: Tri-Met Case Study. *Final Report TNW95-08.* Seattle, WA: Transportation Northwest (TransNow), University of Washington.

Scott, R.A. (2010). Demand responsive passenger transport in low-demand situations. *NZ Transport Agency research report.* 425(116).

Tsygalnitsky, S. 1977. Simplified Methods of Transportation Planning. thesis presented to Massachusetts Institute of Technology, Cambridge, Mass., in partial fulfillment of the requirements for the degree of Master of Science.

Building Tomorrow's Society
Bâtir la Société de Demain

Fredericton, Canada
June 13 to 16, 2018 / *13 au 16 juin, 2018*

CALIBRATION AND VALIDATION OF MICRO-SIMULATION MODELS USING MEASURABLE VARIABLES

Tawfeek, Mostafa H.[1,4], El Esawey, Mohamed[2], El-Araby, Khaled[3] and Abdel-Latif, Hatem[3]
[1] Ph.D. Candidate, University of Alberta, Canada
[2] Associate Professor, Ain Shams University, Egypt
[3] Professor, Ain Shams University, Egypt
[4] mostafa.h.tawfeek@ualberta.ca

Abstract: Traffic micro-simulation models have been extensively used to evaluate the impacts of traffic design alternatives. These models help stakeholders make informed decisions about any new change in the transport network. Calibration and validation of micro-simulation models are essential steps to ensure the reliability of the models. Most of the existing calibration efforts focus on experimental designs of driver behavior and lane changing parameters that have no physical meaning, or at least, cannot be easily measured. This paper suggests an approach for calibrating micro-simulation models using measurable variables. We advocate using an easy-to-understand, yet appropriate measures of the traffic stream such as spot speed as a calibration parameter. The approach was applied to a suburban corridor of about 12 km located in Greater Cairo Region. Video data were collected for about 9 hours on the corridor for both directions and were used to obtain traffic volumes and composition. Spot speeds were collected using a speed gun for a sample that comprised more than 2,000 vehicles belonging to different vehicle classes. VISSIM was used to model the corridor for only one direction that encompassed uninterrupted flow conditions. Spot speeds distributions were edited extensively in VISSIM to replicate the empirical distributions of the collected data. Furthermore, the distributions were developed for each vehicle type to achieve a higher level of accuracy. Statistical tests were carried out to ensure that the modeled speed distributions match the observed distributions. Corridor travel time was used as a validation parameter. Error measurements were computed for different analysis intervals (e.g., average travel time during a particular range), and it was shown that the validation errors decrease as the analysis period increases.
Key Words: Calibration and Validation, VISSIM, micro-simulation models

1 INTRODUCTION

Micro-simulation models have been widely used by transportation experts and professionals to evaluate the effects of new geometric and traffic modifications. A comparison can hence be carried out between the current and the future situation in terms of the operational performance and potential gains. Simulation models should be reliable in a fashion that replicates real-life conditions so that the simulation results would enable making informed decisions. Model calibration and validation are two necessary steps to ensure the reliability of the developed model. The accuracy of the model outputs mainly depends on the quality of these two steps. This research explores the feasibility of using measurable variables for the calibration and validation of simulation models. In this approach, one or more variables are selected and measured with a high level of detail to enable a reliable calibration. One of the issues that are also explored in this research is the effect of the aggregation/measurement interval on the value of the error measurements used in evaluating the validity of the developed model. To elaborate, in the validation of micro-simulation models, data on one or more measures of effectiveness (MOEs) are collected and used for model validation. Examples of these measures include travel time, queue length, etc. In addition, this paper assessed the validity of the simulation model based on measures during different time periods. We hypothesize that the length of the aggregation interval could affect the estimation accuracy of the model and hence lead to a wrong assumption of a validated model.

2 LITERATURE REVIEW

Micro-simulation has been a powerful tool for testing transportation-related solutions and designs before implementation. Applications of these models include studying route choice behavior (Talaat et al. 2007), analysis of unconventional intersection designs (El Esawey and Sayed 2007), testing and evaluating new signal optimization strategies (Kesur 2009), and, recently, assessing the impacts of intelligent vehicles applications (Tawfeek and El-Basyouny 2018). Recognizing the importance of calibrating and validating micro-simulation models, a considerable body of literature exists to address several issues related to the development, calibration, and validation of different micro-simulation models. In the calibration efforts of small network model (e.g., an isolated intersection, a corridor of a series of intersections, or a section of a corridor), the focus is usually on driver behavior and lane change parameters. The FHWA Guide (Dowling et al. 2004) defines calibration as the adjustment of a set of parameters in the simulation model to predict real-life conditions. The guide further recommends keeping this set of parameters as small as possible to minimize the calibration effort. Accordingly, the observed data will be non-adjustable variables in the calibration process to reduce the number of adjustable parameters. The FHWA Guide recommends the following three steps for calibration: capacity calibration, route choice calibration and finally system performance calibration.

Most of the current procedures for calibrating small network models start with identifying a set of calibration parameters along with their ranges. Some MOEs (e.g., travel time) are defined and used to compare the simulation-measured values of each scenario to field observations. An optimization algorithm is then utilized to determine the best set of parameters to make the model output matches field observations. Calibration procedures usually differ in terms of the number of calibration parameters and their ranges, the method for conducting the experimental design, and the optimization algorithm used to obtain the best set of parameters. For example, while calibrating a micro-simulation model for a congested freeway using VISSIM, Gomes et al. (2004) used Wiedemann 99 module and focused on three necessary lane change parameters (i.e., look back distance, emergency stop distance and waiting time before diffusion) and four car-following parameters. The selection of parameters was based on iterations with visual evaluation of the results and manual adjustment of the parameters. In another study by Kim et al. (2005), a simulation model of a signalized arterial was calibrated by adjusting Wiedemann 74 car-following and lane changing parameters. The calibrated car following parameters included the number of observed preceding vehicles, look ahead distance, average standstill distance, and desired safety distance. Adjusted lane changing parameters, on the other hand, included only lane change distance. The selected ranges for the calibration parameters were determined based on engineering judgment. A genetic algorithm was applied to choose the best set of calibration parameters.

Lownes and Machemehl (2006) aimed to examine the variations of the capacity of a congested freeway when changing driver behavior parameters. The authors used the same car following parameters presented in (Gomes et al. 2004) to replicate real-life conditions. Capacity, demand, and queue characteristics were used as measures to evaluate the calibration quality. A visual inspection was carried out to compare the queue characteristics of the model and compare it to real-life conditions. A sensitivity analysis was undertaken to study the impact of eleven driver behavior parameters on the estimated capacity. Chitturi and Benekohal (2008) developed a procedure to calibrate simulation models based on the relationship between capacity and two driver behavior parameters in VISSIM; standstill distance and time gap between vehicles. Although speed distribution was used, the focus was on the change in the capacity of freeways or work zones when the values of the parameters were modified. In a recent effort by Manjunatha et al. (2013), a methodology was presented to calibrate simulation models of heterogeneous traffic networks. Although the case study was for two intersections in an urban area, Wiedemann-99 car following model was used for its flexibility. The optimum values of the calibration parameters were chosen by an optimization formulation solved by a genetic algorithm. A sensitivity analysis showed that a combination of some car following parameters had a significant impact on the capacity. These parameters were defined for each vehicle type. The model was calibrated and validated using time delays as a measure of effectiveness.

Although many calibration procedures are currently available in the existing literature, it is noteworthy that these approaches depend mainly on adjusting model parameters that have no physical meaning or are difficult to obtain from field observations. In this paper, an approach is presented to utilize measurable variables such as spot speeds as the calibration parameters. High level of disaggregation is proposed where the measured variable is estimated for each particular vehicle class in the simulation model. The motivation behind this research is to apply a calibration methodology that is based on realistic data that have natural meaning and can be easily captured from the field.

3 CALIBRATION AND VALIDATION METHODOLOGY

Successful utilization of micro-simulation models is mainly dependent on their accuracy and reliability. Accordingly, it is essential to accurately calibrate and validate micro-simulation models before their use to derive critical design decisions. Calibration of micro-simulation models is the process by which model parameters are adjusted to minimize the differences between the model outputs and the observed values of some traffic parameters (Dowling et al. 2004). A similar definition is presented by Hollander and Liu (Hollander and Liu 2008) where the authors suggested that the purpose of calibration is to fine-tune a subset of driver behavior parameters so that the model output matches field observed data.

In the absence of a robust calibration and validation procedure, the accuracy of the model will always be questionable. Micro-simulation software packages often include several modules that interact with one another to generate the simulation visualization and output. Each of these modules has enormous parameters that can be manipulated to change the simulation results. A modeler seeking to calibrate a micro-simulation model has to be careful not to get jammed in an infinite loop trying to fix a problem by tweaking some parameters just to find another issue popping up elsewhere. As a result, it is necessary to develop a strategy that is based on a series of logical and sequential steps to calibrate the model promptly without compromising the required accuracy level. As well, it is also preferable to base the calibration on measurable variables rather than ad-hoc procedures using non-measurable calibration parameters. In this study, we advocate focusing on traffic variables to act as "calibration parameters" instead of manipulating user default parameters. The calibration variables should satisfy the following conditions 1) measurable in the field, 2) have a physical meaning, 3) representative of the traffic state of the modeled network, and 4) editable in the simulation model. Examples of these variables include spot speed distributions, acceleration and deceleration profiles, etc. We further hypothesize that detailed representation of the measured variable could lead to a well-calibrated model without the need to tweak any default parameter. The following steps describe the overall calibration and validation procedure:

1. Specify the dominant traffic conditions on the modeled network during the simulation period (e.g., free flow, congested, near capacity, etc.),

2. Select one or more calibration variables that are measurable, have physical meaning, editable in the simulation model, and most representative of the traffic conditions (e.g., spot speeds for free flow conditions, queue length for congested conditions),

3. Specify the evaluation criteria, based on which, the success of the calibration process will be judged (e.g., the similarity between the statistical distributions of simulated outcomes and field measures of the variable),

4. Recognize different sets of calibration variables (i.e., controllable inputs) that have an impact on the evaluation criteria (e.g., speed distribution of each vehicle class)

5. Set acceptable error margins and calibration targets,

6. Edit the calibration variables in the simulation model so that they become as close as possible to the field-measured values,

7. Run the model initially with the default parameters of the simulation model and evaluate the results,

8. If the results of the run satisfy the calibration targets, the model is considered calibrated. Otherwise, the model needs further calibration,

9. Edit the calibration parameter(s) in the model and repeat steps 6 and 7 to satisfy the calibration targets,

10. Specify the measures of effectiveness (MOEs) which will be used for model validation,

11. Compare the outputs of the calibrated model with the real-life values to validate the model, and

12. If the results were not satisfactory, repeat the calibration using more detailed data or another calibration variable until an appropriate accuracy level is achieved.

3.1 Evaluation Measures

Evaluation criteria used to assess the quality of micro-simulation models' calibration can be categorized into two major groups: traffic demand measures and system performance measures. Traffic demand measures are often represented by link volumes. Intuitively, an analyst cannot accept a micro-simulation model that generates traffic volumes which are not comparable to those measured from the field. In micro-simulation models of medium/large networks, differences in traffic volumes become an issue because these models are based on dynamic traffic assignment where demand is modeled using an Origin-Destination (OD) matrix. System performance measures, on the other hand, incorporate a wide range of traffic-related variables that can be both measured in the field and generated from the simulation model. These measures/variables can be used for both model calibration and validation. The extent (i.e., period and sample size) of data collection for model calibration is constrained by budget and time. The collected dataset should be split into two subsets: one for model calibration and one for model validation. This split can be either temporal or spatial. For example, if the dataset is collected for two days, data from the first day can be used for the calibration, while data from the second day can be used for the validation. Alternatively, if the data were collected from many entities of the network (e.g., a number of locations), data of some locations can be used for the calibration, while data of the other locations can be used for the validation. For a comprehensive literature review of different criteria used in the calibration of micro-simulation models, readers are directed to consult (Hollander and Liu 2008). The current study is concerned only with micro-simulation models with static routing (i.e., demand is modeled using traffic volumes rather than OD matrices). Hence, the calibration is aimed to assess system performance of the simulation model compared to real-life data.

3.2 Calibration Targets

Calibration targets are determined according to the minimum performance requirements for the micro-simulation model keeping in mind the aim of developing the model and the available resources (*18*). The FHWA Guide (Dowling et al. 2004) includes examples of calibration targets. Although these targets were developed and used specifically for freeways, some of them have been widely applied in simulation models of urban areas (Oketch and Carrick 2005, Choi et al. 2008). The FHWA Guide (Dowling et al. 2004) suggests that calibration targets can vary according to the potential use of the model, the available resources of the analyst, and time constraints.

3.3 Model Validation

Validation of a simulation model is the next stage after ensuring that the model is well-calibrated. Validation is defined as the process of matching the output of the calibrated simulation model with some real-life measurements that were not used in the calibration. The evaluation measure(s) used in this process should be different from the measure(s) used in the calibration process. Alternatively, the measurements of the same variable can be utilized but for various locations, time periods, and/or traffic conditions (Park and Schneeberger 2003). The model is deemed "valid" if the chosen MOE from the unused real-life dataset is close enough to the simulation values. Otherwise, the calibration process has to be re-executed until comparable results are achieved.

4 CASE STUDY: VISSIM MODEL OF 26TH OF JULY OF CORRIDOR

4.1 Test Site

The corridor chosen for this study is a 12 km-suburban expressway section of the 26th of July corridor which stretches between the Ring Road East and Cairo–Alexandria Desert road (CAD road) west. Two locations were selected for the data collection where each location is about 1.5 km away from the starting/ending point of the corridor. This ensured that the flow is uninterrupted and unaffected by the bottlenecks and changes of the road cross section at the starting/ending points of the road.

Figure 1: 26th of July corridor alignment and location of data collection points

The road cross section is divided and separated by New Jersey concrete barriers, where each direction consists of four travel lanes. The corridor is a freeway section as it has no U-turns crossovers, no access points, and no at-grade intersections. The average lane width is 3.5 m with no shoulders, and the posted speed limit is 90 km/hr. Figure 1 shows the alignment of the section and the data collection points locations.

4.2 Data Description

Developing a reliable micro-simulation model requires the availability of a vast amount of traffic data. At the basic level, data of traffic volumes/composition and speeds have to be available. For the calibration and validation of the model, more detailed data have to be collected such as travel times and spot speed distributions. Video cameras were installed at the data collection points where the video recording took place between 9:00 a.m. and 6:00 p.m. to cover the peak hour at each direction. The data collection took place on a typical working day in the summer of 2012. Spot speed data were collected via a handheld radar speed gun of an accuracy of ±1.6 km/hr for a sample that included more than 2,000 vehicles.

4.2.1 Traffic Counts and Composition

Volume data and traffic compositions were extracted at 15-minute intervals by manually observing the captured videos. No queues were observed at the data collection point located in the WB direction of the corridor. Nonetheless, long queues were observed at the other location between 3 p.m. and 6 p.m. As the corridor represents a divided roadway, each direction was modeled separately according to the prevailing traffic conditions. As mentioned earlier, the selection of the calibration variables mainly depends on the type of traffic conditions. Accordingly, measures such as spot speeds can be used for the calibration of the WB direction where no queue took place. On the other hand, the model of the EB direction can be calibrated using a measure such as the queue length. In this paper, we focus our analysis only on the WB direction which was operating at uncongested flow conditions. The recorded peak hour flow was approximately 5,400 vehicles for the hour between 9:00 and 10:00 a.m. with a peak hour factor of about 0.963 showing a uniform traffic distribution along the hour. Assuming an average lane capacity of 1800 veh/hr, the v/c ratio during the peak period would be 0.757 showing a reasonable utilization of the corridor. More than 33,000 vehicles were counted during the data collection period. Uniform distribution of the volumes was noticed during the afternoon period; 12:00 p.m. to 5:00 p.m. Corridor traffic was classified into six distinct vehicle classes: passenger cars (P.C.), which included private cars and taxis, microbuses (M.B.), heavy vehicles (H.V), buses, Motorcycles, and other vehicles (e.g., emergency vehicles). The traffic composition of the total traffic volume is 74.8%, 13.5%, 6.5%, 2.2%, 2.5%, and 0.4% for P.C., M.B., H.V., buses, Motorcycles, and other vehicles respectively.

4.2.2 Spot Speeds

Spot speeds were measured for a sample that comprised 2,012 vehicles during the 9 hours of data collection. This sample represented about 6% of the total number of counted vehicles (i.e., 33,291). The sample size distribution per vehicle class is matching the actual traffic composition.

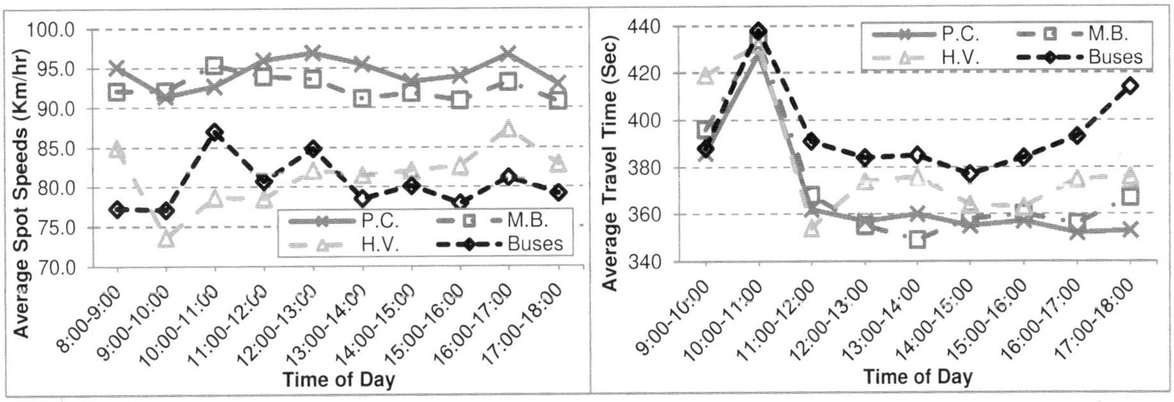

Figure 2: Average spot speeds (left) and average travel times (right) per vehicle class per hour

Figure 2 shows, on the left, the average spot speed per hour for each vehicle class type. Average spot speeds for P.C. varied between 91 km/hr and 97 km/hr while it ranged between 90 km/hr and 95 km/hr for M.B. It is also evident from Figure 2 that no significant differences existed between the recorded average spot speeds of P.C. and M.B. throughout the measurement period (i.e., speed attributes almost remained constant). Noteworthy is that the average spot speeds of buses and H.V. are less than the average spot speeds of P.C. and M.B. just due to the differences in vehicle characteristics.

4.2.3 Travel Times

Travel times were obtained by matching vehicles passages in front of the two cameras and calculating the time elapsed by their passages. This method is similar to the concept of time stamps used to calculate travel times from Automatic Vehicle Identification (AVI) tags. However, the process was carried out manually in this study by human observers. Travel times were determined for each vehicle class and each hour for a total of 783 vehicles. The measured travel times were used to calculate the average values for each vehicle class and each hour of data collection. Figure 2 shows, on the right, the average travel time as estimated from the measured data. The free flow travel time on the corridor is about 360 seconds (i.e., 9 km length and a speed limit of 90 km/hr). As shown in Figure 2, the highest travel times for all vehicle classes were recorded between 10:00 and 11:00 a.m. which was not the peak hour of the traffic volume. This indicated abnormal conditions occurring within the section between the two cameras. Unfortunately, point measurements do not enable clearly identifying the reason for the high travel times during this hour. Nonetheless, it is expected that a number of slow vehicles or maybe H.V. and buses were running together on the corridor at the same time and resulted in temporary delays for all other vehicles.

4.3 Development of the Micro-simulation Model

VISSIM is a stochastic time step behavior-based microscopic simulation software package that uses a Wiedemann psycho-physical car following logic to model traffic on urban streets and freeway environments. VISSIM is versatile and provides the flexibility to model any geometric configuration, a wide range of traffic operations in both the interrupted and uninterrupted traffic environment, and unique driver behaviors encountered within the transportation system (PTV 2011). VISSIM was chosen to simulate the selected section of the corridor where the behavior type was set to be freeway (i.e., free lane selection) which uses Wiedemann 99 car following model. The following sub-sections provide more details on the model development.

4.3.1 Network Coding

Aerial photos were obtained from Google Earth ™, patched together and used as a background for network coding. Photos were scaled using the background scale feature in VISSIM to adjust the dimensions of the section. The number of lanes and lane widths were identified from field visits as mentioned earlier. Hourly traffic volumes were modeled for each of the 9 hours (i.e., from 9:00 to 18:00) of the simulation period. Traffic composition was edited separately for each hour according to the collected data. Transit lines schedules were created according to the observed headways during the data collection period.

4.3.2 Model Calibration

In general, the target of calibration is to ensure that the simulation outputs of the evaluation variable match the field-observed data. Instead of calculating average errors to assess the quality of the calibration, a more robust approach is applied where the "matching" is evaluated using a suitable hypothesis statistical test. The null hypothesis is that both the simulation results and the field values of the calibration variables follow the same distribution. If the simulation output fits the observed data statistically, then the model is calibrated. Otherwise, the calibration variable would need further fine-tuning.

Spot speed per vehicle class was chosen as the calibration variable in the current analysis where the desired speed distribution was selected as the editable calibration parameter. Spot speed distribution is maybe the most influential calibration parameter for uninterrupted flow conditions of a simulation model. Spot speed distribution is usually specified in a simulation model according to the speed limit on the modeled network. The speed limit on the modeled corridor is 90 km/hr. Nevertheless, commuters usually drive at a speed higher than 90 km/hr due to the lack of enforcement as shown in Figure 2. Hence, it became essential to model the speed distributions according to the collected data regardless of the speed limit. The cumulative distribution of field-observed spot speeds was created for each vehicle class according to the collected data. As well, the cumulative distributions of simulated spot speeds were developed. The model will be a calibrated model if a statistical agreement is found between the cumulative distributions of the simulated and field-observed spot speeds.

Several Goodness-of-Fit (GoF) tests can be used to test the null hypothesis that the observed values of a variable are a random sample from a specified distribution. The most widely used test is the Chi-squared test that can be used for testing against discrete and continuous distributions. This test requires the sample of observations to be large enough for the approximations to be valid. Other types of GoF tests are based on the Empirical Distribution Function (EDF). The EDF tests offer advantages over the traditional Chi-squared GoF test, including improved power and invariance with respect to the histogram midpoints (D'Agostino and Stephens 1986). Kolmogorov-Smirnov (K-S) is an example of EDF tests which assumes that the observations come from the same statistical distribution as the null hypothesis. This hypothesis is rejected if the test statistic is higher than the tabulated critical value (D'Agostino and Stephens 1986). The K-S test was used in this study to assess the statistical agreement between the two cumulative distributions. The K-S test is only appropriate for testing data against a continuous distribution. As well, the K-S test is distribution free as the critical values do not depend on the specific distribution being tested. The K-S statistic is computed as the maximum of D+ and D-, where D+ is the largest vertical distance between the EDF and the distribution function when the EDF is higher than the distribution function, and D- is the largest vertical distance when the EDF is less than the distribution function. This is expressed mathematically as (Kim et al. 2005):

[1] $D = max|F_1(x) - F_2(x)|$

Where $F_1(x)$ and $F_2(x)$ are the cumulative probability density functions of x_1 and x_2 (i.e., observed and simulated spot speeds). If the computed statistic D is greater than the critical tabulated value, then the null hypothesis at the chosen level of confidence is rejected.

4.3.3 Model Validation

In this study, corridor travel times were selected as an MOE for model validation. Average travel time was determined for each vehicle class for both the observed and the simulated travel times. Instead of calculating a single value for the average travel time during the entire simulation period, average travel times were computed for different measurement (i.e., aggregation) intervals including 0.5-hour, 1 hour, and 3 hours. The purpose was to examine the effect of the length of aggregation interval on the validation error. It is expected that increasing the aggregation interval will decrease the error and it was desirable to quantify this effect. Different error measurements were used to evaluate the quality of the calibrated model. These included:

The Mean Absolute Percent Error (MAPE), calculated as:

[2] $MAPE = \frac{1}{N}\sum_{i=1}^{N}\left|\frac{Travel\ Time_{RL} - Travel\ Time_{Sim}}{Travel\ Time_{RL}}\right|$

The Root Relative Squared Error (RRSE), calculated as follows:

[3] $RRSE = \sqrt{\frac{1}{\sum_{i=1}^{N} Travel\ Time_{RL}} \cdot \sum_{i=1}^{N}\left(\frac{Travel\ Time_{RL} - Travel\ Time_{Sim}}{Travel\ Time_{RL}}\right)^2 . Travel\ Time_{RL}}$

The Normalized Root Mean Square Error (RMSN), calculated as Balakrishna et al. (*17*):

[4] $RMSN = \frac{\sqrt{N \cdot \sum_{i=1}^{N}(Travel\ Time_{RL} - Travel\ Time_{Sim})^2}}{\sum_{i=1}^{N} Travel\ Time_{RL}}$

Where:

Travel Time $_{Sim}$= Simulated Travel Time,

Travel Time $_{RL}$= Observed real-life Travel Time, and

N = Number of validation observation

Each of these error measures was applied to travel times of each vehicle class and to each time interval to make the validation more trustworthy. The FHWA Guide (*18*) presented calibration criteria which generally suggest that a 15% error margin can be acceptable in similar exercises.

5 RESULTS

5.1 Model Calibration

Initially, ten simulation runs were executed with the default parameters of VISSIM (i.e., with no calibration). The cumulative frequency distributions of the field-observed (i.e., actual) and the simulated spot speeds for different vehicle classes were compared using the K-S test which rejected the null hypothesis of distributions similarity. The results indicated that running the model with the default parameters would not be appropriate as shown in Table 1. Therefore, to calibrate the model, the desired speed distributions were manually edited for all vehicle classes to match the observed spot speeds cumulative frequency distributions closely. The model was re-run, and the simulated speed distributions of the calibrated model were estimated. From K-S test statistics, a close matching was found between the observed and simulated speed distributions as presented in Table 1. As shown in Table 1, the statistics of the calibrated model indicate a significant improvement over the results of the default model. The hypotheses of similarity were accepted, and hence, the model was successfully calibrated. Validation of the simulation model will be the last step to ensure that the model replicates field conditions.

Table 1: Results of K-S Test for Model Calibration

Vehicle class	D_{max}		Improvement (%)
	No calibration	After calibration	
Passenger Cars	0.283	0.12	58%
Microbuses	0.28	0.099	65%
Heavy vehicles	0.352	0.067	81%
Buses	0.434	0.038	91%

5.2 Model Validation

The average travel times of the corridor were used for model validation. The simulation average travel times of the calibrated model were compared to the field-observed travel times. Different error measurements were computed for each vehicle class, and aggregation intervals and the results are presented in Table 2.

Table 2: Error measurements for model validation by vehicle class

Vehicle class	MAPE			RRSE			RMSN		
	0.5 hour	1 hour	3 hours	0.5 hour	1 hour	3 hours	0.5 hour	1 hour	3 hours
Passenger Cars	3%	3%	3%	5%	5%	**3%**	6%	5%	**3%**
Microbuses	5%	4%	3%	7%	5%	4%	7%	6%	4%
Heavy vehicles	11%	10%	9%	12%	10%	9%	12%	10%	9%
Buses	9%	8%	7%	10%	8%	7%	9%	8%	7%

As shown in Table 2, all the error measurements were below 12% indicating a reasonable matching between the simulated and the actual travel times. Comparing the error values of different vehicle classes, it was found that the lowest validation errors were observed for the travel times of P.C. and M.B. Conversely, the travel times of H.V. exhibited the highest differences between the simulation and the field values. The reason is maybe that most of the H.V. observed on the corridor were unloaded, and hence they run faster than the simulated H.V. Another reason could be the small sample size of the measured travel times for the H.V. which might magnify the magnitude of validation errors. As also shown in Table 2, the validation errors decrease with the increase of the aggregation period (e.g., the error calculated from the average travel times during 1 hour is less than the error calculated during 0.5 hours). Nevertheless, the decrease in the amount of error is not significant (i.e., around only 2%). Overall, all validation results were satisfactory with minimal errors. Therefore, it can be concluded that the model is successfully calibrated and validated.

6 SUMMARY AND CONCLUSIONS

A simple-yet-powerful procedure for the calibration and validation of micro-simulation models based on measurable variables was presented. The proposed method starts with running the model with the default parameters and comparing the simulated values and the field-observed ones for the calibration variable. If close matching is found between the two datasets, the model is deemed suitable and of no need for further calibration. Otherwise, the default parameters of the model would require fine-tuning. Unlike other calibration procedures that heavily depend on experimental designs and optimization algorithms for the selection of the best set of parameters, the method presented in this study relies on traffic variables that are measurable and have natural meaning. The procedure starts with identifying one or more measurable variables that are most representative of the traffic conditions on the network. The variable(s) also has to be editable in the simulation model and easy to capture in real-life. Examples of these variables include spot speeds for uninterrupted free flow environments and queue length for uninterrupted congested flow conditions. Detailed data have to be collected about that variable to enable a more reliable calibration. Such detailed data could include a larger sample, disaggregate data for different vehicle classes, or data for different times of the day. A case study was presented using a VISSIM model of a suburban freeway in Greater Cairo. Extensive traffic volume and spot speed data were collected for 9 hours on the corridor. Desired speed distribution per vehicle class was chosen as a measurable calibration parameter instead of using driver behavior parameters which are difficult to measure in real-life. All cumulative speed distributions were edited in VISSIM to closely match the observed spot speed distribution of each vehicle class. Statistical tests were performed to compare the simulated and the field values. The results showed the superiority of the calibrated model compared to the model with default parameters in terms of improved test statistics. The model was finally validated using average corridor travel time per vehicle class. Average travel times were calculated for different aggregation periods for both the simulated and the real-life values. Different error measurements were computed to assess the validity of the model. In general, all the errors were below 12% showing a reasonable matching between the observed and the simulated travel times. It was also revealed that the error decreases as the measurement interval increases, but the drops in error were insignificant. This conclusion should be treated with caution as the studied corridor exhibits uninterrupted flow conditions and hence the differences in error for different measurement intervals might not be large. In summary, the model could be calibrated and validated with reasonable trade-offs between accuracy and modeling efforts. The approach is logical as it depends only on measurable variables for calibrating simulation models. However, the method requires an extensive amount of data to ensure reliable calibration. The paper discussed some practical and theoretical issues that are essential for the development and application of micro-simulation models. These concepts are beneficial for both practitioners and researchers. As an extension of this paper, the calibration process using measurable and tangible variables from the field with interrupted/congested conditions will be explored.

References

Chitturi, M. V., and Benekohal, R.F. 2008. Calibration of Vissim for Freeways. Transportation Research

Board 87th Annual Meeting,.
Choi, K., Jayakrishnan, R., Kim, H., Yang, I., and Lee, J. 2008. Dynamic OD Estimation using Dynamic Traffic Simulation Model in an Urban Arterial Corridor. *In* Transportation Research Board 88th Annual Meeting.
D'Agostino, R.B., and Stephens, M.A. 1986. Goodness-of-Fit Techniques.
Dowling, R., Skabardonis, A., and Alexiadis, V. 2004. Traffic Analysis Toolbox Volume III : Guidelines for Applying Traffic Microsimulation Modeling Software. Rep. No. FHWA-HRT-04-040, U.S. DOT, Federal Highway Administration, Washington, D.C, **III**(July): 146.
El Esawey, M., and Sayed, T.A. 2007. Comparison of Two Unconventional Intersection Schemes: Crossover Displaced Left-Turn and Upstream Signalized Crossover Intersections. Transportation Research Record: Journal of the Transportation Research Board, (2023): 10–19.
Gomes, G., May, A., and Horowitz, R. 2004. Congested Freeway Microsimulation Model Using VISSIM. Transportation Research Record, **1876**(1): 71–81.
Hollander, Y., and Liu, R. 2008. The principles of calibrating traffic microsimulation models. Transportation, **35**(3): 347–362.
Kesur, K.B. 2009. Advances in Genetic Algorithm Optimization of Traffic Signals. Journal of Transportation Engineering, **135**(4): 160.
Kim, S.-J., Kim, W., and Rilett., L.R. 2005. Calibration of microsimulation models using nonparametric statistical techniques. Journal of the Transportation Research Board 1935, (1): 111–119.
Lownes, N., and Machemehl, R. 2006. Sensitivity of Simulated Capacity to Modification of VISSIM Driver Behavior Parameters. Transportation Research Record: Journal of the Transportation Research Board, **1988**: 102–110.
Manjunatha, P., Vortisch, P., and Mathew, T. 2013. Methodology for the Calibration of VISSIM in Mixed Traffic. Transportation Research Board 92nd Annual Meeting,: 11. Available from http://docs.trb.org/prp/13-3677.pdf.
Oketch, T., and Carrick, M. (2005). 2005. Calibration and validation of a micro-simulation model in network analysis. In: TRB annual meeting, Washington, DC,.
Park, B., and Schneeberger, J. 2003. Microscopic Simulation Model Calibration and Validation: Case Study of VISSIM Simulation Model for a Coordinated Actuated Signal System. Transportation Research Record: Journal of the Transportation Research Board, **1856**: 185–192.
PTV, A.G. 2011. VISSIM 5.40 user manual. *In* Karlsruhe, Germany. Available from http://cgi.ptvgroup.com/cgi-bin/en/traffic/vissim_download.pl.
Talaat, H., Masoud, M., and Abdulhai, B. 2007. A Simple Mixed Reality Infrastructure for Experimental Analysis of Route Choice Behaviour under ITS Applications. *In* Transportation Research Board 86th Annual Meeting.
Tawfeek, M.H., and El-Basyouny, K. Forthcoming 2018. Network-level Comparison of Various Forward Collision Warning Algorithms. Simulation: Transactions of the Society for Modeling and Simulation International.

Building Tomorrow's Society
Bâtir la Société de Demain

Fredericton, Canada
June 13 to 16, 2018 / *13 au 16 juin, 2018*

CAN WE BETTER LEVERAGE NATIONAL DATA COLLECTION INITIATIVES TO SUPPORT TRANSPORTATION PLANNING IN SMALLER CITIES AND RURAL AREAS IN CANADA? A VIEW FROM NEW BRUNSWICK

Hanson, Trevor R.[1,2]
[1] University of New Brunswick, Canada
[2] thanson@unb.ca

Abstract: As Canada's population ages, there will be growing calls for transportation solutions that address the challenges associated with transportation and aging, in particular, for those who are unable to drive. Effective transportation solutions require a clear understanding of travel behaviour, yet the most sophisticated transportation studies are undertaken in the largest urban centres where many alternatives to driving already exist. Smaller cities and rural areas are among the most rapidly aging in the country, yet may not have the resources to conduct studies to understand the travel needs of their residents, necessitating reliance on national datasets. National datasets, such as the Census, can be used to gain some insight into broad travel needs, but are limited in terms of specific conclusions that could be used to support transportation solutions for non-driving older adults. This paper profiles national Canadian data collection efforts in terms of their suitability and limitations to be used for transportation planning in smaller communities, with a New Brunswick focus. The paper identifies opportunities to consider new questions or wording among these national surveys that could provide an additional resource for smaller cities and rural areas to help support their transportation planning for an aging population.

1 INTRODUCTION

Statistics Canada anticipates that over 20% of the Canadian population will be over the age of 65 years by the year 2036 (Statistics Canada 2015d), and in some jurisdictions, like New Brunswick, this percentage may be nearly 30% (Statistics Canada 2015c). The vast majority of Canadians depend on the automobile to meet their transportation needs, yet the health effects of aging can make driving difficult or impossible over time. This demographic shift is of such significant national importance that numerous federal agencies, such as Statistics Canada, have devoted resources to understanding the issues and opportunities associated with an aging population. While the issues are generally well articulated at the national level, responding to the transportation needs of an aging population will require deploying solutions at the local level.

Effective transportation solutions require a clear understanding of travel behaviour, yet the most sophisticated transportation studies are undertaken in the largest urban centres where many alternatives to driving already exist. Smaller cities and rural areas, such as those in New Brunswick, are among the most rapidly aging in the country, are the most car-dependent, and yet may not have the resources to conduct studies to understand the travel needs of their residents, necessitating reliance on national datasets. Any transportation issues that manifest among an aging driving population will likely be more pronounced in smaller cities and rural communities due to higher proportions of older adults and higher automobile dependence. Conversely, smaller cities and rural communities may also be the incubators for new ideas that address transportation issues for an aging population. The challenge is ensuring that smaller cities and rural areas have the tools to understand the issues and to mobilize in response.

2 APPROACHES FOR TRAVEL DATA FOR TRANSPORTATION PLANNING

"Transportation planning studies are designed to collect and analyze information pertaining to present and future transportation needs" (Institute of Transportation Engineers 1999), and for passenger transportation, this information can come from personal and household travel data collected through surveys. There are three main approaches to personal and household travel data collection in Canada:

- Commuting travel data associated with the census of population
- National surveys that capture travel behaviour as part of a wider study (e.g. General Social Survey on Time Use in Canada) for understanding regional and national trends
- Metropolitan surveys on travel behaviour (e.g. Transportation Tomorrow Survey, Toronto, Destination 2040, Greater Moncton) for service and infrastructure planning

Each of these approaches are explored below in terms of their relevance to rural communities and small cities.

2.1 Commuting travel data associated with the census of population

The most relevant (and accessible) national data for smaller and rural communities is the Statistics Canada "Journey to Work" data, which are open data available on the Statistics Canada website and taken from the Census of Population. "Journey to Work" is also the only transportation component of the U.S. Census (U.S. Census Bureau 2017) and reports data in a comparable way to Canadian data, a fact noted by Pucher and Buehler (2006) when looking at cycling data. The data in Figure 1 is organized by Census Agglomeration, but data are also available by Census Subdivision, which would typically include any incorporated municipality (regardless of size).

The information presented in tables like Figure 1 provides valuable insight on travel behaviour and commuting patterns, including mode choice, and is available to even the smallest communities, provided they are a recognized Census Subdivision. The challenge with these data is that they have fairly limited utility for transportation planning; the issue of seeing a Toronto to Fredericton driving commute (1300 km) is described in this footnote from the Statistics Canada data table:

"The census assumes that the commute to work originates from the usual place of residence, but this may not always be the case. Sometimes, respondents may be on a business trip and may have reported their place of work or main mode of commuting based on where they were working during the trip. Some persons maintain a residence close to work and commute to their home on weekends. Students often work after school at a location near their school. As a result, the data may show unusual commutes or unusual main modes of commuting" (Statistics Canada 2016c).

Data table

Select data categories for this table

Geography: Fredericton

Geography type: Place of work

Commuting duration (6): Total - Commuting duration

Sex (3): Total - Sex

Submit

		Main mode of commuting (10)									
Place of residence	Place of work	Total - Main mode of commuting [1]	Car, truck or van	Driver, alone	2 or more persons shared the ride to work	Driver, with 1 or more passengers	Passenger, 2 or more persons in the vehicle	Sustainable transportation	Public transit	Active transport	Other method
Fredericton	Fredericton	37,570	33,185	26,590	6,590	3,325	3,265	4,060	1,045	3,020	330
Saint John	Fredericton	295	250	205	45	30	20	40	10	30	0
Moncton	Fredericton	155	140	125	15	10	10	15	0	10	0
Bathurst	Fredericton	45	45	35	0	0	0	0	0	0	0
Miramichi	Fredericton	45	45	40	0	0	0	0	10	0	0
Toronto	Fredericton	25	15	15	0	0	0	0	0	0	0
Halifax	Fredericton	20	15	15	0	0	0	10	0	10	0

Figure 1: Commuting data to Fredericton by place of residence (Statistics Canada 2016b)

These data are not associated with a timeframe or frequency (e.g. how many trips per day, how many days per week) or modal split for the trips taken (e.g. what percentage of the trips are taken as a driver). The caveat provided by Statistics Canada above suggests that the farther away the place of residence is from the place of work, the less likely that the data represent a "usual" daily commute. This can present further complications for interpreting data from rural areas as commuters may need to travel farther for work than in urban areas, and the distinction between what makes an "unusual" commute and a long-distance commuter is unclear.

2.2 Challenges with the 2011 National Household Survey commuting data for Rural New Brunswick

Given the small numbers associated with commuter flows from small communities, data issues (such as non-response) or errors that may be minor on a metropolitan scale can have major impacts on conclusions drawn from data from small communities. The New Brunswick experience with the 2011 National Household Survey (NHS) provides some insight into some of the challenges with this survey for understanding rural needs. The NHS was introduced as a voluntary replacement for the mandatory long-form census which previously provided information that could be used for understanding commuting behaviour. While Smith (2015) highlights that data from 1100 communities or 3% of the Canadian population were not released from the 2011 NHS due to "because of unacceptably low response rates by Statistics Canada's standards", this actually accounted for 82 New Brunswick Census Subdivisions representing 11% of the New Brunswick population (calculated from cross-referencing communities profiled in Statistics Canada (2015b) and their population from the 2011 Census). The majority of communities included in Table 1 were unincorporated rural areas accounting for 64,451 New Brunswickers, with the largest unincorporated area having 5032 people.

Table 1: Breakdown of New Brunswick communities excluded from NHS data release in 2011

Parishes (unincorporated)	61
Villages	14
Rural Community (incorporated)	3
Town	2
First Nation	2
Total number of communities where data not released (2011)	82
Total population of communities where data not released (2011)	83,794
Total population NB (2011)	751,171
% population where data not released	11%

In addition, coding errors among communities with similar names can result in commuting flow data that appears unrealistic. Commuting flow data for one New Brunswick community was found to be entirely incorrect due to assigning people working at Canadian Forces Base Gagetown (in the Town of Oromocto) to the Village of Gagetown (pop. 698), 40 km away (Hanratty 2015). The 2011 commuter flow data showed 660 commuters to the Village of Gagetown, with 410 from Oromocto; the 2016 data show only 85 total commuters destined for the Village of Gagetown and none from Oromocto (Statistics Canada 2016a). As of March 2018 the values have not been corrected online.

The return of the long-form census and mandatory completion has had positive impacts on Global Non-Response Rates (GNR) for a number of small communities in New Brunswick, as some had high GNR for the 2011 National Household Survey (Table 2), which was the source for "Journey to Work" data in 2011.

Table 2: Select New Brunswick communities and Global Non-Response to the 2011 NHS and 2016 long-form census (retrieved from community profiles from both surveys)

Community	Non-response to 2011 NHS	Non-response to 2016 long-form census
Minto	48.6%	5.8%
Chipman	48.1%	6.6%
Canning	39.4%	9.7%
Chipman Parish	40.6%	3.8%

Even though "Journey to Work" is only one aspect of travel in a community, these data may represent the only origin-destination information available at a small community level, but as evidenced from New Brunswick examples for the 2011 NHS, the data may not always be available, and when it is, it may not be correct. The improved response rate for the latest census can be expected to increase the reliability and value of this information for planning in small communities.

2.3 National surveys that include travel behaviour

The Census of Canada is the only national data source that collects and disseminates data at the community level, and the only travel behaviour characteristics presented are commuting-based. There are additional national data sources that, with appropriate analysis, can provide considerable insight into aggregate individual and household travel behaviour.

The national data source that is likely most relevant for understanding travel behaviour is the General Social Survey on Time Use. These data are available upon request from Statistics Canada or can be accessed through the Data Liberation Initiative at participating universities, such as UNB, and can be analyzed through such tools as CHASS from the University of Toronto. The data are available by province (and include samples from each province, but no territories) and by urban (CMA/CA) and rural (non CMA/CA) areas, and include demographic variables, such as age. They are based on a 24hr time-use diary obtained through a telephone interview, which is coded to provide information on the time spent on various activities, including transportation between activities, and the type of transportation. Data from earlier surveys (Cycle 19 (2005) and Cycle 24 (2010)) go much further in depth on access to transportation than the most recent Cycle 29 (2015), asking whether respondents had a driver's license (or had to give it up) whether others depend on them for transportation, and why people do not use public transportation. Cycle 29 (2015) represented a considerable change compared to previous Cycles, with only one question on access to transportation ("How often do you have a vehicle at your disposal?") and a question on mode to commute to work in the previous week (Statistics Canada 2015a).

These data are valuable for understanding how Canadians use transportation, within the confines of overall how they spend their time, and can form the basis for modelling exercises to identify important variables influencing transportation use (see Young and Lachapelle (2017)) and earlier versions of the survey have been used explore travel behaviours of older Canadians (Newbold et al. 2005). Cycle 29 drew from a sample of 17,390, of which 1,058 were from New Brunswick. These data could be used to understand how rural older adults in New Brunswick spend time on transportation and earlier Cycles could provide insight on older adults' perspectives on transportation access and preferences. It may be possible to develop mathematical models relating demographic and socio-economic variables with travel behaviour, permitting some estimation of local needs based upon census attributes.

The Canadian Community Health Survey (CCHS) is another national survey that asks transportation questions that are relevant to an aging population. In a study of access to health care services and waiting times, participants were asked about difficulties is accessing care (e.g. specialist) and one of the options was "Transportation problems" (Statistics Canada 2013). These survey data were the basis for the compilation of 33 New Brunswick community profiles for the New Brunswick Health Council, with results that indicate the percentage of the population in that community reporting transportation problems to healthcare (New Brunswick Health Council 2018). The benefit of these data is that they do give a finer resolution to transportation issues by geography, and the entire province is covered, but the geographic boundaries are generally associated with health catchments areas rather than municipal or taxation boundaries which would expected to be the service delivery mechanisms. The challenge remains from the transportation planning perspective that "transportation problems" are not defined. The 2008-2009 Health Aging Canadian Community Health Survey has numerous questions about someone's ability to meet their own transportation needs, including how much they drive, whether they rely on others for transportation, and the data are organized by province and CMA and Non-CMA, providing another valuable dataset.

2.4 Metropolitan (urban) surveys on travel behaviour

Data collected through metropolitan or urban travel surveys explicitly focus on measuring, understanding , and forecasting transportation, and do so in a way that align with the planning and forecasting tools typically available (e.g. four-step modelling). While the Census "Journey to Work" data looks at the number of workers and where they live and work most often, an urban travel survey will look at all elements of a journey, and the one-way trips generated from Home and Non-home, and include more categories than "Work", since the goal is often to understand travel impacts on infrastructure capacity and service levels. The population census data can serve to help with the validation of model estimates.

These data can be collected at the local (metropolitan) level involving participation from municipalities, regions, transit agencies and can involve provincial governments. Canada's two largest metropolitan areas, Toronto and Montréal, for example, conduct regular (and extensive) household travel surveys. The Transportation Tomorrow Survey (TTS) is a detailed household travel survey conducted by the University of Toronto Transportation Research Institute every five years for urban transportation planning in Greater Toronto & Hamilton (GTHA) (University of Toronto 2018b). In 2011 the TTS included a random sample of 159,200 households, representing information on 8.5 million people (University of Toronto 2018a). The Réseau de transport metropolitan (RTM) in Montréal in 2013 conducted an "extensive Origin-Destination survey throughout the entire Greater Montréal Area" involving 80,000 households, "in order to adapt transportation policies and projects to the real needs of the population"

(Société de transport de Montréal 2018). By contrast, the U.S. National Household Travel Survey in 2017 relied on participation from 129,000 households (U.S. Department of Transportation 2018). Both the Toronto and Montréal travel studies have expanded their scopes to include outlying communities and regions.

Smaller metropolitan areas also collect similar household travel data, though the regularity may be more ad hoc. Hanson (2017) profiled the progression of transportation planning in New Brunswick from 1946 to 2017 and found many of the metropolitan transportation plans for New Brunswick cities did involve household travel surveys, but the plans were focused on understanding the impacts of highway infrastructure projects and typically did not cover rural or suburban areas. More recent approaches, including the Destination 2040 plan for Greater Moncton (Cities of Moncton and Dieppe, Town of Riverview) (HDR Inc. 2015), have starting moving towards modal split targets involving transit and active transportation instead of exclusively forecasting automobile usage. The challenge is that in the case of Destination 2040, 20% of the metropolitan population lives outside the boundaries of the three communities in the study area; by 2041 this is forecast to be 18% or over 37,000 people (calculated from data in HDR Inc. (2015)). These 37,000 people will likely be highly automobile dependent, and primarily living in smaller, rural communities, yet fall outside any transportation planning framework that could assist them planning for a future where access to the automobile is reduced.

3 THE CHALLENGE AND OPPORTUNITIES FOR BUILDING TOMORROW'S SOCIETY BY DEPENDING ON EXISTING DATA

The recently released "From Surfaces to Services" 20 year (2017-2037) transportation strategy from the Rural and Urban Transportation Advisory Committee of the New Brunswick Economic and Social Inclusion Corporation (ESIC, an anti-poverty Crown Corporation) is predicated on the need to address the economic and social inclusion needs of tomorrow's society by focusing on transportation today. "New Brunswick has largely achieved its previous vision for automobile-based mobility, but this mobility now makes populations dependent on the ability to own and drive a car." (RUTAC 2017). The strategy highlights the challenges with this automobile dependence for New Brunswick where it may have up to 1/3 of its population over the age of 65, with "no plan to help prepare for the number of people in this group who will look to transition from being drivers to passengers due to health effects of aging and the onset of disability". The strategy also indicates that it is unclear what agency has the responsibility to develop such a plan.

Many of the most rapidly aging areas of New Brunswick are in rural areas, many of which are outside the reach of any transportation planning and service domains. The two largest urban areas in New Brunswick, Greater Moncton and Greater Saint John have involved outlying or adjacent municipalities in their transportation plans, but "there is no obligation for coordination of transportation service planning among smaller municipalities and unincorporated areas." (RUTAC 2017). Unincorporated areas are home to 37% of the New Brunswick population (Government of New Brunswick 2017).

The existing national datasets, such as the Census, can provide insight for some rural travel behaviour, but the focus on "commuting" for understanding travel behaviour can be expected to have a decreasing relevance for an aging population, many of whom will be retired. Hanson and Hildebrand (2011) found in their sample of 60 drivers rural drivers aged 54-92 that only 6% of one-way trips were for "Work" purposes, with "Shopping" accounting for 20% and "Social Visits" for 16%. Hanson and Hildebrand's sample was primarily retired from work; therefore may not be representative of the employment status of the entire 54-92 year old population. The findings did suggest that the greatest need for transportation for older adults who do not drive will be for non-work related transportation, yet it is not possible to quantify this need from the census of population and rural communities are likely not well positioned to conduct travel studies at the same scale as urban areas. The challenge is that rural communities are or will be on the leading edge of demographic changes that will make driving a less realistic option for some over time; these communities can be the best positioned to mobilize in response to these changes, yet have the fewest resources to do so.

There would be value in extending the long-form census questions to include such trip purposes as "Journey to Shop", "Journey to Health", and "Journey for Social Visits". While these are covered to some extent in the General Social Survey and Canadian Community Health Survey, it is not at the same resolution as "Journey to Work" where origin and destination communities are identified. A rural area that is looking to develop a transportation solution for those who cannot drive to the doctor, for example, would be much better positioned to respond to this need if it had easily accessible data on the level of need. The census would provide a mechanism for that data collection, analysis and dissemination.

Looking to the future, the ubiquity of data-enabled smartphones and the "Big Data" they generate remains an opportunity for obtaining travel data in rural areas at a level not previously available. Traffic congestion data, for example, is already available for small cities and rural areas in New Brunswick on Google Maps. Furthermore, autonomous vehicles appear poised to be realistic options for non-drivers in the near future, yet more work is needed to understand the deployment implications for serving older adults in rural areas, from a policy, technology, infrastructure and user preference standpoint. "Big Data" may have a role in preparing for future technological changes, such as autonomous vehicles, that could support rural transportation needs.

4 CONCLUSIONS

Statistics Canada anticipates that over 20% of the Canadian population will be over the age of 65 years by the year 2036 (Statistics Canada 2015d), and in some jurisdictions, like New Brunswick, this percentage may be nearly 30% (Statistics Canada 2015c). Rural areas and small cities are among the most rapidly aging areas of Canada, yet in provinces like New Brunswick, they face challenges with respect to having the resources to obtain sufficient data to understand the magnitude and breadth of transportation needs posed by an aging population. Census data are the most comprehensive community data available, yet the only information transportation-related data focuses on commuting to work, which may not be relevant for understanding needs of a predominantly retired population. General Social Survey data and Canadian Community Health Survey data provide valuable insight into demographic and socio-economic factors influencing transportation use, but is not readily translated into understanding local transportation needs that can facilitate an operational response. Urban transportation studies can fill in data gaps between the Census and other surveys, but may not cover rural areas. Expanding the long-form census to include transportation-related questions beyond "Journey to Work", such as "Journey to Health", could provide valuable insight to rural and small communities as they look at addressing present and future needs in their communities. "Big Data" may also have a role to play in understanding rural transportation needs and preparing for a future than may include autonomous vehicles.

Acknowledgements

The author wishes to acknowledge the Computing in the Humanities and Social Sciences (CHASS) at the University of Toronto through the Statistics Canada Data Liberation Initiative (DLI) for access to the General Social Survey data and Censuses of Canada. The author's research in this area is supported through the Natural Sciences and Engineering Research Council of Canada (NSERC).

References

Government of New Brunswick. 2017. Local Government Resource Manual. edited by Local Government. Fredericton.

Hanratty, Siobhan. 2015. "Fwd: Re: Commuting flows table - interpretation issue or error?", July 20, 2015.

Hanson, T.R, and E.D Hildebrand. 2011. "Can rural older drivers meet their needs without a car? Stated adaptation responses from a GPS travel diary survey." *Transportation* 38 (6):975-992. doi: 10.1007/s11116-011-9323-3.

Hanson, Trevor R;. 2017. "Exploring the past 70 years of transportation planning tools and practices among select New Brunswick cities: are we ready for the future?" 52nd Annual Conference of the Canadian Transportation Research Forum, Winnipeg.

HDR Inc. 2015. Regional Sustainable Transportation Master Plan
Institute of Transportation Engineers. 1999. *Transportation Planning Handbook*. 2 ed: Institute of Transportation Engineers,.
New Brunswick Health Council. 2018. "Community Profiles 2017." accessed March 12. https://www.nbhc.ca/community-profiles#.WqbNcnxG1QI.
Newbold, K Bruce, Darren M Scott, Jamie EL Spinney, Pavlos Kanaroglou, and Antonio Páez. 2005. "Travel behavior within Canada's older population: a cohort analysis." *Journal of Transport Geography* 13 (4):340-351.
Pucher, John, and Ralph Buehler. 2006. "Why Canadians cycle more than Americans: a comparative analysis of bicycling trends and policies." *Transport Policy* 13 (3):265-279.
RUTAC. 2017. From Surfaces to Services: an inclusive and sustainable transportation strategy for the Province of New Brunswick. edited by Rural and Urban Transportation Advisory Committee. Fredericton: Economic and Social Inclusion Corporation,.
Smith, Wayne R. 2015. "The 2011 National Household Survey: the complete statistical story." Statistics Canada, accessed March 15. http://www.statcan.ca/eng/blog-blogue/cs-sc/2011NHSstory.
Société de transport de Montréal. 2018. "2013 origin-destination survey - Launch of the largest travel behaviour survey." accessed March 9. http://www.stm.info/en/press/press-releases/2013/2013-origin-destination-survey---launch-of-the-largest-travel-behaviour-survey.
Statistics Canada. 2013. Access to Health Care Services and Waiting Times Canadian Community Health Survey Rapid Response Questionnaire.
Statistics Canada. 2015a. "General Social Survey on Time Use, 2015." accessed March 15. http://www23.statcan.gc.ca/imdb/p3Instr.pl?Function=assembleInstr&lang=en&Item_Id=217656#qb218258.
Statistics Canada. 2015b. "NHS Census subdivisions (CSD) not released." accessed March 15. http://www12.statcan.gc.ca/nhs-enm/2011/ref/sup_CSD-SDR-eng.cfm#NB.
Statistics Canada. 2015c. "Population by age group and sex, medium-growth - historical trends (1981 to 2008) scenario (M1), July 1st — New Brunswick, 2010 to 2036 ", accessed March 15. https://www.statcan.gc.ca/pub/91-520-x/2010001/t367-eng.htm.
Statistics Canada. 2015d. "Population Projections for Canada, Provinces and Territories, Highlights." accessed March 15. https://www.statcan.gc.ca/pub/91-520-x/2010001/aftertoc-aprestdm1-eng.htm.
Statistics Canada. 2016a. 2016 Census of Population, Catalogue no. 98-400-X2016325.
Statistics Canada. 2016b. 2016 Census of Population, Statistics Canada Catalogue no. 98-400-X2016327 Government of Canada,.
Statistics Canada. 2016c. Commuting Destination (5), Main Mode of Commuting (10), Sex (3) and Age (5) for the Employed Labour Force Aged 15 Years and Over Having a Usual Place of Work, in Private Households of Canada, Provinces and Territories, Census Divisions and Census Subdivisions, 2016 Census - 25% Sample Data. edited by Statistics Canada: Government of Canada,.
U.S. Census Bureau. 2017. "2009-2013 5-Year American Community Survey Commuting Flows." accessed March 15. https://www.census.gov/data/tables/time-series/demo/commuting/commuting-flows.html.
U.S. Department of Transportation. 2018. "National Household Travel Survey." accessed March 8. http://nhts.ornl.gov/.
University of Toronto. 2018a. "2011, 2006, 1996 & 1986 Travel Survey Summaries." accessed March 9. http://dmg.utoronto.ca/transportation-tomorrow-survey/tts-demographic-and-travel-summaries/2011regional-travel-summaries.
University of Toronto. 2018b. "About TTS 2.0." University of Toronto, accessed March 9. https://tts2.ca/welcome-to-tts-2-0/.
Young, Mischa, and Ugo Lachapelle. 2017. "Transportation behaviours of the growing Canadian single-person households." *Transport Policy* 57:41-50. doi: https://doi.org/10.1016/j.tranpol.2017.03.022.

Building Tomorrow's Society
Bâtir la Société de Demain

Fredericton, Canada
June 13 to 16, 2018 / *13 au 16 juin, 2018*

DEMANDING CHANGE / CHANGING DEMAND

Lanigan, Adam[1,5], Kostyniuk, Jon[2], Bello, Mariam[3], and Probert, Jeffrey[4]
[1] Dillon Consulting Limited, Canada
[2] City of London (formerly of Dillon Consulting), Canada
[3] Dillon Consulting Limited, Canada
[4] Dillon Consulting Limited, Canada
[5] alanigan@dillon.ca

Abstract: It has become clear in recent years that traditional methods of forecasting and analyzing future demands for transportation infrastructure are largely out of step with the progression of many provincial and municipal agencies towards a more equitable, sustainable, and integrated transportation network that seeks to curb auto use and increase the use of transit and active modes. Estimating 'demand' to solely be the number of cars on the road and equating success with pushing more cars down that road is guaranteed to maintain the car as the dominant – and in some cases, only – choice. As the goals of the transportation system evolve to meet changes in policy, so too must our methods of forecasting and analyzing the demand for this system. This paper examines the changes that will be required to reposition transportation demand forecasts to serve the desired outcomes within the framework of new data sources, new transportation technologies and services, and an increasingly knowledgeable, nimble, and flexible travelling public. Important in all of this will be the evolution of practical and cost-effective solutions that are not restricted to academic research, but can be applied on real-world projects to help municipalities and provinces to achieve their policy goals.

1 INTRODUCTION

The forecasting of future demand for transportation infrastructure is undoubtedly one of the most important aspects of the transportation planning profession. This is what guides our investments in infrastructure and allows us to look beyond today's problems, peer into the future that we would like to see for the places where we live, and determine the path to achieving that future. As the principles of sustainable transportation and integrated mobility have taken hold in municipalities across Canada, it has become obvious that how we have traditionally gone about forecasting transportation demand – rooted very much in an auto-focused or auto-only approach – must evolve to provide a more complex, equitable, and realistic examination of the true drivers behind our transportation choices. As the options for travel continue to expand into the realms of the sharing economy, automated vehicles, smartphones, and other technological advances, our society is also experiencing cultural shifts away from a car-based society towards one that recognises the health benefits of active transportation, towards one where young people no longer want to live where they need a car, and one where higher order transit service is becoming more prevalent by the day. All of these aspects need to be better understood to allow us to provide our best estimate of what the future will bring, and create the multi-modal transportation system to meet it.

This paper is separated into five separate sections to allow us to discuss the major elements of such a wide ranging topic, these are:

- Input: Investigation of the changing needs and availability of travel data in light of the advent of GPS, smartphones, and other technologies

- Processes: Traditional modelling processes treat complex daily journeys as a series of disconnected events, but each trip is linked and relates to other choices made along the way.

- Detail: As the principles of sustainable transportation and integrated mobility have come to the fore, it is increasingly important to understand more about the people making the trip.

- Changes: Trends related to how people get around need more consideration in light of changing priorities, technologies, and disruptors like automated vehicles, on-demand ride and bike sharing, and technologies not yet considered.

- Output: The measures we apply in our analyses should be directly linked to the policy goals to allow for the city to evolve appropriately and cost-effectively.

2 INPUT: DATA NEEDS TO BE BETTER, DATA NEEDS TO BE BETTER UNDERSTOOD

As the rate of technological development increases, so too do its potential applications. The advent of smartphones, GPS technologies, and increasingly small and powerful computing hardware has created a new realm and scale of data collection not yet seen in the transportation planning industry.

Traditionally, transportation analysis has been limited by the amount of data on hand with which to understand the world. In large part, data was traditionally collected manually in the field, via surveys, or via the use of expensive hardware, all of which would typically be focused on the issue at hand for a limited time period and would also likely require significant time and effort to decode and understand the data. This is very quickly changing and, as the sheer amount of data available has increased, other issues have risen to the forefront such as: privacy, data access, data management, and data analytics. The catch-all term for this new ocean of data is *Big Data*.

The main difference between traditional data sources and Big Data can be summarised conceptually as: volume, velocity, and variety. That is, the sheer amount of data available (volume), the rate at which is collected, made available, and processed (velocity), and the range of sources and types of data (variety) all far outstrip traditional approaches. The difference is so significant that whole new professions have emerged in recent years related to processing and analysing large mountains of transportation data. No longer is a transportation engineer or planner someone who simply stands at the roadside with a clipboard and counts the cars as they pass; these days the terms ‘data scientist’, ‘programmer’, and the like are becoming more prevalent in the transportation industry.

2.1 The Inflection Point

It is interesting to consider why things are changing now. In part, this is due to the natural progression of technology, efficiency of communication, and increased integration of data collection. This has been an evolving process where several technologies have created a turning point in the realm of data collection.

The prevalence, pervasiveness, and constancy of smartphones and similar technologies in our lives have enabled data collection in a broader, more precise way than previously possible. This new data has resulted in an inflection point - data is now more real-time and passive in its collection and analysis, rather than deferred and active. In other words, it is now possible for massive quantities of data to be available more or less instantly rather than more traditional data collection means where data is actively collected, processed, and available months or even years later.

In some ways, the traditional problem with transportation data has been flipped on its head. Whereas before the problem was creating a focused data collection program to collect as much data as possible on your current issue and *expanding* it outward to represent the whole, now the issue is often *reducing* and understanding the information you need from the vast ocean.

Data collection can now be done virtually anywhere through anonymous GPS and smartphone tracking, rather than the traditional approach of utilizing human observation or more isolated ITS infrastructure, such as sensors on a highway for traffic monitoring. The capital and labour cost to the agency or consultant can be significantly reduced in the collection of the data. The investment comes, however, in the form of time and effort to understand the data and convert it to a useful and meaningful form. This has had the effect of creating a more robust data collection industry, where commercial firms develop ways to add value and services related to their data sources. This is an exciting and quickly expanding market in the transportation industry, as firms race to create better algorithms and even Artificial Intelligence (AI) routines to mine the data for more and more useful conclusions.

2.2 Cross-Platform Data Collection and Integration

An important consideration regarding big data sources, databases, and interactive user interfaces is that the data be 'open'. Open data is data that is available in a portable digital format without restrictions on the ability to use, consume, or share the information. This does not necessarily mean "free" data, but open in the sense that data flows to where it's needed - either for a fee or for free. This is critical to transportation model analysis and the need to increase of efficiency throughout transportation networks.

Although gathered data is largely proprietary, it must also be kept open and available to promote the integration of different data sources. The value of data is not the data itself. Data gains value in how it is aggregated, analyzed, and applied. Systems must be in place to perform data analytics and results must be communicated effectively. A non-open or proprietary system will limit the value of the data collected.

Encouraging the development and maintenance of open data is of utmost importance. But, how can open data sources be integrated cross-platform to provide the best value to end users? In contrast, if each big data company were to collected limited data in a siloed, gate-kept fashion, it would be of more limited benefit to end users. Through data standards and pipelines to integrate data sources, this enables a broader range of potential use cases for data application by end users.

2.3 Privacy Concerns and Data Ownership

To properly integrate data, standards need to be promoted and used, so that the data becomes useful to other parties. Data standards should encompass data formats, ethics, and privacy considerations.

Another consideration is data consolidation and reconciliation. For example, adjacent municipalities with the need to integrate and share data could address this at a base level with an agreed-upon data framework and/or a private consortium of member organizations. This could apply distributed consensus or blockchain technologies with a database being held by each consortium member.

A distributed, common database could be developed where everyone and nobody owns the data. In this way, a distributed and integrated network could be created to anonymize data based on specified rules at the source (i.e. smart contract algorithms), rather than navigate through each company's own ethics and data sharing policies. A common ledger provides immutable proof of who created individual data records and could be maintained to settle payment and provide incentives for data collection.

Caution should be applied when exploring use cases for blockchain technology. For each potential use case, risk assessments and proof-of-concepts will need to be applied prior to any practical investment.

2.4 Importance of Detail and Data Management for a Bigger Picture

Smartphones have provided a level of location-based accuracy previously unattainable on such a large scale and have become a leading factor in the facilitation of big data collection. This has provided companies with a wealth of data and has made the management of data critical.

The level of detail in the data is vital when it comes to its distribution and analysis down the line. If appropriate detail is not collected at the source, this same data cannot be analyzed in depth at a later time. Big data collection has brought us into an age where it is finally possible to see a bigger picture of how people and goods are interacting and how best to promote their efficiency. This will provide a framework for greater value extraction by individual data source companies and enable the provision of a more detailed analysis for transportation engineers, planners, and modellers.

3 PROCESSES: DOING THINGS DIFFERENTLY

The traditional form of transportation demand modelling is typically referred to as Four-Step Modelling, as it has four general 'steps' - trip generation, distribution, mode choice, and assignment. This basic approach has been in practice for decades and is useful for examination of an auto-focused society. More recently, however, we have seen an increase in the use of tour-based, journey-based, and activity-based models, which look at how individual trips connect together. However, despite the increasing sophistication and complexities represented by newer types of models, how people travel has largely remained unchanged for the past century - walk, bicycle, transit, automobile, and trucks.

The fundamental question to consider within our model processes is whether or not they are still useful. For decades, the means by which people travel have fundamentally been the same and, thus, observing the trends over time was typically useful. However, with the imminent and ongoing paradigm shifts due to new technology (e.g., automated vehicles, ride sharing, bike sharing) and changing attitudes surrounding the use and design of urban space, the question is whether traditional processes are still useful.

To consider the new paradigm of models, a more useful term than "traffic model" or "transportation demand model" may be: *Mobility Model.* In essence, this is a blending of our traditional processes and experiences as professionals with new and emerging trends in transportation such as Mobility-as-a-Service (MaaS) and Integrated Mobility along with non-conventional modes of travel and services. This process should consider the mobility of *all people* who need to move through the city no matter their choice and represent the increasing fluidity between modes and choices as technologies and other transportation services become available. A Mobility Model will be a tool to help shape the future based on behavioural economics, demographics, technology, and policy.

3.1 Limitations of Traditional Four-Step Modelling

More traditional four-step models will continue to be useful in situations with less population density and fewer transportation options. However, for regions and municipalities that have more options, the usefulness of traditional four-step models is likely to be more limited. The approach taken to model processes will continue to be at the judgement and discretion of the professional in most cases.

We are at a crossroads in our approach to transportation models, which presents a bit of a "Catch 22" under current circumstances. The question is: when do forecasted trends from current models get dampened by changing attitudes and technologies? The answer to this is imprecise and not very predictable. On one hand, we are observing the emergence of new technologies and mobility options that are not yet fully commonplace which raises questions about the usefulness of current models and processes. Yet, we do not currently have the empirical data required to gain a full understanding or extent of the impacts these new technologies and mobility options will have.

Ultimately, while models will likely continue to be "wrong", we must still adapt to ensure their usefulness over time. With increased real-life complexity and changing needs, our models must adapt to approximate this behaviour as simply as possible in order to optimize the effort required to develop those models.

3.2 Mobility Modelling Defined

A mobility model is defined as a model with the primary focus of moving people, regardless of travel mode, and has the ability to forecast infrastructure needs and impacts. The key assumption here is that fewer people will use a single mode choice in their daily travel.

Mobility models will include algorithms to capture travel elements and metrics, such as (but not limited to):

- Implementation of Mobility-as-a-Service or similar approaches to integrate travel modes;

- Implementation and effects of flex-route transit, mini-transit, and/or rider sharing for first/last mile trips; and

- Positive and negative effects of autonomous vehicle (AV) implementation (e.g. additional accessibility benefits vs. increased deadhead trip disbenefits).

In the current modelling landscape, there are many agencies that have already implemented more sophisticated modelling approaches, such as activity-based, journey-based, and tour-based modelling. To be sure, these approaches will continue to be utilized, especially by larger agencies that have invested in developing them. For these cases, it is a matter of implementing new and emerging algorithms that encapsulate mobility modelling concepts into their current model systems. Perhaps the greater challenge exists for small and medium sized agencies that have more limited resources, but wish to better understand mobility modelling concepts in the context of their transportation planning needs.

3.3 Forecasting with Mobility Models

The mobility modelling approach brings up a classic modelling problem - modelling and forecasting the behaviour for emergent or non-existent travel modes. For these emerging or theoretical travel modes, we likely have little to no empirical data to calibrate against. Therefore, in the short term, we need some creative approaches to forecast future conditions.

In this case, a sensitivity testing approach can be taken to model more theoretical travel choices using processes such as ensemble modelling. This is analogous to the modelling the future path range of a hurricane looking at the worst to best cases based on defined inputs and output metrics and selecting the most likely path from a large range of results.

While there is a need to address short term challenges in mobility modelling processes, we need to primarily look to longer term solutions and prepare for the time when empirical data for the mobility modelling approaches begin to flow. The intent is to iterate and refine our mobility modelling processes so that we have a solid foundation in place as theoretical assumptions transition to real-world data over time.

4 DETAILS: MOVING PEOPLE, NOT JUST CARS

Canada's urban, suburban, and rural areas are made up of a fascinating variety of people, each of whom have their own thoughts, needs, preferences, and abilities related to moving around. So, it's interesting to think about how that variety has been historically distilled in our transportation planning efforts to essentially look at only one type of person: the motorist. As a country – and planet – that is very rapidly urbanising and becoming more sophisticated and flexible in how we choose to get around, this myopic focus on one generic person moving via a single generic mode of travel to and from their single destination, is truly working ourselves towards failure and neglecting to recognize not only the variety of people, but also the variety of solutions available to us to keep our cities operating and growing smoothly.

It follows pretty directly, then, that we need to open up our thinking when trying to forecast future travel demand to not only represent a greater variety of *people*, but also better represent the increasing variety of *options* available to those people to get around.

4.1 Understanding the Population

Looking at the people first, we need to dig deeper into understanding who they are from a wider variety of elements that influence how, why, and when they travel. Traditionally, our knowledge would come from the Census or a manual survey and would be boiled down to a population number for a model zone and an understanding of typical household size and auto ownership. In reality, personal characteristics that determine our travel behaviour could be related directly or indirectly to a laundry list of criteria, such as: age, marital status, in-house dependents, education level, income level, occupation, and dwelling type.

Newer approaches to forecasting are pushing towards this with concepts such as agent-based simulation, which uses demographic information at the household and individual level to gain a better understanding of the individual. To do this, an agent-based simulation needs to rely on some form of population synthesis; that is, creating a more detailed understanding about the population from a range of sources, as data available from the Census and other surveys are necessarily aggregated for privacy or of limited sample size. This aggregated understanding needs to be fleshed out for the entire community.

To date, population synthesis (and thereby agent-based simulation) has typically been reserved for academic research and application, or for larger municipalities with the resources to tackle the coding necessary to create their own population synthesis approach. Some options are beginning to emerge for more general use, with some commercial and open source tools being released. Development of more practical, plug-and-play tools or implementation by commercial software vendors is necessary to allow agent-based methods to become more prominent and give us a better picture of the population.

Other approaches to population synthesis, at perhaps a less detailed level, are possible through the combination of census data with other data sources. There are vendors, for example, that have broken the demographic characteristics for all of Canada at a postal code level into many distinct groups based on their life stage, income, education, number of dependents, and other factors. Consideration of these categories with relation to trip-making behaviour and allowing the categorisations to shift over time could be a powerful tool in better understanding the shifting sands of travel behaviour over time.

An approach like this would reduce the level of effort required for municipalities to understand their population now and how that population will change over time. In essence, if you don't understand who the people are, their needs, and their reasons for travelling, you will not be able to adequately plan your system to address or change that behaviour for the betterment of the city.

4.2 Understanding the Options

And likely just as important as who is moving around, we also need to do a better job at understanding how they move around their community amongst an increasing range of choices. As above, we have historically looked at this almost exclusively through a cars first (or cars only) lens. At best, the typical transportation model examines the use of cars, transit, and commercial traffic to determine our future transportation needs, which purposely ignores, or at best pays lip service to, the reality of an urbanising population that wants to walk and bike in increasing numbers, and altogether omits a rapidly changing market for alternative transportation service delivery through new sharing concepts enabled by smartphones and other technologies.

Not only are the options for travel increasing via newer concepts such as ride sharing, bike sharing, and car sharing, but the flexibility of moving between different modes on a single trip or between trips through the day is also increasing. The concepts behind Mobility as a Service (MaaS) are an attempt to bring together this increasing granularity and dexterity for travellers as they move through their community. MaaS is that umbrella that recognises that mobility needs can and should be able to change throughout a day, week, month, or season. The provision of seamless mobility 'packages' via MaaS that tie together elements like transit, biking, and car sharing demonstrate the evolution of mobility from a static one-time choice to a fluid reasoning of current needs. Some major *car* companies, for example, are currently investing heavily into becoming *mobility* companies. Our approaches to modelling travel behaviour need to evolve along with these concepts.

All of this means that our forecasting models need to better understand the mode choices available to people, build upon our increased knowledge of who they are and what they need to accomplish in a day, and give them the flexibility to move between modes as they would in reality. Each mode has certain characteristics that drive how they are used and how convenient they are for travellers.

5 CHANGES: TRENDS, DISRUPTORS, AND LESSER BULLIES

One of the main drivers of change in transportation is vehicle congestion. As the urbanization of areas continues to increase and intensify, new technologies, efficiencies, and modes of transportation are becoming necessary to accommodate them. Urbanization is one of the main drivers for Smart Cities, integrated mobility, and ITS, for example.

To promote the adoption of ITS and the coming disruptors (i.e. changes) to the current transportation system, services are being altered across the globe. Education campaigns to promote connected and autonomous vehicles (CAVs) are ongoing, and supporting infrastructure for these technologies is being constructed in order to accommodate the disruptive changes to transportation as they arrive.

5.1 Areas of Disruption

The premise and underlying tether for many areas of disruption in the transportation system lies in connectivity. Mobility-as-a-Service (MaaS), for instance, relies on the connectivity of various platforms to work together seamlessly. The concept of MaaS makes the case for shared mobility: ride-sourcing, micro-transit, shared parking, etc. Similarly with CAVs, vehicle-to-infrastructure (V2I) and vehicle-to-vehicle (V2V) communications rely on communication protocols and standards, such as bluetooth, cellular, dedicated short range communications (DSRC), or wireless access in vehicular environments (WAVE).

In order to provide MaaS seamlessly, it will be important for authorities to ensure that the base infrastructure to support these services is there. For example, DSRC technology may need to be present in the right of way if that is the chosen means of communication. Educational campaigns will also have to be carried out to help abate natural distrust of these coming technologies by the public. Concerns with CAVs, for instance, center on the risk of possible technology failures, legal liability, and a vehicle's systems being hacked.

This connectivity will improve the efficiency of the transportation system. Dedicated CAV lanes, reserved solely for CAVs in platoon, would be especially efficient for trucking companies. Autonomous vehicles are classified by a 0-5 level of automation, where 0 is no automation and 5 is full. It is important to recognize that these changes to the transportation system will not happen overnight. There has been and will continue to be an incremental shift from manually operated cars, to CAVs, and everything in between.

Infrastructure to promote these changes is already being implemented and there will be an incremental shift toward CAVs, education campaigns, development of policy and legislation, and adjustment time for the public. The degree of adoption will likely increase over time as public trust is gained.

5.2 Modelling Non-Existent Travel Modes

One central issue that has arisen in the development of mobility models is the speed at which mobility changes are occurring. Forecasting methods have traditionally been based on historical data with little differences in specific modes of travel. However, it is becoming more necessary to represent behavioral changes that may not yet be present in empirical data. For example, there is little empirical data to calibrate these models against as CAVs and MaaS are still largely theoretical.

One potential way to approach this problem may be ensemble modeling. Ensembling is a technique of combining several algorithms to create a more robust system which incorporates the predictions from each algorithm. The resulting final decision is a more accurate, less biased solution. There are also disadvantages to this approach, for example, as conclusions can be more difficult to draw depending on the models chosen to create the ensemble and it can be time consuming to develop. Depending on the potential use cases a reusable framework may be a better approach.

Another approach may be a form of sensitivity-based modelling. A sensitivity analysis is a systematic method for examining how the outcome of an analysis changes with variations in inputs, assumptions, or methods. This could be applied in mobility models by combining various possible outcomes (i.e. best and worst cases) using multiple algorithms. An example of this is how weather patterns are predicted; or, how multiple models attempt to forecast a range of potential hurricane paths based on current conditions. The range of paths and likelihoods are analyzed to determine one of "best fit".

Future modelling will likely involve more sensitivity testing of desirable and undesirable trends, rather than simply past historical data, to rapidly test a range of scenarios and potential outcomes. To provide the capacity to conduct these analyses, it will be necessary to generate base assumptions from a variety of sources, and improve these assumptions as more empirical data is available over time. A solid evaluation framework will need to be developed to report on likely ranges rather than definitive conclusions.

5.3 Looking Back to Look Ahead

In the process of forecasting, it may be challenging to utilize historical trends for modes of transportation that have yet to be observed, but it is the best information available at this time. By making educated assumptions as inputs to our models, they can still be useful in the interim. It is vital to determine how much weight current trends carry in making assumptions and monitor how these trends evolve over time.

Working from the end to the beginning, backcasting is a planning method that begins with defining a desirable outcome and then looking backward to define interim goals. This can be useful in arriving at a futuristic destination, but has obvious predictive limitations, especially when interim goals are not realized. A mix of forecasting and backcasting may be desirable in order to attain our future transportation visions, depending how these models are most useful to practitioners.

5.4 Integration

The mobility models needed to anticipate changes in our transportation systems will have to take into account inputs such as population segments (e.g. age, income, lifestyle, behaviours, etc.), generational shifts, and the impacts of improved travel choices and technology. The eras of data used to implement travel demand forecasting should be delineated so that future trends carry more weight in certain areas, such as pre/post smartphone, pre/post ridesharing, pre/post CAVs, etc.

Different jurisdictions will adopt travel options and shifts at different rates, based on any number of factors, such as income, political will, urban structure, regional policy, etc. For example, Montreal has an excellent bike network in place; therefore, preparation for the adoption of CAVs may be less of a focus.

Transportation demands will evolve at different paces in highly urbanized areas than in more rural areas. In order to facilitate the use of CAVs, the concern of where they are able to function will need to be addressed. Travel demands are also evolving at a different pace across the country. For example, the magnitude of Toronto traffic and congestion is hardly comparable to more rural areas.

As disruptors become part of the new normal, adaptable modeling processes that account for these integrations will need to be developed for use now and for when more empirical data begins to emerge. Individual jurisdictions will need to evaluate their needs and determine how mobility models can best be useful to them to plan for infrastructure investments.

The evolution of mobility modelling will likely be some balance between focusing on preferred outcomes (backcasting) and anticipating future trends from a wide range of unknowns (forecasting). Practitioners cannot simply ignore changing travel behaviour and may have the ability to influence the direction of CAV implementation through a more proactive approach, so long as they align themselves with current trends.

6 OUTPUTS: MEASURING PAINS AND GAINS

As discussed above, conventional transportation engineering philosophy steered transportation professionals towards design and operational decisions that prioritized the comfort of personal motorized vehicle users while lowering the comfort for other travel modes or discouraging their use altogether.

This prioritization affected many levels of transportation planning and design. We planned networks to "optimize" roads by maximizing auto capacity and minimizing auto travel time. We evaluated the Levels of Service (LOS) with measures such as delay, queue, and volume-to-capacity ratio, all of which are most applicable to vehicular traffic. And we also often planned and designed land use and transportation corridors independently, resulting in failed attempts at redevelopment and wide roads that are barriers between communities rather than parts of them. The impacts of decades of this thinking are present-day traffic congestion, primacy of the automobile, and entire cities built around driving.

However, the growing awareness of the link between health and physical activity, building support for sustainable cities, and increasing demand for downtown living has caused many stakeholders to challenge the priorities for street planning and design. This changing societal landscape has resulted in wide ranging municipal support for Complete Streets, a school of thought for street design that advocates for streets serving all modes of transportation rather than cars exclusively.

In this time of change, two major questions arise – how do we adequately measure changes to performance in the evolving landscapes of urban centres and how can we better examine street performance for all modes of transportation?

6.1 Multi-Modal Level of Service

Multi-Modal Level of Service (MMLOS) is an evaluation tool for measuring the level of service present on streets for *all modes* of transportation. Performance measures are assigned to each mode for safety, system efficiency, mobility, sustainability, etc. These measures are mode-specific – for example, a measure of mobility for automobiles might be a midblock volume-to-capacity ratio or travel time reliability, whereas a measure of mobility for pedestrians may be width of the sidewalk.

Of course, transportation professionals are constantly faced with the reality that the rights-of-way of streets in urban areas are limited. Few streets are capable of delivering a high level of service to every mode of transportation due to physical and/or monetary constraints. Therefore, we need trade-offs among level of service objectives for each transportation mode within an established right-of-way. With trade-offs, the prioritization of the modes depends on the context of the intersection, corridor, or network. We can make decisions on a case by case basis to prioritize certain modes and make compromises with others.

As of time of writing, the concept of MMLOS is still relatively new and there is no agreed-upon standard for performance measures, targets, or preferred trade-offs. On the one hand, this lack of precedent may present itself as a challenge to policymakers. However, not yet having an established industry standard also allows us to create a custom approach tailored to a municipality's unique goals. Ideally this leads to a sort of "municipal soul searching" where the community is able to work through a reasoned process towards an MMLOS system that helps them accomplish their policy goals.

MMLOS can allow a city to tailor their frameworks, measures, targets, and tradeoff decisions to be locally sensitive and reflect unique goals for all modes of transportation. That's why Canadian cities like Toronto, Ottawa, London, Edmonton, and Calgary have already made efforts to incorporate MMLOS into their Official Plans, Transportation Master Plans, and/or Complete Street guideline documents.

6.2 Evaluating Street Performance

For certain scenarios, mode priority is relatively obvious: you probably wouldn't make plans for extensive cycling infrastructure and a 3.0m wide sidewalk for a city bypass freeway. For others, such as for a bustling main street, the priority will depend on the city's vision for the street. But once priorities are established, how can we logistically evaluate the street's performance across the modes?

Transportation professionals can determine, tweak, and/or evaluate different performance measures at different stages: planning, design, and operation/maintenance. For instance, the decision to turn a corridor into a designated truck route, which could be a performance measure of a truck's environment, would be determined at the planning stage. Meanwhile, the design speed limit on an adjacent road can be utilized as a measure of the pedestrian mode's environment, a measure that can be changed as needed during design. And at the operation stage, providing transit signal priority at intersections could be a measure of safety for the transit mode.

Simply determining performance measures and then getting values for each measure is meaningless; we need to compare them to some form of rubric that can give an LOS "score" to each mode based on the values for each measure. And in fact, the aforementioned municipalities that already use an MMLOS framework do just that, though not all in the same way. And that's not necessarily a bad thing - a municipality's "rubric" can also be customized to reflect its unique objectives, goals, and vision. But having that "rubric" to consistently and efficiently screen planned transportation projects gives transportation planners and engineers the necessary targets for the best possible design.

6.3 Person Capacity

As we focus on building more progressive and inclusive cities, we should remain cognisant that our population is growing and more people are moving towards urban centres. Therefore, the efficiency of urban space is becoming more crucial and transportation professionals should be considering person capacity on streets. But with limited room between building faces and many cities already reaching or coming close to automobile capacity, how can we design streets to allow for this continued growth?

Person capacity is a rejection of the traditional definition of capacity as being the number of automobiles that can fit on a road and a move towards streets being designed for people. Person capacity measures the number of users of various modes that can fit on a given street. And when using person capacity to evaluate our streets, something quickly becomes obvious - a single street can fit more cyclists per square metre of space (and even more pedestrians) than drivers in cars since automobiles take up significantly more space than bicycles and shoes. And these comparisons vary by orders of magnitude.

The reality is that single-user vehicles are not an efficient use of space. Even buses, though larger than a single automobile, can fit more people per square metre than automobiles. Therefore, more people can be on a single bus at any given time than in cars. By focusing on person capacity and reducing delays for transit, we can create less person-delay per lane of buses than would have been possible with cars.

With the fixed amount of street space combined with the growing number of people, the need to shift to shared and active transportation modes becomes more prominent. Switching out a car lane for a transit lane on an existing street would reduce traditional automobile capacity, which we have been trained to think is undesirable and poor road design. However, with person capacity, while automobile capacity may be decreasing, by focusing on alternative modes, the overall capacity of the street would increase. And there is evidence that this is not just possible - it works. Toronto has added minimal significant auto capacity moving into the downtown core in the last 30 to 40 years but this has not hindered its growth - the city managed to triple the number of commuters in that same time period by increasing the transit (largely by rail), cycling, and pedestrian mode shares.

Of course, as a performance measure in an MMLOS framework, the concept of person capacity still requires refinement. We should also remain cognisant of designing for people. A utilitarian mindset of simply measuring throughput for a corridor may work for automobiles, trucks, and transit, but perhaps not pedestrians. And even the idea of just measuring number of people per square metre may be flawed. A number for capacity obtained through this method on its own does not consider the user's abilities or street grades, for instance, which would affect the requirements of the pedestrian environment, let alone the presence of a mix of land uses and urban design considerations that make the space itself useful and attractive.

7 CONCLUSION

The above discussion presents an overview of some of the ways that the modelling of existing and future demands on our transportation networks will need to evolve to better serve the development of our cities. Just as it will be a challenge to help our cities evolve as they need to meet the demands of changing culture and technology in the coming years, it will be equally challenging to stay ahead of that curve and predict and direct where we are headed next.

Building Tomorrow's Society
Bâtir la Société de Demain

Fredericton, Canada
June 13 to 16, 2018 / *13 au 16 juin, 2018*

DEVELOPMENT OF A PASSING COLLISION WARNING SYSTEM PROTOTYPE FOR OVERTAKING TRUCKS ON TWO-LANE HIGHWAYS

Hassein, Udai[1,2], Diachuk, Maksym[1] and Easa, Said[1]
[1] Department of Civil Engineering, Ryerson University, Toronto, Canada
[2] uhassein@gmail.com

Abstract: Passing collisions are one of the most serious traffic safety problems on two-lane highways. These collisions occur when the driver cannot correctly assess the situation. This paper provides the framework design for a passing collision warning system (PCWS) that assists drivers, avoiding passing collisions on two-lane highways by reducing the probability of human error. The system uses sensors to identify the impeding vehicle type and to detect opposing vehicles travelling in on the oncoming lane. The objectives of this research are: (1) to design driving simulator experiments for data collection to calculate passing parameters, (2) to develop the PCWS algorithm based on camera and radar sensors' signals for the detection of a truck (impeding vehicle), (3) to develop a the driver handling model that describes situations, in which steering wheel action is necessary for lane changes and maneuver interruptions, and (4) to develop SUV and articulated vehicle dynamic models which are represented as universal Simulink blocks combined in one simulation scenario. The simulation series were carried out to confirm the effectiveness of the PCWS algorithm. This study involved the investigation of the effect of driver behaviour on passing maneuvers, and the development of the algorithms for a PCWS that assists drivers in the selection of proper passing gaps during passing maneuvers. Certain techniques, such as mathematical and imitation models, were enhanced in order to replicate real-life driving situations in Simulink. The different factors that affect system accuracy were also examined.

1 Introduction

Passing collisions occur when a fast-moving vehicle tries to pass a slow-moving vehicle by travelling in the opposing lane on a two-lane highway. These collisions are often caused by driving errors and lead to the loss of human lives and costly property damages. The development of a Collision Warning System (CWS) makes a significant contribution to the field of vehicle transportation. These systems help to prevent accidents and save human lives, while avoiding the potential expenses involved in collisions and road infrastructure damage, benefiting the economy. The most important outcome from the implementation of these systems is expected to be autonomous vehicles and intelligent traffic, which help optimize vehicle flow and reduce erroneous human influences.

The evolution of CWS technology began with the study of the overtaking process. Using statistical estimation, researchers tried to distinguish certain zones (according to AASHTO, 2011) in the overtaking path and mathematically describe regressions for the corresponding times (Granet et al., 2003). The bicycle model (in the form of a state-space system) was often used to represent vehicles (Kretschmer et al., 2005). Using this approach, driver behaviour, which depends on different parameters such as age, gender, experience, etc., can be studied to reflect deviations in various time components (Hassein et al., 2017, Mehmood, 2010). The next step in the evolution of CWS technology included the study of vehicle handling and acceleration models using simulations. When automotive radar sensing technology was developed, CWS systems, which assist in the prevention of collisions, were considered in studies

(Dabbour et al., 2014). In approximately 2005, the new computer vision technology became affordable for implementation. By means of installed video camera and special image analysis techniques, the measurement of distances and object dimensions became an option for road object recognition (Isermann et al., 2012). The latest technology combination provides multi-sensor fusion and fully automatic handling to reach a zero-tolerance accident strategy.

The purpose of this paper was to develop a Passing Collision Warning System (PCWS) for driving on two-lane highways to avoid passing collisions and to improve road safety. Regression models were provided to determine the initial time, passing time, and acceleration needed by the driver of the passing vehicle using driving simulator data collected from 63 different drivers (Hassein, 2017). Mathworks Simulink was used to replicate real-life passing situations. The following section presents the proposed PCWS, driver model, and mathematical basis of the handling model. The Simulink model and application example are then presented, followed by the conclusion.

2 Proposed Collision Warning System

The algorithm is capable of detecting various opposing vehicles in the passing maneuver and the same procedures are followed for each impeding vehicle. Fig. 1 shows the procedure of the passing collision warning algorithm. When the system detects a vehicle in the opposing lane, a message regarding passing safety is displayed for the driver. The message can be in text format or an arrow symbol with different colours for each message (e.g. green for "Safe" and red for "Not-safe"). The message stays on until the algorithm confirms that a safe passing maneuver is available.

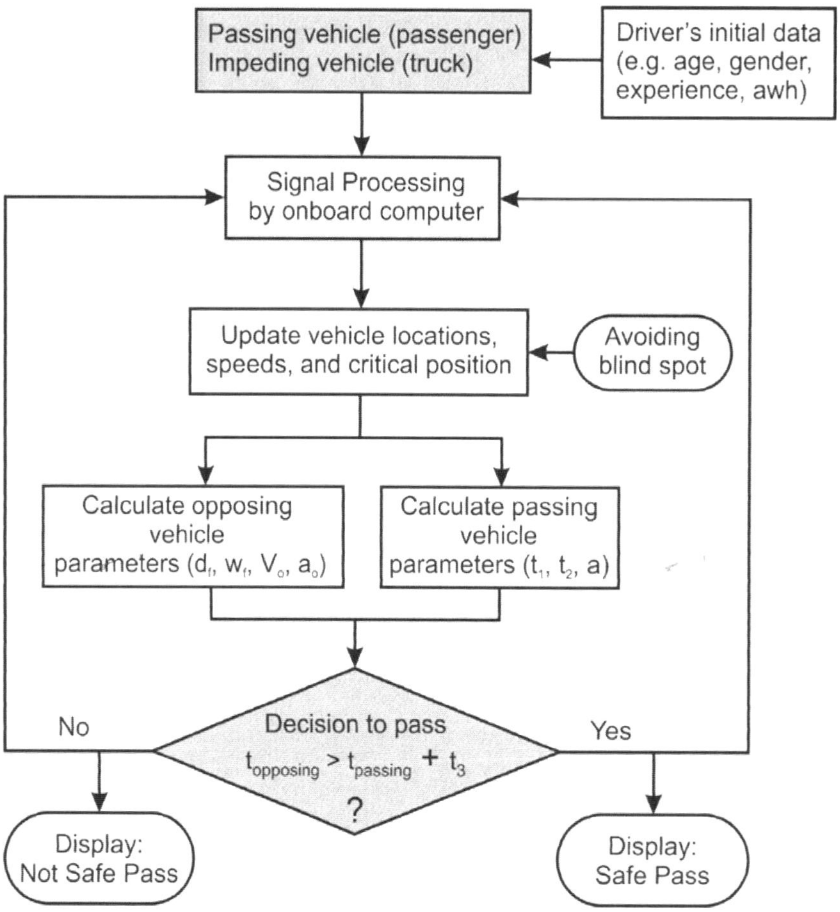

Figure 1: Passing collision warning algorithm

2.1 General Algorithm Description

1. The driver characteristics are entered in the system (e.g., age, gender, experience).
2. The system takes two successive measurements of both the location and speed of the nearest vehicle detected in the closest lane on the left side and the time interval between the two measurements equals the inverse of the detector's frequency.
3. Using these measurements, the system estimates the acceleration rate of the vehicle that has been detected.
4. The system then determines the time needed for the opposing vehicle to reach the safe point, $t_{opposing}$.
5. This information is then used to determine the time ($t_{passing}$) needed for the passing vehicle to complete the pass which is equal to the sum of t_1 (the perception-reaction time and initial acceleration of the passing driver) and t_2 (passing time), as shown in Fig. 2.a.
6. The system compares times $t_{passing}$, t_o (the sum of $t_{opposing}$ and t_{sm}) where t_{sm} represents a safety margin of 2 s and makes a decision according to the following criteria: (a) If $t_{passing}$ is greater than t_o, the "Not safe to pass" message stays on and the system replicates the algorithm to find a safe time gap; or (b) If $t_{passing}$ is smaller than t_o, the "Safe to pass" message is turned on and the driver is free to begin the passing maneuver.

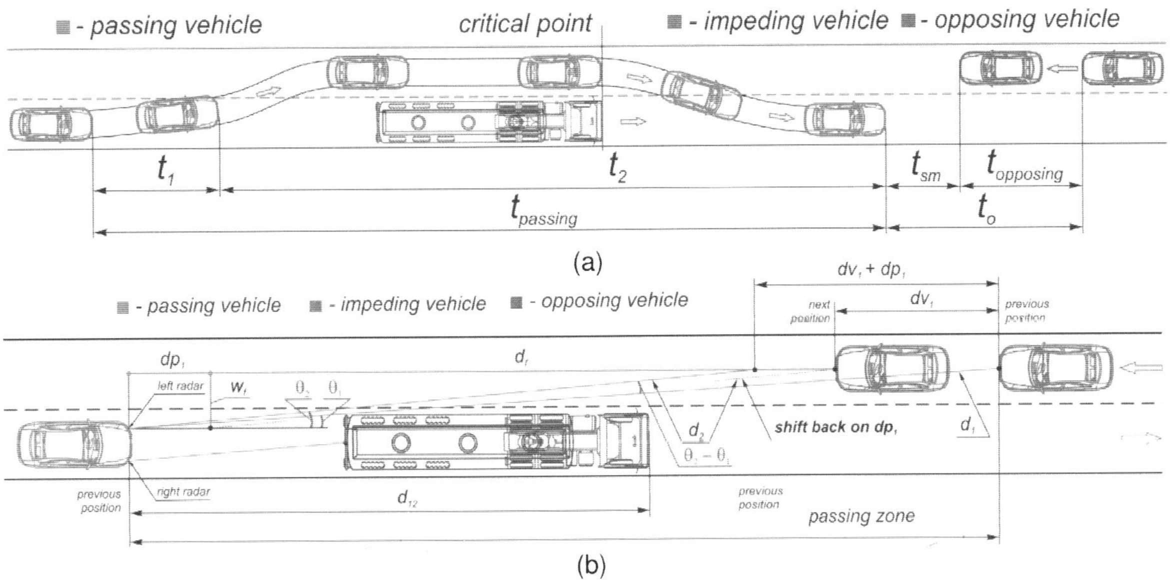

Figure 2: (a) Time definitions; (b) Calculating distance and angle at each time interval

2.2 Passing Vehicle Location

The total time required for the driver of a passing vehicle to complete the pass, $t_{passing}$, is the sum of the following:

1. The driver's perception-reaction time and initial acceleration (t_1), which is the time needed by the driver to notice the "safe" signal and then take the necessary action (to activate throttle); and

2. The vehicle's travel time (t_2), which is the time needed to accelerate the vehicle and clear the path for the oncoming vehicle. This involves the time needed to cross the offset distance between the passing and opposing vehicles, along with the length of the passing vehicle itself.

Finally, the total time required for the passing vehicle to complete the pass, $t_{passing}$, is calculated using the following formula:

[1] $t_{passing} = t_1 + t_2$

Linear regression models for initiating the passing time (t_1), and passing time (t_2) were developed (see Fig. 2.a) using the field data collected using SAS software (SAS, 2015). The new parameter developments are presented in the following sections.

2.3 Opposing Vehicle Location

Fig. 2.b provides an illustration of the distance and angle calculations at each time interval. The radar sensor detectors produce the detection beams at time T in order to scan the left lane of the two-lane highway. If nothing is detected, a "Safe" message is displayed. Otherwise, the closest object detected is measured at distance d_1 and azimuth angle θ_1, where the polar coordinates used with the origin point coexist with the detector's location. The algorithm (see Fig. 1) assumes that the opposing vehicle is approaching from the east and the azimuth angle is measured counter clockwise from the north meridian located at the detector's location. At time $T + \Delta t$, the detector produces another radar beam. The Δt represents the detector's time interval (the inverse of the frequency = 0.01). The new location of vehicle A is recorded at distance d_2 and azimuth angle θ_2 (Fig. 2.b).

The speeds v_1 and v_2 of the opposing vehicle are also measured at time interval Δt. If d_1 and d_2 are equal (v_1 and v_2 are both zero), the algorithm infers that the object is stationary (e.g., a tree or a post) and the "Safe" message is displayed. If d_2 is greater than d_1, the algorithm infers that the object is moving in the opposite direction (away) and a "Safe" message is displayed. If d_2 is smaller than d_1, the distance crossed by the vehicle during the time interval at time $T+\Delta t$ (dv_1) is obtained using:

[2] $dv_1 = \sqrt{d_1^2 + d_2^2 - 2 \cdot d_1 \cdot d_2 \cos(\theta_2 - \theta_i)} - dp_1$

A third radar signal is produced at time $T+2\Delta t$ and records the information for vehicle A at distance d_3 and azimuth angle θ_3. The distance that is crossed by the vehicle in the second time interval, dv_2, can be calculated in a similar way as dv_1 using the following formula:

[3] $dv_2 = \sqrt{d_2^2 + d_3^2 - 2 \cdot d_2 \cdot d_3 \cos(\theta_3 - \theta_2)} - dp_2$

The speed, $v_{T+2\Delta t}$, and acceleration, $a_{T+2\Delta t}$, were determined using the following formulas:

[4] $V_{T+2 \cdot \Delta t} = (3 \cdot dv_2 - dv_1)/(2 \cdot \Delta t)$

[5] $a_{T+2 \cdot \Delta t} = (dv_2 - dv_1)/\Delta t^2$

The side offset between the opposing vehicle and the passing vehicle, w_f is obtained using:

[6] $w_f = s_l + d_{i+1} \cdot \sin(\theta_{i+1})$

The offset may be used as a gauge in order to ascertain whether the opposing vehicle is travelling in the left lane. This is accomplished by comparing the calculated value with the sum of the setback distance (between the detector and the edge of the left lane) and the width of the left lane. Where s_l = represents the location of the left radar sensor on the passing vehicle, which is 0.618 (m). The distance between the opposing vehicle and the detector, d_f, at time $T+2\Delta t$ is obtained using:

[7] $d_f = d_3 \cdot cos(\theta_3)$

Based on conflict types involving the opposing vehicle and the passing vehicle (travelling in the opposing lane or along the same lane), an opposing time, $t_{opposing}$, with a passing time, $t_{passing}$, are calculated in one of the two ways described below. The formula obtained using regression analysis to determine the time t_2 only contains passing vehicle speed as the main unknown variable. This means that the influence of the acceleration mode is neglected in this case. In a real-life situation, the acceleration at the start of overtaking reduces the estimated runtime of the maneuver. In this case, some of the imperfections of the model represented by the formulas for t_2 may be considered as a factor for the tightening of conditions.

Obviously, the formula is more suitable for the flying pass, when the speed is reached in advance and is kept constant while overtaking the impeding vehicle. The necessary linear space for the complete passage is therefore calculated using $v_p \cdot t_{passing}$, where $t_{passing} = t_1 + t_2 - t_C$. Here, t_C represents the current time from the start of sensor data processing, providing the measurement of the distance between the opposing vehicles. Since t_1 and t_2 were determined during the initiation of the warning system, t_C allows for the calculation of the remaining time for the maneuver and ensures the continued monitoring of the safety of the overtaking maneuver. Thus, assuming that the probability of a significant positive change in the speed of an opposing car is small (leading to increased danger), the time required for the opposing vehicle to overcome its portion of the distance to the critical point is calculated using the following formula:

[8] $t_{opposing} = d_f / (v_{opp} + v_p)$

where d_f can be calculated from Eq. (7); v_p represents the passing vehicle speed; v_{opp} represents the instantaneous velocity of an opposing vehicle, which is determined for each value of the time interval Δt, and v_{opp} can be calculated from the speed of the opposing vehicle Eq. (5).

2.4 Decision-Making Process

The last step in the algorithm compares the time needed for the opposing vehicle to reach a conflict point, $t_{opposing}$, with the time needed for the passing vehicle to complete the passing maneuver, $t_{passing}$. If the time needed to complete the passing maneuver, $t_{passing}$, is less than the time needed for the opposing vehicle to reach a conflict point, t_o, a "Safe" message is displayed for the driver. Otherwise, a "Not safe" message will be displayed (see Fig. 1).

3 Proposed Steering Control Model

The development of adequate driver models is critical for vehicle dynamics simulations and for the development of control algorithms that can be added to autonomous vehicle technologies. It should be noted that a driver has a much broader range of parameters characterizing his/her behaviour than a robotic analogue. The various vehicle handling models can be categorized into the following two approaches, which have been successfully used in researchs of overtaking research. Isermann et al. (2012) described the coincidence between the vehicle's mass center trajectory and a hypothetical trajectory. Kretschmer et al. (2005) described a steering controller, which used a current lateral vehicle offset from a determined overtaking trajectory. The controller uses the virtual guidance point, when the look-ahead distance depends on many vehicle properties, including the vehicle speed.

The proposed driver model is illustrated in the Fig. 3.a. A hypothetical curve is shown by the blue line that represents a driver's guide for visual reference starting from point ***0*** to point ***G***, used when it is necessary to change direction. That is, the guidance point in the process of vehicle motion slips along this conditional trajectory. Consider a generalized variant, when the vehicle mass center has current displacements ΔX, ΔY in this curve's fixated coordinate system, and a current course angle ϕ (Yaw Angle) is determined by the rotation of the local coordinate system ***ξCμ*** around the coordinate system ***x'Cy'***, which slides parallel to the global coordinate system (Fig. 3.a). In addition, assume that the driver intuitively holds a certain area in his/her field of vision, which is limited by the sight angles and radius ***R***, thus forming the circle sector. If the hypothetical trace of the guidance point ***G*** is given by function *y = f(x)*, then taking into account the equation of circle, the required value *x* can be derived from the following equation:

[9] $\sqrt{R^2-(x-\Delta X)^2}-\Delta Y-f(x)=0$

The line segment that combines the mass center ***C*** and the point of intersection of an arc with a hypothetical track of the guidance point sets the angle $\boldsymbol{\varphi}$ relative to the ***x'*** axis, which is defined as:

[10] $\varphi = arcsin\left(\left(y-\Delta Y\right)/R\right)$

The difference between the angles $\boldsymbol{\phi}$ and $\boldsymbol{\varphi}$ is $\boldsymbol{\Delta\phi}$, which must be compensated by a feedback circuit with a PID (proportional-integral-derivative) controller. Fig. 3.b depicts the main functions reduced to a normalized scale, which are proposed for the guidance spot trace. The *x*-axis is a longitudinal coordinate, which varies from ***0*** to ***L***, and the *y*-axis is a side offset of point ***G*** from ***0*** to ***B***. The primary requirement for these functions is a continuous derivative. The simplest option in this regard is the inverted and vertically shifted trigonometric cosine function, which has a zero *y* value and a zero derivative at the zero point.

[11] $y(x) = B\cdot\left(1-cos\left(x\cdot\pi/L\right)\right)/2$

Another function, which provides smoothness and smooth curvature changes, is a cubic polynomial with zero derivatives at the boundary points. Function y(x) in Eqs. (11) and (12) are closely adjoined.

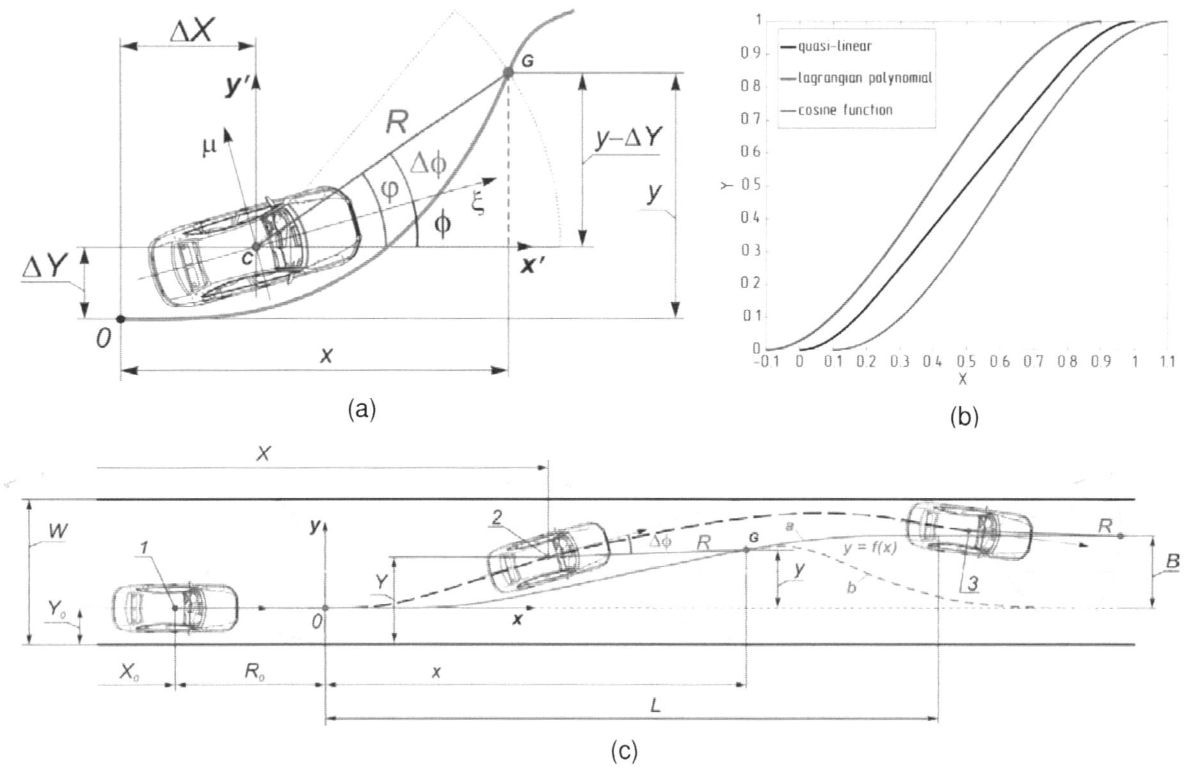

Figure 3: Elements of the proposed driving strategy: (a) Scheme for creating the desired course relative to the current direction of movement; (b) Typical normalized tracking curves for guidance spot; (c) Scheme of the formation of the vehicle mass center trajectory during the lane change

$$[12]\ y(x) = B\cdot\left(3\cdot L\cdot x^2 - 2\cdot x^3\right)/L^3$$

The last function is composed of three parts: a cubic polynomial, Eq. (12), with a zero-value function and a zero derivative for the left boundary point; a value of the function, and a derivative at the right boundary point matching the corresponding left boundary parameters of the following linear component function. The steps are then repeated in the opposite direction using polynomial Eq. (13) instead of Eq. (12).

$$[13]\ y(x) = B\cdot\left(2\cdot x^3 - 3\cdot L\cdot x^2 + L^3\right)/L^3$$

We then obtain a quasi-linear function with smoothed conjugation. The description of the formation of the vehicle mass center trajectory is based on Fig. 3.c, where the lane change process is shown according to the handling strategy on a two-lane highway. The proposed glide function $y = f(x)$ of the virtual guidance point ***G*** has a longitudinal basis ***L*** and a transversal basis ***B***. In order to immediately change the direction of spot ***G*** without a sharp transition between the virtual curves ***a*** and ***b***, the condition of derivative continuity in the current ***G*** point must be satisfied. Consequently, both the guidance track curve $y = f(x)$ and its derivative *dy/dx* should be calculated simultaneously. Thus, the driver's decision to continue or to abort an overtaking maneuver would be possible at any moment within curve basis ***L***. The expression for curve b is as follows:

$$[14]\ y(x) = y_0\cdot\left(2\cdot x^3 - 3\cdot L\cdot x^2 + L^3\right)/L^3 + dy_0\cdot\left(x^3 - 2\cdot L\cdot x^2 + L^2\cdot x\right)/L^2$$

where y_0 and dy_0 represent the function and its derivative for the left end of the base L, respectively. It is assumed that the function and derivative of the right end are equal to zero. This indicates a return to the initial location within the lane.

4 Simulink Model Description

4.1 Collision Warning System: Simulink Model for Overtaking Cases Involving Trucks

The general Simulink model for overtaking scenarios involving a truck as the impeding vehicle consists of three components which are combined by data exchange buses (Fig. 4.a). Blocks 1, 5 (Single Unit Vehicle) and 9 (Articulated vehicle (truck + semitrailer)) represent the 2D-dynamic steerability models for passenger cars and articulated vehicles. These blocks implement the mathematical models that describe the vehicle plane dynamics. Each vehicle block-model has an extended inner hierarchy of subsystems that are based on the Simulink library elements. This can be adapted for any vehicle configuration, taking into account geometric, physical, and technical characteristics. Each unit has one input and one output complex port for more compact data transfer. The input data are generated using blocks 4, 8, 13 - bus creator. Each vehicle model has a block with initial conditions (positions 2, 6, 11), which specify the initial kinematic parameters (displacements, velocities) of the vehicle movement.

The control models (positions 3, 7, 12) form the signals corresponding to the vehicle speed modes and the steering wheel angles, allow for independent control of the vehicle models. In Block 10 - Road Cond, the general parameters of the road conditions, such as the rolling resistance coefficient and the maximum coefficient of the road adhesion, can be specified. All the kinematic output data from each particular vehicle block (displacements, velocities, accelerations) are combined in block 14 - Results. From this block, the parameters involved in the functioning of the proposed collision avoidance system are directly selected along with data that affect the control algorithm (driver behaviour) for a passing vehicle. The output variables associated with the activation buttons for creating visualizations or filming can be found in block 15 - Vis log.

4.2 Control-model Development

Fig. 4.b shows a block structure of the control signal formation, which sets the direction of vehicle movement. Longitudinal displacement signals are received in input port 1, marked by block *1 - X*; Transverse displacement signals are received in port 2 of block *2 - Y*; The longitudinal velocity signal of the local vehicle coordinates is received in input port 3 of block *3 - Vx*. Port 4 (block *4 - Yaw*) transmits the course angle.

All of these parameters are the numerical solution results of the SUV's motion differential equation system. In block 5 - Driving strategy, the lane changing algorithm is implemented using Stateflow. This algorithm also contains the scheme in Fig. 4.d. The output parameter of the Stateflow chart is the desired direction angle DesYaw (output port 2). The error between the angle that was predicted by the driving strategy and the actual yaw angle is evaluated in block 6 - Sum. The PID-controller (block *7 - PID Controller*, see Fig. 4.c) generates a deviation compensative signal. Block 8 - Transport Delay, allows for the avoidance of the occurrence of the detrimental algebraic circuit, and simulates the delay associated with the driver's reaction time. The Limitation 9 - Saturation restricts the steering wheel rotation angle within rational boundaries. Output port 1 (block *10 - StAng*) transmits the steering wheel angle to outside, where it is converted into driver wheel rotational angles by corresponding subsystems in the vehicle model structure.

4.3 Simulink-function solver to search the guidance track point

Fig. 4.d illustrates the Simulink-function, which is part of the Stateflow chart, for the determination of the current guidance point coordinates. The input parameters are as follows: ports 1, 2 (blocks 1 - DX, 2 - DY) receive the vehicle mass center's relative displacements in the guidance point function coordinate system; input ports 3 - L, 4 - B – receive the corresponding values that parameterize the virtual track function in the space (on the highway lanes); the input ports 5 - R_{min}, 6 - R_{max} receive the values of the minimum and maximum radii, which form the changing sight distance range; input port 7 – Vmax, is the signal of a vehicle's maximum conditional speed corresponding to the most outlying guidance point in the space ahead of the vehicle.

Input block 8 - V is the vehicle's current longitudinal velocity. Block 9 - Variant function, offers one of the three guidance curves defined in Fig. 3.b, and block 10 - Circle function, determines the arc of a ring with a moving center coinciding with the vehicle's mass center. In block 11 – Subtract, the difference between these two functions is computed. Block 12 - Algebraic Constraint, uses iterations to find the value of the output parameter (which is the current *x* to be found), which provides a zero function subtract at the input of this block. The current distance to the guidance spot is recalculated repeatedly in block 13 - Current Radius. The determination of the desired angular position value is carried out in block 14 - Required Angle, and is transmitted to the output port 1 of the 15 - ang block. Lastly, the output port 2 of block 16 - D transmits the derivative of the virtual curve track to the outside, and the output port 3 of block 17 - r sends the current radius R.

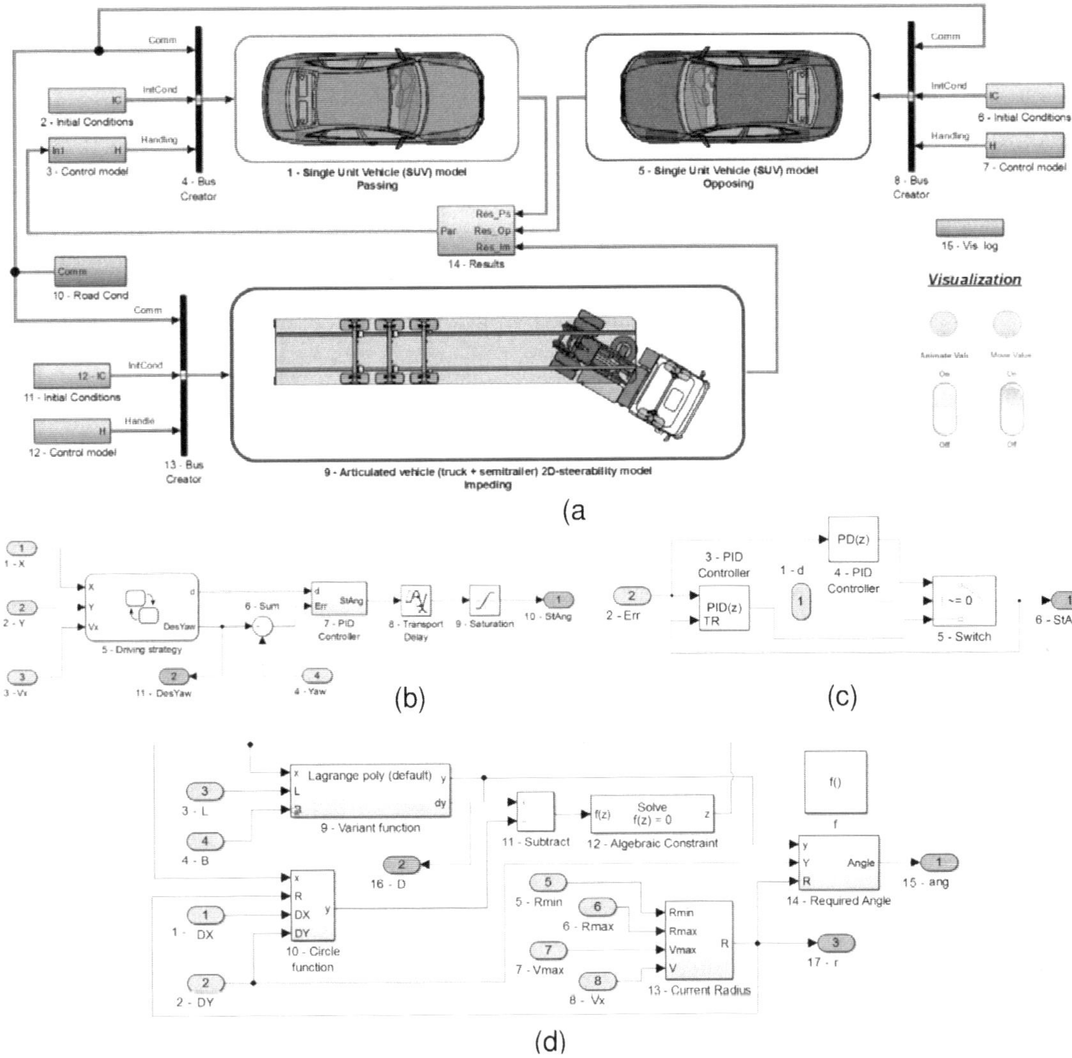

Figure 4: Simulink-model of overtaking scenario: (a) Simulink model structure of a truck overtaking scenario on a two-lane highway; (b) Structural Simulink-model of automatic steering angle control; (c) Internal structure of the PID-controller; (d) The structure of the Simulink-function solver for intersection point search between the arc and the guidance point functions

5 Application Example

An application example is provided in this section in order to illustrate the methodology used in this study. The input data for the passing vehicle from field studies for Passing Collision Warning Systems (PCWSs) included the following: (1) initial speed of 20.83 (m/s); (2) initial distance of 480 (m); (3) initial acceleration rate of the opposing vehicle of 0.0 (m/s^2); (4) detection sensor angle of 90°; (5) gender 0 (male = 0), age of 27 (years), driving experience of 10 (years) and average weekly driving hours of 30 (hours) for the passing vehicle driver; and (6) the number of lanes is 2, lane width of 3.75 (m), and design speed of 22.22 (m/s), which is equal to 80 (km/h) in highway geometric design.

The example assumes that the advisory system is using a detector with a frequency of 100 Hz and precision levels of 0.1 m and 0.1 for the reading distance and azimuth angle, respectively. The first detector was installed on the right side of the front bumper of the passing vehicle. The setback distance (d_{12}) was approximately 31.5 m from the front bumper of the passing vehicle to the front bumper of the impeding vehicle. The second detector was installed on the left side of the front bumper of the passing vehicle and the setback distance was approximately 475 m from the front bumper of the passing vehicle to the front bumper of the opposing vehicle.

The initial detection measurements for the opposing vehicle were obtained at a distance (d_1) of 475.3 m, a sensor ray angle (θ_1) of 0.375°, and a speed of approximately 20.83 m/s. The second detection measurements: d_2 = 474.9 m, θ_2 = 0.3752°, speed = 20.82 m/s. The third detection measurements: d_3 = 474.5 m, θ_3 = 0.3755°, speed = 20.8 m/s. The final distance between the opposing vehicle and the detector: d_f = 474.1 m, θ_4 = 0.3759°, and the speed was 20.79 m/s. The distance between the opposing vehicle and the conflict point was 220.3 m. Since the impeding vehicle was travelling at a near-constant speed, the time required for the passing vehicle to overtake the impeding vehicle ($t_{passing} = t_1 + t_2$), computed using Eq. (1), was 9.6 s. The time required for the opposing vehicle to reach the conflict point ($t_{opposing}$), computed using Eq. (8), was 11.4 s. The difference between $t_{opposing}$ and $t_{passing}$, the 1.8 s safety margin, was larger than the critical 1.4 s (95% confidence interval). This safety margin was significantly larger than the measurement error associated with the precision level (0.002 s). The warning message displayed to the passing driver can therefore be deactivated.

It is important to note that the warning message would not have been deactivated if the opposing vehicle was detected at a shorter distance (less than 460 m) from the passing vehicle or if it was travelling at a higher speed (higher than 22.22 m/s). It is also important to note that the warning message would not have been deactivated if a higher confidence level had been used (higher than 99%) or if the precision levels of the detector were above 0.1 m and 0.1 when measuring the distance and the angle, respectively. Fig. 5 shows the example of numerical safety margin time processing.

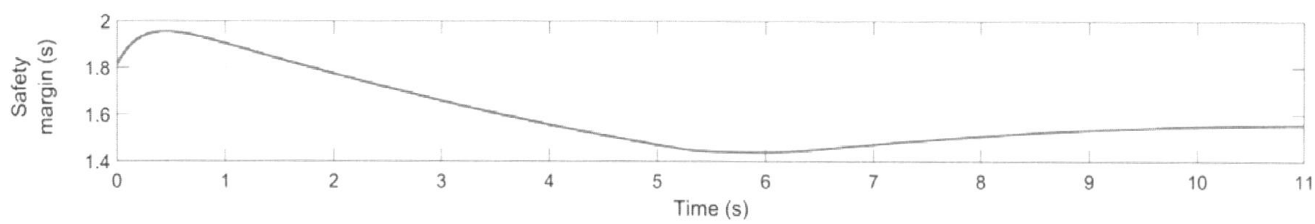

Figure 5: Example of numerical safety margin time processing

6 Conclusions

This paper presents the framework and algorithm design of a collision warning system intended to help drivers conduct safe passing maneuvers on two-lane highways.

1. The system makes use of a detector (radar sensor or laser scanner) to measure the speed and location of the closest opposing vehicle on a two-lane highway at two successive time intervals in order to discern the distance to the conflict point, as well as the speed and rate of acceleration.

2. The algorithm used in this system calculates the time required for the passing vehicle to complete the pass. The algorithm then ascertains whether or not there is any possible conflict between the opposing and passing vehicles and displays a warning if a conflict is detected. An in-vehicle detector (placed in the passing vehicle) is used to activate the system and launch the algorithm.

3. The algorithm considers the time required for the driver to notice the warning displayed by the system and react to it (to start or abort the passing maneuver).

4. Mathematical models were developed in order to estimate the perception reaction time. The initial acceleration, passing time, and acceleration rate of passing drivers were obtained using data collected from field experiments in which a GPS data logger device recorded the position and speed of various vehicles at 1s intervals.

5. The results revealed that 95% of drivers have a safety margin time of 1.5 s or more. The inclusion of the safety margin time in the system's algorithm may increase the reliability of the system by decreasing the amount of mistimed warnings.

This study presents the development of a Matlab Simulink model that simulates passing scenarios in order to evaluate the performance of the system when different reading error levels are present. This mathematical model can be used in any similar passing collision warning systems as they are independent of technology. The results of this study are also applicable to detectors with reading errors that exceed those suggested in this study, although the proper safety margin threshold should be utilized (see Fig. 5). Further research will be required in order to identify various hardware components that meet the minimum requirements of the proposed system.

Acknowledgements

The financial support of the Natural Sciences and Engineering Research Council of Canada (NSERC) is acknowledged.

References

American Association of State Highway and Transportation Officials. (2011). *A Policy on Geometric Design for Highways and Streets.* AASHTO, Washington, D. C.

Dabbour, Essam and Easa, Said M. (2014). *"Proposed collision warning system for right-turning vehicles at two-way stop-controlled rural intersections"*, Elsevier Journal for Transportation Research, Part C 42. Vol. 121-131.

Granet, F., Picado, R., and Smith, L. (2003). *Longitudinal Collision Avoidance.* Institute of Transportation Studies at the University of California at Berkeley, CA, USA.

Hassein, U. (2017). *Deveopment of Framework for In-Vehicle Passing Collison Warning System Considering Driver Characteristics.* Ph.D thesis, Ryerson University, Toronto, Canada

Hassein, U., Diachuk, M., Easa, S. (2017). *"Evaluation of Dynamic Passing Gap Acceptance on Two-Lane Highways Using Field Data"*, Canadian Journal of Civil Engineering, Accepted for Publication. https://doi.org/10.1139/cjce-2016-0572

Isermann, R., Mannale, R., Schmitt, K. (2012). *"Collision-avoidance systems PRORETA: Situation analysis and intervention control"*, Control Eng. Pract., Vol. 20, pp. 1236-1246.

Kretschmer, M., Neubeck, J., Wiedemann, J. (2005). *"Combining Vehicle Dynamic Control and Decision Making Procedures toward Safe Overtaking Maneuvers".* SAE International, 01-0412.

Mehmood, Atif. (2010). *Development of framework for in-vehicle rear-end collision warning system considering driver characteristics.* Ph.D thesis, Ryerson University, Toronto, Canada.

SAS. (2015). SAS: *Enterprise Guide*. Retrieved from http://www.sas.com/technologies/bi/query_reporting/guide/

Building Tomorrow's Society
Bâtir la Société de Demain

Fredericton, Canada
June 13 to 16, 2018 / *13 au 16 juin, 2018*

EFFECT OF RAINY DARKNESS INTENSITY AT TWO-LANE HIGHWAY ON TRAVEL TIME DIFFERENTIALS

Makinde, Opeyemi Oluyemisi[1,3] and Ben-Edigbe, Johnnie[2]
[1] University of KwaZulu-Natal, South Africa
[2] University of KwaZulu-Natal, South Africa
[3] 216073398@stu.ukzn.ac.za

Abstract: In this paper travel time differentials prompted by rainy darkness on Two-Lane Highway were investigated. The investigation raises the issue of night travel on roads without lights. Even though drivers react to rainfall and darkness simultaneously, studies have not been carried out insufficiently. Hence the paper wants to provoke debate on this issue. Based on the hypothesis that rainy darkness irrespective of intensity will cause travel time increase, with and without rainy darkness studies were carried out at three selected sites in Nigeria during the rainy season (June-August). Traffic and rainfall data were collected continuously for eight weeks. Rainfall intensity was classified according to the conventional rate of precipitation: Light rain — when the precipitation rate is < 2.5 mm (0.098 in) per hour. Moderate rain — when the precipitation rate is between 2.5 mm (0.098 in) - 7.6 mm (0.30 in) or 10 mm (0.39 in) per hour. Collated data on which the study was based were analyzed. Travel times for with and without rainy darkness were estimated and compared. Results show that there is speed reduction of 4.1% due to light rain effect, 8.6% and 14.8% due to moderate and heavy rain effect respectively. Further, travel time increase relative to speed decrease during rainy darkness, thus suggesting that motorists reduce vehicle speed when it is dark and rainy irrespective of intensity. The paper concluded that rainy darkness has a significant effect on travel time.

1 Introduction

The highway is an important element of transportation because it connects all other forms of transportation. Traffic flows on the highway are expected to be safe, efficient, give good driving comfort and with minimum delay. Vehicle performance on highway depends on the road geometry, traffic volume, traffic composition and weather condition like fog, snow, wind, darkness and rainfall. These weather conditions can be mild and could be harsh depending on the intensity of rain and fog, wind speed and their duration. Harsh weather condition can bring about total traffic flow disruption on the highway. The weather impacts on traffic flows affect delays, safety, travel demand, and road accident (FHWA, 2008 and Cools *et al.* 2010). Weather impact on traffic flow may increase accident, delay, speed reduction and traffic flow contraction (Alhassan and Ben-Edigbe, 2014). Rainfall disrupts traffic flow by reducing driver's visibility, thereby leading to speed reduction, and increase in headways (Ben-Edigbe, 2010). Other studies by Chung et al, (2005) and Mukhlas et al (2016) showed that increase in rainfall intensity causes speed and capacity decrease and increase in travel time. Under darkness condition where there is no road lightning, there is poor or no illumination, drivers are dependent on the performance of their vehicle headlights. This condition affects the driver's visibility because of deterioration in colour vision which could lead to a reduction in sight distance. If rainfall is thrown in the mix, then driving conditions deteriorate even further. Rainy darkness depicts rainfall and darkness conditions. When driving under rainy darkness, the driver will be subjected to the combined effect of rainfall and darkness. This might make the reaction of the driver to rainfall under daylight to be different from under darkness. No research has been carried out on the effect of rainy-darkness on the travel time. Hence, the need for this study. This study is not about comparing the difference in the rain effect under daylight and darkness but the rainfall effect on travel time under darkness. Based on the hypothesis that travel time increases irrespective of rain intensity under dark weather condition, the remainder of this paper is divided four sections; the immediate section is the literature review, section 3.0 is on materials and methodology, and section 4.0 is the findings and discussion. In section 5, conclusions are drawn.

2 Literature Review

Rainfall as a major cause of road hazard occurs more frequently than other weather conditions (Alhassan and Ben-Edigbe, 2011a). The spatiotemporal nature of rainfall accounts for its effects on traffic flow. The effects of rainfall on road traffic majorly are poor visibility and aquaplaning which may result in flow contraction, speed reduction and increase in headways. The extent of the rainfall effect is a function of the variability in rainfall intensity. These rainfall problems affect the behaviour of traffic on road. The behaviour of traffic flow is crucial and important in planning, designing and managing of highway and its facilities. The major traffic flow parameters are speed, flow, density and headway. Travel time is a function of speed and traffic flow with optimum interest for highways planners and designers. Traffic parameters are used to analyse traffic flow processes and highway performances with respect to changes in any or all the traffic parameters. As contained in previous studies the fundamental relationship traffic stream parameters is expressed in equations 1 and 2.

[1] $q = uk \Rightarrow u = \frac{k}{q} \Rightarrow k = \frac{q}{u}$

[2] $u = u_f - \varphi k$

where: q = flow, u = mean speed, k = density, u_f= free-flow speed, and φ = a coefficient.

The parameters speed and density are empirically related and are used to evaluate the quantity of flow on a roadway and to assess the congestion level of a freeway. These parameters though inter-related are dependent on various factors such as vehicle type, size and composition, weather conditions such as rainfall, snow, ice, fog, daylight and darkness. The individual effect of the various factors varies in magnitude and time of occurrence.

2.1 Travel Time

Travel time described as the average time spent by vehicles in moving through a highway segment (HCM,2000). Travel time is important to both road users and highway designers in measuring road performance. Travel time used together with delay and capacity, is used for measuring the effectiveness of roadway (Mashros et al 2012). The studies showed that travel time effectiveness is affected by road, traffic and rainfall under daylight conditions.

Travel time equations are shown in 3 and 4 below:

[3] $u = \frac{d}{t_f} \Rightarrow t_f = \frac{d}{u}$

where: u = speed, d = distance, t_f = travel time at free flow.

t_f occurs when density is zero i.e no vehicle on the road and/or operating speed are determined and controlled by drivers themselves. In general, prediction of travel time over a length of roadway is calculated using the equation postulated by US Bureau of Public Roads (US-BPR).

[4] $T = t_f\{1 + \rho(x)^{\beta}\}$

where T = predicted travel time over the length of roadway

t_f = travel time at free flow speed

x = degree of saturation (v/Q) (defined as 115% of speed at capacity)

v = demand flow ($q = uk$)

Q = capacity (80% of actual capacity)

ρ = ratio of free-flow speed to the speed at capacity

β = abrupt drop of curve from the free flow speed

Equation 4 can be re-written as:

$$[5]\ T = t_f\left\{1 + \rho\left(\frac{v}{Q}\right)^{\beta}\right\}$$

BPR further stated that a high value of β causes speed to be unresponsive to "x" but as "x" tends to 1, the speed drops suddenly. BPR 1965 version initially proposed the values 0.15 and 4 for "ρ" and "β" respectively in the travel time equation. However, studies conducted on the BPR curve by Dowling and Skabardonis (1993) reported an underestimation of speed and re-plotted the BPR equation by proposing 0.2 for ρ and 10 for β. The new proposed values were later adopted and updated by BPR 1994.

For this study, our focus is on off-peak traffic period which implies that the degree of saturation v/c < 0.9, the values ρ = 0.2 and β = 10 are acceptable. The choice of off-peak is premised on the reasoning that peak travel data is biased and skewed towards unstable traffic flow. Equation 5 is appropriate for dry and rainy darkness conditions and can be re-written as

$$[6]\ T = t_f\{1 + 0.2(x)^{10}\} \Rightarrow T = t_f\left\{1 + 0.2\left(\frac{v}{Q}\right)^{10}\right\}$$

From equation 6, the parameter Q known as capacity is very important in road design both in terms of planning and management. It plays an indispensable part in defining the performance of a road. Estimation of road capacity is used in assessing the conditions of the road relative to congestion, delay and the number of traffic expected to be involved. Capacity is the maximum hourly flow rate by which vehicles or persons are expected to pass a certain point or uniform section of a roadway for a certain period of the specific roadway, traffic, geometric, control and environmental conditions (HCM, 2000).

Estimation methods of roadway capacity using traffic headway, volume, speed and density data include headway, product limit, bimodal distribution, selected maxima and fundamental diagram (Minderhound, 1996). Nevertheless, the use of any method is a function of traffic data available. Fundamental diagram employs the relationship existing between the three macroscopic flow parameters which are volume (q), speed (u) and density estimate capacity. The fundamental relationships constructed using any two variables are speed-density, flow-density and flow-speed diagrams. The speed-density is fitted to a linear function while a quadratic equation represents the flow-density.

With the flow-density relationship, it is possible to estimate capacity, critical density (density at which capacity is reached) with its corresponding traffic mean speed and jam density (density at which traffic volume and mean speed is zero). The capacity value can be obtained from the fundamental diagram or by finding the derivative value of the quadratic function. The capacity equation is written as:

$$[7]\ Q = -c + (v_f)\frac{v_f}{2\left(\frac{v_f}{k_j}\right)} - \frac{v_f}{k_j}\left(\frac{v_f}{2\left(\frac{v_f}{k_j}\right)}\right)^2$$

The capacity equation above has been shown in past studies (Ben-Edigbe and Ferguson, 2005, Anais et al., 2010) to be reliable for off-peak traffic stream characteristics.

Now plug equation 7 into equation 6.

$$[8]\ T = t_f\{1 + 0.2(x)^{10}\} \Longrightarrow T = t_f \left\{ 1 + 0.2 \left(\frac{v}{-c + (v_f)\frac{v_f}{2\left(\frac{v_f}{k_j}\right)} - \frac{v_f}{k_j}\left(\frac{v_f}{2\left(\frac{v_f}{k_j}\right)}\right)^2} \right)^{10} \right\}$$

Recall that $t_f = \frac{d}{v}$ and $u = u_f - \varphi k$, therefore equation 8 can be re-written as:

$$[9]\ T = \frac{d}{u_f - \varphi k} \left\{ 1 + 0.2 \left(\frac{uk}{-c + (v_f)\frac{v_f}{2\left(\frac{v_f}{k_j}\right)} - \frac{v_f}{k_j}\left(\frac{v_f}{2\left(\frac{v_f}{k_j}\right)}\right)^2} \right)^{10} \right\}$$

Equation 9 above can then be used to determine the travel time for both dry and rainy conditions based on the advantage of the fundamental diagram which allows traffic state to be determined at any point along the road section irrespective of whether capacity is reached or not.

2.1.1 Impact of Rainy Darkness on Travel Time

Rainy darkness can affect the travel time of a vehicle on a roadway. Based on past research work, darkness is known to affect visibility which impacts negatively on travel time while rainfall irrespective of its intensity leads to speed reduction. Alhassan and Ben-Edigbe (2011b) described the impact of rainfall as "clouding effect" that leads to speed reduction. Often drivers tend to reduce their speed once darkness set in due to poor and limited visibility, but a situation of darkness coupled with rainfall i.e. rainy darkness calls for an in-depth knowledge. The extent of the rainy-darkness effect on travel time could be argued to depend on the intensity of the rainfall at the period of darkness. Vehicles travelling in darkness irrespective of the fact that the degree of saturation < 0.9 (i.e off-peak) road are bound to reduce their speed which consequently leads to travel time increase.

3 Methodology

Three sites were selected in Nigeria for this rainy-darkness impact study. The selected road sections were characterised with a good drainage system, straight and flat terrain, void of attractions and free of structural pavement deterioration. The selected roads were without road lightning. Rainfall precipitation was collected with a rain gauge and the precipitation amount was converted to rain intensity. Rainfall was divided into three intensity classes viz light (i < 2.5mm/hr), moderate (2.5 ≤ i < 10 mm/hr) and heavy (10 ≤ i < 50 mm/hr) in accordance with World Meteorological Organisation (WMO) rainfall classification. Very heavy rainfall (i > 50mm/hr) was not considered for the study because of drag force, aquaplaning and water splash on vehicles which might cause aggression and anxiety in driver's reaction. These effects might be difficult to separate from the rainy-darkness effect on travel time. As shown in figure 1, automatic traffic counter (ATC) was used to collect traffic volume, headway, type of vehicle, speed, date and time of vehicle tires hit on ATC pneumatic tubes continuously for eight weeks during the rainy season at the selected sites. The continuous traffic data collection allows traffic under darkness to be captured. Off-peak traffic data were used for the study in order to separate the effect of peak traffic flow from rainy-darkness effect. The separated off-peak traffic data was synchronised with equivalent rainfall data obtained.

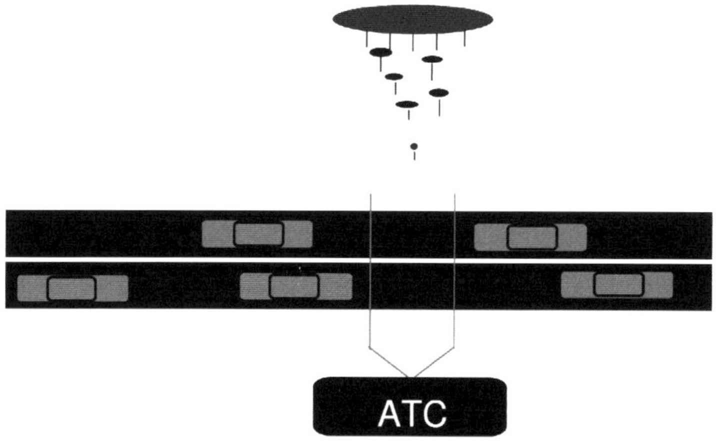

Figure 1: Set up of rainy darkness impact study

4 Results and Discussions

The stepwise method of analysis is used for simplicity and clarity.

Step 1: The traffic and rainfall data collected were separated into daylight and darkness. The traffic data under darkness falls into the off-peak period which eliminates the effect of the peal period. The rainfall data under darkness was classified into light rainfall ($i < 2.5$mm/hr), moderate rainfall ($2.5 \le i \le 10$mm/hr) and heavy rainfall ($10 \le i \le 50$mm/hr) based on its intensity. The traffic data under darkness were separated into traffic data with and without rainfall. The traffic data under darkness was synchronized with the rainfall data to generate rainy-darkness traffic data. The rainy darkness traffic data was divided into a passenger car, light vans, and heavy trucks.

Step 2: The passenger car equivalent was modified under rainy darkness. The modified passenger car was used in the conversion of traffic volume to traffic flow.

Step 3: The speed and traffic flow were used in computing the density using equation 10.

[10] $q = uk$

As an example, when the speed = 71km/h, and flow = 540pce/h,

Then density (k) = $\frac{q}{u}$ = $\frac{540}{71}$ = 8pce/km

This was estimated for all conditions under dry, light, moderate and heavy rainy conditions under darkness.

Step 4: The estimated density was related to traffic flow to determine the speed - density flow quadratic relationship as shown in figures 2 to 5 under dry and rainy darkness conditions. The model equations were statistically tested. The coefficient of determinant (R^2) was more than 0,5 which shows that the variables are significant, the two-tail t-test result if more than 2.2 which shows that the variables are significant, the F-test is more than 4.0 which suggests that the equation did not occur by chance. This show that the model equations are statistically satisfactory and could be used for prediction.

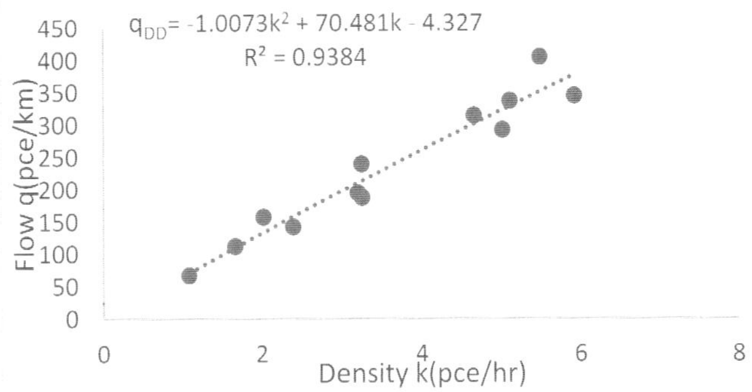

Figure 2: Flow-density relationship of dry darkness condition

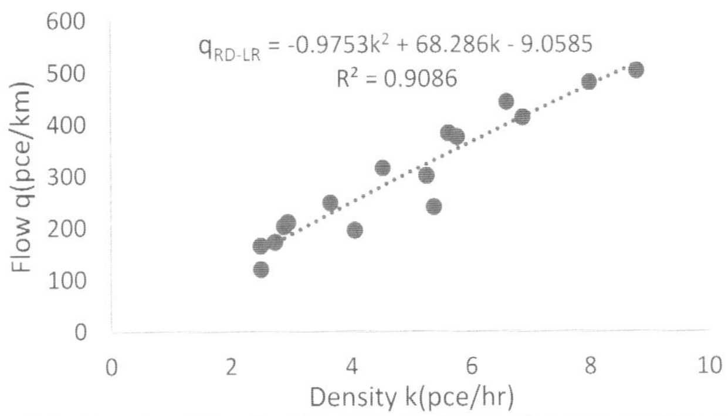

Figure 3: Flow-density relationship of rainy darkness under light rainfall

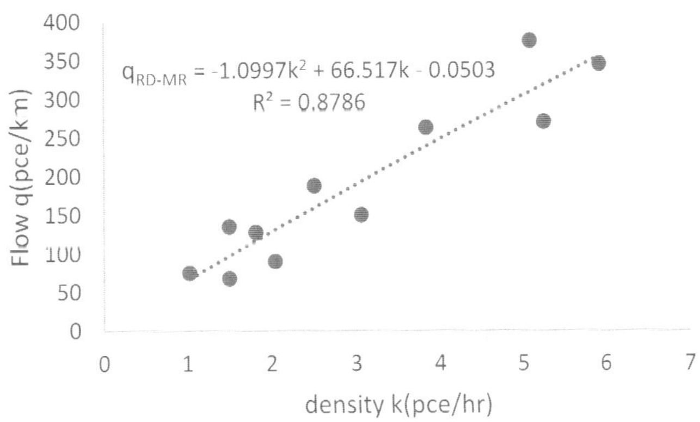

Figure 4: Flow-density relationship of rainy darkness under moderate rainfall

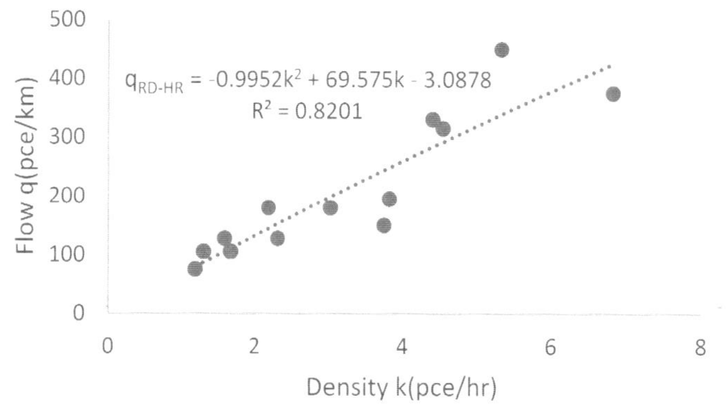

Figure 5: Flow-density relationship of rainy darkness under heavy rainfall

Step 5: The capacity was computed using the model equation. Under dry darkness, the model equation is

[11] $q_{DD} = -1.0073k^2 + 70.481k - 4.3270$

Equation 11 was differentiated as shown in equation 12 and the result is set to 0 to determine the critical density because the maximum flow occurs when the critical density is reached.

[12] $\frac{dq}{dk} = -2\ x\ 1.0073k + 70.481$

$$-2.0146k + 70.481 = 0$$

$$k = \frac{70.481}{2.0146}$$

$$k = 34.99\ pce/km$$

$$q_{DD} = -1.0073(34.99^2) + 70.481(35.01) - 4.327$$

$$= 1228.91\text{pce/h}$$

The capacity is estimated under dry and rainy darkness at the three selected sites. The summary of the capacity result is presented in table 1.

Step 6: The maximum speed was estimated using equation 10. Under dry darkness, the speed is calculated as

$$S = \frac{1228.91}{34.99}$$

= 35.12km/hr

The maximum speed is estimated under dry and rainy darkness for the three sites. The summary of speed result is presented in table 1.

Step 7: Travel time for dry and rainy darkness conditions were calculated using equation 13

[13] $T = \frac{d}{u_f - \varphi k}\left\{1 + 0.2\left(\frac{uk}{-c + (v_f)\frac{v_f}{2\left(\frac{v_f}{k_j}\right)} - \frac{v_f}{k_j}\left(\frac{v_f}{2\left(\frac{v_f}{k_j}\right)}\right)^2}\right)^{10}\right\} \Rightarrow \frac{d}{v}\left\{1 + 0.2\left(\frac{v}{Q}\right)^{10}\right\}$

$$T_{DD} = \frac{60}{35.12}\left\{1 + 0.2\left(\frac{1228.91}{1228.91}\right)^{10}\right\}$$

$= 2.05\text{min}$

Typical highway maximum flow calculations from the model coefficients in Table 3 are shown below and the remainder results are shown in Table 4: Where flow and density quadratic function is used:

The travel times for dry and rainy conditions for the selected sites were computed. Table 1 shows the summary of the model equations, maximum flow, speed, travel time and travel time differentials for with and without rainfall under conditions for the selected sites.

Table 1: Estimated flow-density model equations and travel time for rainy-darkness

Site	Weather condition	Model Equation	*Q (Pce/h)	*U (km/h)	*T (min)	*ΔT (min)
1	Dry darkness	$q = -1.0073k^2 + 70.481k - 4.3270$	1229	35	2.05	N/A
	Light rain (Darkness)	$q = -0.9753k^2 + 62.286k - 9.0585$	1186	34	2.28	0.23
	Moderate rain (Darkness)	$q = -1.0997k^2 + 66.512k - 0.0503$	1006	33	4.49	2.24
	Heavy rain (Darkness)	$q = -0.9952k^2 + 60.575k - 3.0878$	918	30	8.45	6.40
2	Dry darkness	$q = -0.9872k^2 + 71.257k - 4.3292$	1281	36	2.03	N/A
	Light rain (Darkness)	$q = -0.9895k^2 + 65.753k - 7.8339$	1085	33	3.79	1.76
	Moderate rain (Darkness)	$q = -0.9939k^2 + 62.844k - 6.3467$	987	31	7.13	5.10
	Heavy rain (Darkness)	$q = -0.9150k^2 + 60.443k - 0.9252$	997	30	6.86	4.83
3	Dry darkness	$q = -1.024k^2 + 75.176k - 1.0089$	1379	38	1.92	N/A
	Light rain (Darkness)	$q = -1.1001k^2 + 74.753k - 4.7826$	1265	37	2.37	0.45
	Moderate rain (Darkness)	$q = -1.0420k^2 + 68.809k - 0.057$	1136	34	4.17	2.25
	Heavy rain (Darkness)	$q = -0.9953k^2 + 63.575k - 4.0848$	1015	32	9.98	8.06

*Q ~ Capacity, U ~ optimum speed, T ~ Travel time, ΔT ~ Travel time differential.

The result in table 1.0 shows that the capacity and speed reduce with the increase in rain intensity under darkness. The reason is that as the rain intensity increases, the drivers become more cautious due to increase in impaired visibility and reduction in frictional coefficient between tires and road pavement. As the rain intensity increases, the travel time increase as shown in table 1.0. This is because as the drivers become more cautious by reducing the speed and keep the gap from the leading vehicle. It is expected that as the gap increases with reduction in speed, the travel time might increase. Under light rainy darkness, the travel time increases from 2.05min to 2.28min with an increase of 0.23min at site 01. It increases from 2.03min to 3.79min with an increase of 1.76min at site 02. At site 03 the increase is from 1.92min to 2.37min with an increase of 0.45min. Under moderate rain the increase in travel time is 2.24min, 5.10min, 2.25min at sites 01 to 03 respectively. Under heavy rainy darkness the travel time increase by 6.40min, 4.83min, 8.06min at sites 01 to 03 respectively. The results follow the same trend at all the sites as the travel time increases with increase in rain intensity under darkness. At site 02 the travel time under moderate and heavy rain are close in value. This is because the heavy and moderate rain at this site might be at the borderline of the moderate and heavy rain. This rain intensity at the borderline cannot be taken as fixed because of the rain fluctuation. The moderate and heavy rain intensity at this site might be fluctuating between the moderate and heavy rain. At all the sites, the travel time differential under light rainy darkness has an insignificant travel time differential, this is because light rain with the intensity of less 2.5mm/h has little effect on the driver's visibility and at times the rain drizzle is taking as light rain and drivers react less to light rain.

Taking the average, the speed reduces from 36.06km/h to 34.59km/h with reduction of 1.47km/h or 4.1% under light rain. Under moderate rain it reduces to 32.96km/h with reduction of 8.6% and it reduces to 30.72km/h under heavy rain with reduction of 5.34km/h or 14.8%. The average travel time under dry darkness is 1.59min, 2.07min under light rainy darkness, 3.93min and 5.58min under moderate and heavy rainy darkness respectively. The travel time differential is 0.49min, 2.34min and 4.00min under the light, moderate and heavy rainy darkness conditions respectively. This shows that as the driver responds to rainfall effect by speed reduction, travel time increases with increase in rain intensity under darkness.

5 Conclusion

Based on the synthesis evidence obtained from travel time changes under rainy-darkness, drivers reduce speed as the rain intensity increases. The travel time increases in response to the speed reduction. The light rain has the lowest effect and heavy rain has the highest effect on travel time under darkness. It is therefore correct to conclude that as driver reduces speed, travel time increases with increase in rain intensity under darkness.

Acknowledgements

We are grateful to the University of KwaZulu-Natal Centre for Transportation Studies for proving an enabling data processing environment even though the traffic data were collected at sites in Nigeria, We are thankful to the university for funding the research studies and many others for their support especially during the challenging traffic data collection period.

References

Alhassan, H. M. and Ben-Edigbe, J. (2011a). Effect of Rainfall Intensity Variability on Highway Capacity. *European Journal of Scientific Research*, **49**(1): 18-27.

Alhassan, H. M. and Ben-Edigbe, J. (2011b). Highway Capacity Prediction in Adverse Weather. *Journal of Applied Sciences*, **11**(12): 2193-2199.

Alhassan, H. M. and Ben-Edigbe, J. (2014). Effect of Rainfall on Traffic Flow Shock Wave Propagation. *Journal of Applied Sciences*, **14**(1): 54-59.

Ben-Edigbe, J (2010). Assessment of Speed-Flow-Density Functions under Adverse Pavement Condition. *International Journal of Sustainable Development and Planning*, **5**(3): 238-252.

Ben-Edigbe, J. and Ferguson, N. (2005). The extent of Capacity Loss Resulting from Pavement Distress. Proceedings of the Institution of Civil Engineers -Transport **158**(1): 27-32.

Cools, M., Moons, E and Wets, G. (2010). Assessing the Impact of Weather on Traffic Intensity. *Weather, Climate and Society, American Meteorological Society*, **2**(1): 60-68.

Chung, E., Ohtani, O. and Kuwahara, M. (2005). Effect of Rainfall on Travel Time and Travel Demand. 5th ITS European Congress, Hannover, Germany.

Dowling, R. and Skabardonis, A. (1993). Improving Average Travel Speeds Estimated by Planning Models. *Transportation Research Board*, National Research Council, 68-74

FHWA (2008). "Road Weather Management Overview." http://ops.fhwa.gov./weather/overview. Assessed November 17, 2017.

Highway Capacity Manual (HCM) (2000). *Transportation Research Board*, National Research Council, Washington, DC.

Minderhound, M.M., Botma, H. and Bovy, P.H.L. (1996). An Assessment of Roadway Capacity Estimation Methods. Transportation and Traffic Engineering Section, Report No. VK 2201, 321/LVV 0920-0592, Delft University of Technology, Delft.

Mayeur A., Bremond R. and Christian Bastien, J. M. (2010). Effect of the Viewing Context on Target Detection, Implications for Road Lightning Design. *Journal of Applied Ergonomics*, 41(3): 461-468.

Mashros M., Ben-Edigbe J. and Rahman R. (2012). The extent of Highway Travel Time Differentials Resulting from Rainfall Intensities. *British Journal of Science and Technology*, 2(3): 254-265.

Mukhlas N. A., Mashros N., Puan O. C., Hassan S. A., Hassan N. A., Abdullah R. A. and Rahman R (2016). Effect of Rainfall on Traffic Flow Characteristics During Night Time. *Jurnal Teknologi (Science and Engineering)* **78**, 1-7.

Building Tomorrow's Society
Bâtir la Société de Demain

Fredericton, Canada
June 13 to 16, 2018 / *13 au 16 juin, 2018*

ANALYSIS AND MODELING OF NEW YORK AND CALIFORNIA SECTIONS IN LTPP DATA

Raffaniello, Alexa[1], Bauer, Matthew[2] and El-Hakim, Mohab[3,4]
[1] Graduate Student, Manhattan College, USA
[2] Former Undergraduate Student, Manhattan College, USA
[3] Assistant Professor, Manhattan College, USA
[4] mohab.elhakim@manhattan.edu

Abstract: This paper analyzed four Long-Term Pavement Performance (LTPP) sections located in California and New York. Two selected sections are located within 1 mile on I-5 in northern California and two sections located within 1 mile on US-4 in northern New York. Sections located in the same state were subjected to identical traffic spectrum but exhibited variation in structural capacity. Two levels of analysis were performed in this study: a) in-state analysis and b) cross state analysis. The in-state analysis focused on comparing the two sections located in each state. Sections located in California exhibited insignificant variation in thermal crack resistance regardless of the presence of additional 1.5" Asphalt Concrete (AC) layer in one section. On the other hand, the additional 1.5" AC layer resulted in significant reduction of rutting depth. The in-state analysis of sections located in New York concluded that chip seal layers has significantly low resistance to thermal cracking and rutting. The cross state analysis examined two sections of comparable structural capacity. The two structurally weak sections (sections 06-0606 in California and 36-A350 in New York) proved that chip seal layers exhibit significantly high rutting depth even under high traffic loads in freezing climate compared to low traffic sections with regular HMA in warmer climate. The analysis of strong sections in California and New York concluded that properly designed and constructed HMA in freeze climate exhibits equal thermal cracking resistance as sections in no-freeze zones. Furthermore, the paper presents pavement deterioration models for LTPP sections using linear regression.

1 Introduction

This paper investigates the deterioration of selective Long-Term Pavement Performance (LTPP) test sections. The selected test sections are located in New York and California. The selected sections were characterized by similarities in traffic spectrum and pavement layer's thickness. This paper investigated the deterioration rates of these sections to evaluate the impact of climate and pavement maintenance on the pavement functional and structural performance.

Two selected LTPP sections from California were part of the Specific Pavement Studies-6 (SPS-6). These sections were utilized to study the rehabilitation of Jointed Portland Cement Concrete (PCC) Pavement. The two LTPP sections selected from New York State were part of General Pavement Studies-6B (GPS-6B) and SPS-3. The two studies are designed to evaluate Asphalt Concrete (AC) on bound base and AC overlay on AC pavement, respectively. Further details about the four LTPP test sections are stated in subsequent segments of this paper.

The pavement functional and structural condition of the pavement sections were well-documented through frequent in-situ testing. The authors focused on transverse cracking and rutting as indicators of the pavement deterioration within pavement's service life. The analysis concluded that application of chip sealing significantly reduces the pavement's resistance to thermal cracking and has low rutting resistance compared to thin HMA overlay. In addition, the authors developed deterioration models for transverse cracking and rutting in four LTPP sections using multiple linear regression.

2 Background

The Federal Highway Administration (FHWA) manages the LTPP program since 1992. The program was initiated through the Strategic Highway Research Program (NCHRP) in 1986 (FHWA 2009). The program included several experiments on flexible and rigid pavements. The background section of this paper focuses on the SPS-6, SPS-3 and GPS-6B experiments as the analyzed sections were included in these studies.

2.1 SPS-6

The SPS-6 experiment was designed to study rehabilitation methods of Jointed PCC pavements. The test sections were spread to include two climatic regions (wet-freeze and wet no-freeze), pavement types (plain and reinforced concrete) and traffic loads. The rehabilitation techniques included Asphalt Concrete (AC) overlays using 4" and 8" layers. Limited slab preparation or full slab restoration were applied prior to AC overlays (FHWA 1992).

The experiment identified the most effective rehabilitation techniques to restore IRI, rutting and cracking. The treatments (ranked from the most effective to least effective) are (Hall 2002):

1. 8" AC overlay on cracked/broken and seated pavement
2. 4" AC overlay of either intact or cracked/broken and seated pavement with or without sawing and sealing of transverse joints and with minimal or intensive preoverlay repair.
3. concrete pavement restoration with diamond grinding, full-depth repair, and joint and crack sealing
4. concrete pavement restoration without diamond grinding but with full-depth repair and joint and crack sealing.

2.2 SPS-3

The SPS-3 experiment evaluates the effectiveness of preventive maintenance of flexible pavements. The experiment was designed to study several parameters including climatic zones, subgrade type (fine or coarse), traffic loading (greater or less than 85,000 ESALs/year), initial condition (good, fair, or poor), and structural adequacy (high or low). The preventive maintenance treatments included slurry seal, chip seal, crack seal and thin overlay (FHWA 2011).

The experiment concluded that chip seal and thin overlay are the most effective treatments for fatigue cracking. This overlay was noted as the most effective preventive maintenance for pavement rutting and delaying the roughness progression (FHWA 2011). The structural adequacy of pavement sections had no significant impact on the performance of maintenance treatments (Morian 1998).

2.3 GPS-6B

The GPS-6B experiment studied sections originally included in GPS-1 and GPS-2 experiments. The test section studied in this paper originally belonged to GPS-2. The GPS-2 examined dense-graded HMAC surface layer with or without other HMAC layers, placed over a bound base layer. The test sections included bituminous and nonbituminous bound base layers. Bituminous bases included asphalt cements, cutbacks, emulsions, and road tars. Nonbituminous bases included all hydraulic cements, lime, fly ashes, and natural pozzolans, or combinations thereof. The test sections were constructed to examine variations among climatic zones, subgrade types, traffic spectrum, and surface and base thicknesses (Elkins 2012). The GPS-6B sections received 1" overlay on the original pavement sections (Hall 2002).

The GPS-6B experiment concluded that about 0.24" of rutting are developed on the first year following AC overlay on top of AC pavement. This could be resulting from traffic compaction of new AC layer. The initial rutting is independent of the overlay thickness, mixture type, pre-overlay preparation, and preoverlay rutting level (Hall 2002).

3 Test Sections

3.1 Location

Two LTPP sections from California were studied and analyzed. Sections 06-0603 and 06-0606 located on Interstate 5 in Siskiyou County belonged to SPS-6 experiment. The test sections are located in Northern California climatic region which is wet, non-freeze. The two test sections are located in close proximity on I-5. The sections are located 60 miles north of City of Redding.

The two LTPP sections from New York State are sections 36-1643 and 36-A350. The two sections are located on US-4 highway within Washington county. The two sections from New York State are 1 mile apart and located in 66 miles north of Albany, NY. The two sections are located in a wet, freeze climatic zone.

3.2 Structural Cross-Section

3.2.1 California Sections

The pavement cross-section on LTPP 06-0603 was determined as 4.8" of AC on top of 8.2" of Jointed PCC and 4.4" cement-treated aggregate base. LTPP section 06-0606 was formed through 3.3" of AC and 8.4" of Jointed PCC and 4.7" cement-treated granular base layers. The subgrade in both LTPP sections is poorly graded gravel with sand (Program n.d.). The main difference between the two sections is the thickness of the asphalt concrete layer. The two sections are located within close proximity on I-5. Therefore, limited variation is noted in the traffic loads applied on both sections. Variation in pavement distresses are driven by the 1.5" difference in AC layer's thickness. Figures 1-A and 1-B present the pavement cross-sections in LTPP sections 06-0603 and 06-0606 respectively.

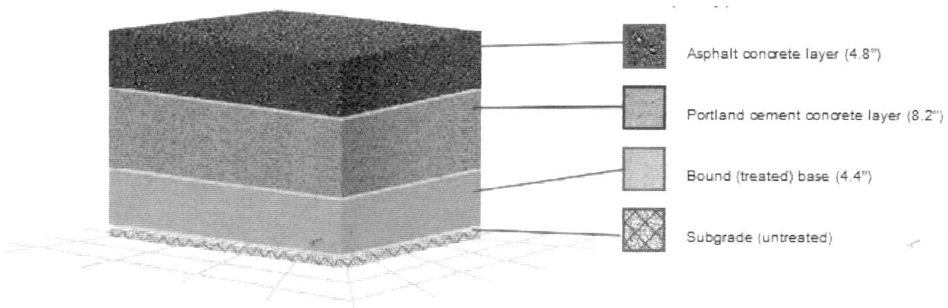

Figure 1-A: Pavement Cross-Section on LTPP section 06-0603 (LTPP 2017)

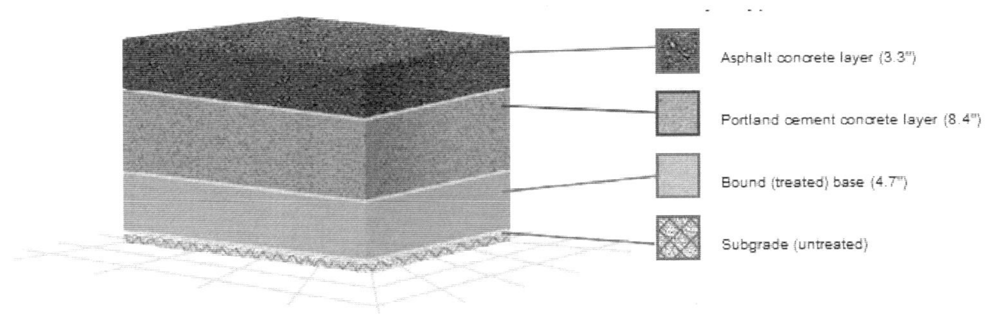

Figure 1-B: Pavement Cross-Section on LTPP section 06-0606 (LTPP 2017)

3.2.2 New York Sections

The pavement cross-section on LTPP 36-1643 is constructed through 5.1" of dense-graded AC layers followed by 8.2" Hot Mix Asphalt (HMA) Base and 7.2" unbound granular subbase . Section 36-A350 consists of 0.8" of chip sealing, 2.6" of dense-graded AC, 8" HMA base and 7.2" unbound granular subbase. The subgrade in both test sections is well graded sand with silt and gravel. Figures 2-A and 2-B present the pavement cross-sections of 36-1643 and 36-A350 respectively.

The sections selected from New York State and California are structurally comparable. Section 06-0603 (California) and section 36-1643 (New York) consist of 4.8" and 5.1" of AC layers respectively. The base layers in both sections are 8.2" jointed PCC and 8.2" HMA base.

Sections 06-0606 (California) and 36-A350 (New York) consist of 3.3" AC and 3.4" AC layers respectively. The base layers in both sections are 8.4" jointed PCC and 8" HMA base.

The authors analyzed multiple alternatives and the presented sections were the best alternatives to perform the comparison between pavement performance assuming structurally equivalent pavement sections.

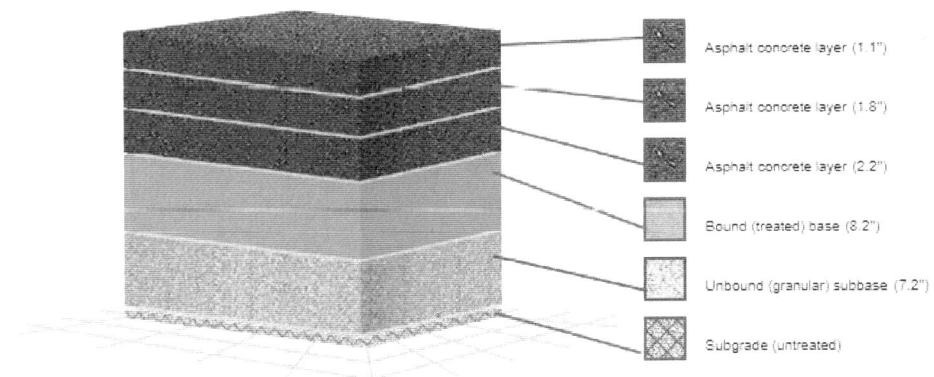

Figure 2-A: Pavement Cross-Section on LTPP section 36-1643 (LTPP 2017)

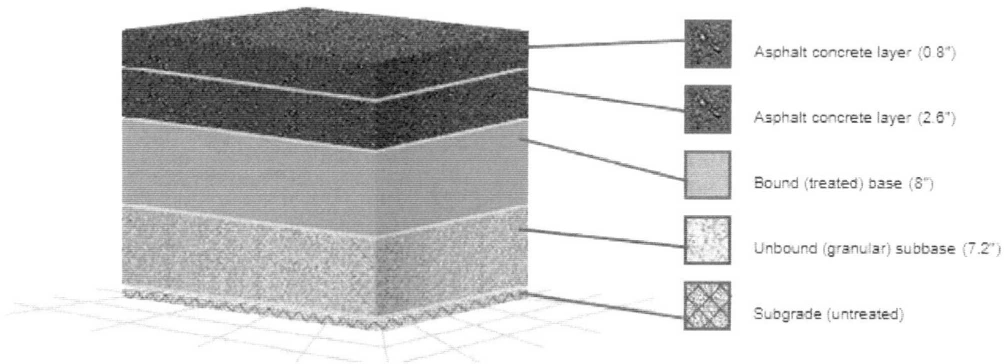

Figure 2-B: Pavement Cross-Section on LTPP section 36-A350 (LTPP 2017)

3.3 Traffic

In general, the Average Annual Daily Truck Traffic (AADTT) in both test sections located in California falls within the range of 2,000 to 3,000. The two sections are characterized by identical traffic spectrum due to their close proximity on the Interstate-5 freeway. Figure 3 presents the AADTT on sections 06-0603 and 06-0606 between 1976 and 2015. The selected sections located in New York State are located on US-4 within 1 mile stretch. The traffic loading on both sections is identical and the AADTT ranges between 600 and 1,000. Figure 4 presents the AADTT on sections 36-1643 and 36-A350 between 1980 and 2006.

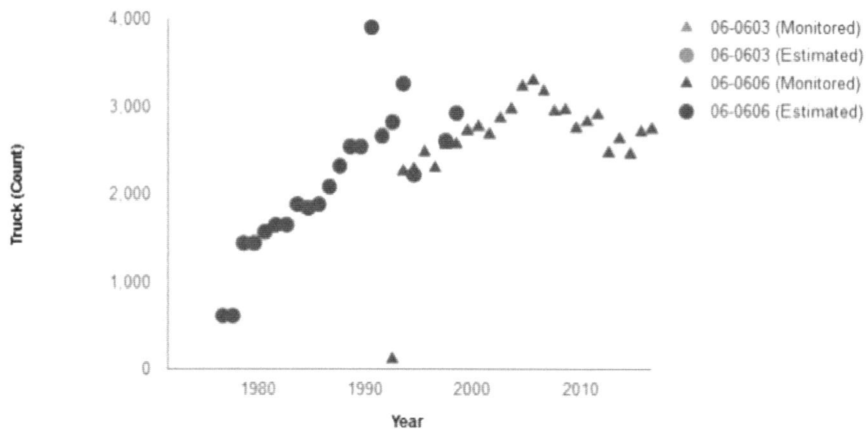

Figure 3: AADTT in Sections 06-0603 and 06-0606

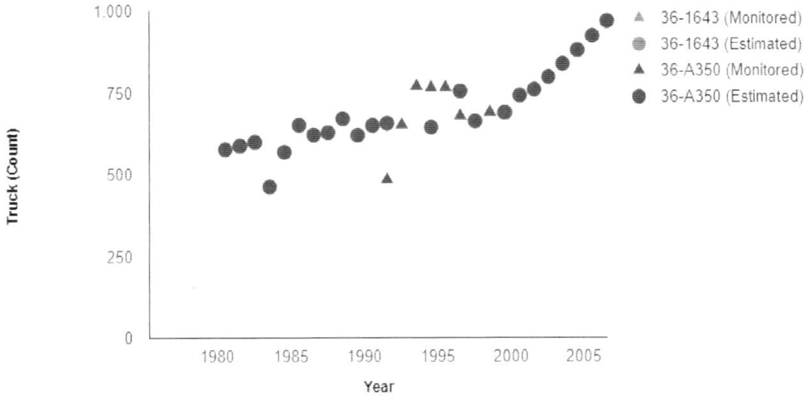

Figure 4: AADTT in Sections 36-1643 and 36-A350

3.4 Maintenance History

The LTPP sections located in California received a full depth patch and shoulder restoration in 1992. The authors identified a 7-year analysis period for this study between 1992 and 1999. Crack sealing was performed on the AC pavement surface in 1999 which impacted the distress records thereafter. Therefore, the authors performed the presented analysis based on data collected from 1992 to 1999 from sections 06-0603 and 06-0606. The sections located in New York State received patching and shoulder restoration in 1995. The following patching was recorded in 2002. The authors used the distress data between 1995 to 2002 (7 years) to perform the analysis presented herein.

4 In-State Analysis

This section presents findings by comparing the performance of the adjacent sections located on the same highway with variation in pavement thickness. The in-state analysis evaluates the impact of variation in structural strength of pavement section.

4.1 California Sections

Limited variation is noted in the transverse cracking count by comparing the performance of sections 06-0603 and 06-0606. The two sections are characterized by identical climate impact, traffic load and subgrade type. Performance variation between these two sections results from variation in AC layer's thickness. Section 06-0603 is characterized by additional 1.5" of AC on the surface pavement layer compared to section 06-0606. However, the additional pavement thickness had insignificant impact on the transverse (thermal) crack count. Figure 5 presents the progression of transverse crack counts over the analysis period for both pavement sections.

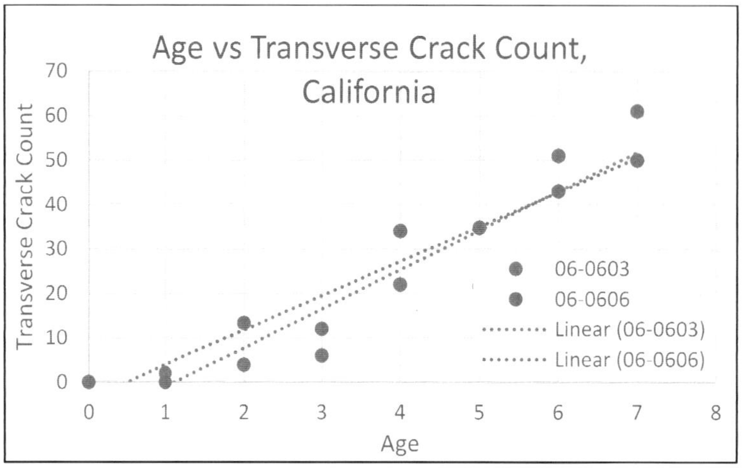

Figure 5: Transverse Crack Counts on Sections Located in California

The variation in AC thickness among sections 06-0603 and 06-0606 had significant impact on the pavement rutting. Performing a statistical T-test, the P-value comparing mean rutting in both sections resulted in a P-value of 0.01. Therefore, significant improvement (at 95% confidence level) is noted in rutting through the additional 1.5" AC constructed on section 06-0603. Figure 6 presents the rutting progression of both sections throughout the analysis period.

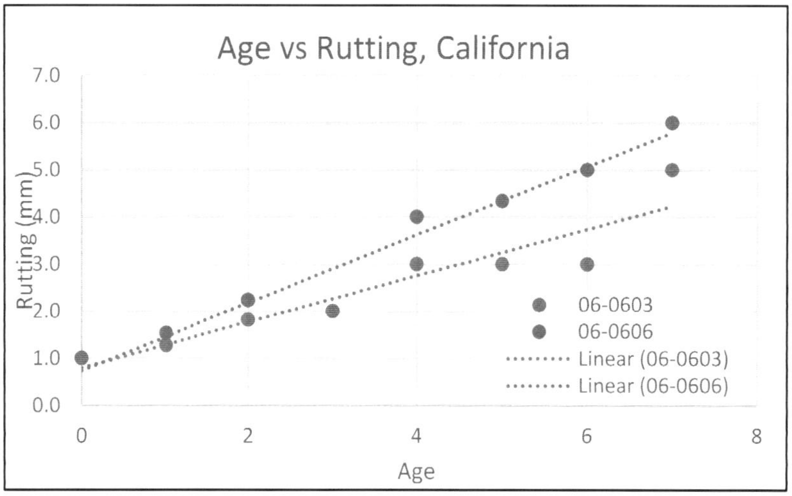

Figure 6: Rutting Progression on Sections Located in California

4.2 New York Sections

Significant variation in the performance of sections 36-1643 and 36-A350 is noted by analyzing the count of transverse cracks and rutting. The principal difference between the two sections located in New York State is the AC surface material and thickness. Section 36-A350 is characterized by 0.8" Chip sealing layer followed by thin 2.6" AC surface layer. Section 36-1643 includes 5.1" AC layers. The material properties of Chip sealing or the relatively thin layer resulted in a significant increase in thermal cracks on section 36-A350. Performing statistical T-test, the count of transverse cracks in section 36-A350 is significantly more than those on section 36-1643. The P-value for that test is 0.002. Figure 7 presents the progression of transverse crack counts on sections located in New York State.

Rutting depth in section 36-A350 is significantly higher than that in section 36-1643. The P-value determined from the T-test is 0.002. It should be noted that section 36-1643 was treated by thin overlay at year 0 (1995) and exhibited 0.07" rutting on the 1st year. This behavior matches the conclusion determined by NCHRP Project 20-50(3/4) (Hall 2002).

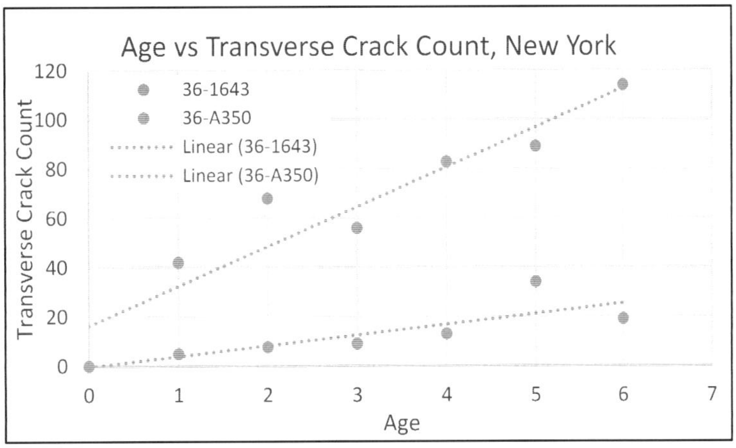

Figure 7: Transverse Crack Counts on Sections Located in New York State

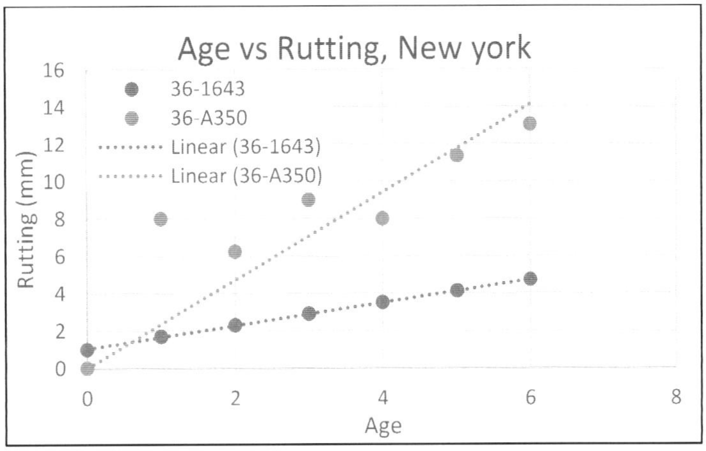

Figure 8: Rutting Progression on Sections Located in New York State

5 Cross-State Analysis

The analysis of pavement performance in the California sections and the New York State sections would be influenced by several parameters. Variations between the test sections would include traffic spectrum, structural strength and climate zones. However, that level of analysis is essential to develop a comprehensive investigation of pavement performance. The cross-state level of analysis compares the performance of structurally weak sections from both states. The structurally weak sections are 06-0606 (California) and 36-A350 (New York). Section 06-0606 consist of 3.3" AC layers and 8.4" jointed PCC base. On the other hand, section 36-A350 is formed through 3.4" AC layers (0.8" chip sealing and 2.6" AC layer) followed by 8" HMA base. The structurally strong sections in both states are compared in this level of analysis. The strong sections are 06-0603 (California) and 36-1643 (New York). Section 06-0603 consist of 4.8" AC and 8.2" jointed PCC base. Section 36-1643 includes 5.1" of AC layers and 8.2" HMA base.

The AADTT on California sections ranges from 2,000 to 3,000 while that on New York State sections ranges from 600 to 1,000. It should be noted as well that the California sections are located in a wet, no-freeze zone while the New York State sections are located in a wet,freeze zone.

5.1 Weak Sections

The transverse crack counts in New York's weak section (36-A350) is significantly higher than the count in California's section 06-0606. The P-value for this T-test is 0.002. This result is expected as New York section is located in a wet, freeze climate zone while the climate zone in California is wet, no-freeze. However, this comparison proves that traffic loads and AADTT would have no impact on the transverse crack progression. The AADTT in section 36-A350 is much lower than that in 06-0606. However, the thermal cracks show rapid deterioration even when subjected to low AADTT. Figure 9 presents the transverse crack count on both weak sections.

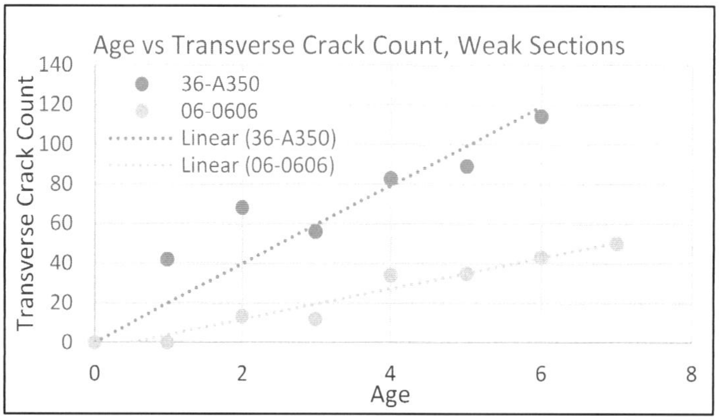

Figure 9: Transverse Crack Count on Weak Sections

The comparison between rutting in sections located in California and New York State illustrates significantly high rutting depth in New York's section 36-A350 compared to section 06-0606 in California. The P-value for this T-test is 0.003. The authors should note that AADTT on section 36-A350 is almost 67% lower than that on section 06-0606. In addition, the presence of section 36-A350 in a wet-freeze climate zone would offer it a relative advantage compared to section 06-0606 as it is located in a wet, no-freeze zone. Therefore, pavement temperature is relatively lower in the New York's test section location. The presence of rutting is facilitated through high pavement surface temperature (Kandhal 2003). The authors believe the significantly high rutting occurring on section 36-A350 is resulting from the 0.8" chip seal layer. This layer showed weak performance in rutting and thermal cracking. Figure 10 shows progressions of rutting on weak sections.

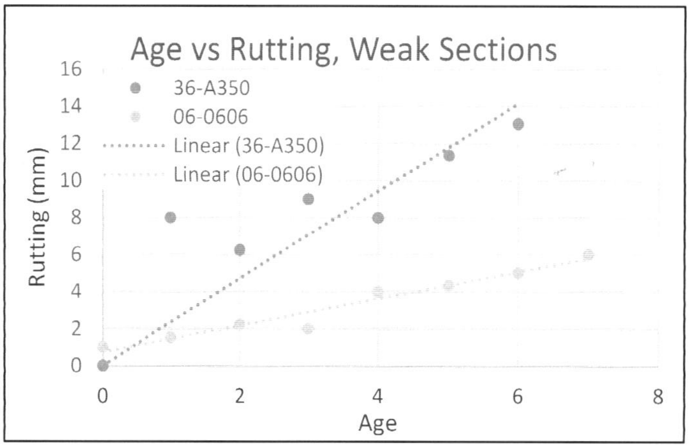

Figure 10: Rutting Progression on Weak Sections

5.2 Strong Sections

The transverse crack count in strong sections indicated weak statistical evidence the section in New York (36-1643) has slight advantage compared to California's section (06-0603). The P-value for this T-test is 0.048 (slightly lower than the 0.05 threshold). The thermal cracking count was pretty close during the first 5 years. The authors wish to note that the transverse cracking count at year 6 in New York's section (36-1643) could be an outlier resulting from operator's error. Thermal crack counts would not decrease unless maintenance treatment was performed but not documented on the LTPP database. As well, the thermal crack count at year 7 was not documented in the New York's section. Therefore, the authors would consider the two sections to have comparable performance in transverse cracking. This result as well confirms the traffic load (AADTT) has no significant influence on transverse cracking. Figure 11 presents the transverse crack count on the strong pavement sections.

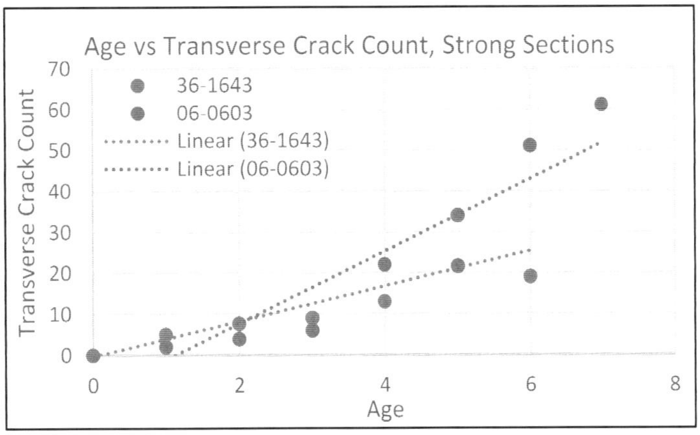

Figure 11: Transverse Crack Count on Strong Sections

The analysis of rutting depth in both sections showed no statistical difference among rutting depth. The P-value for this T-test is 0.22. This result gives slight advantage to the California's section (06-0603) as it performed slightly better even though it is subjected to high AADTT compared to the New York Section (36-1643) and it is located in a warmer climate zone.

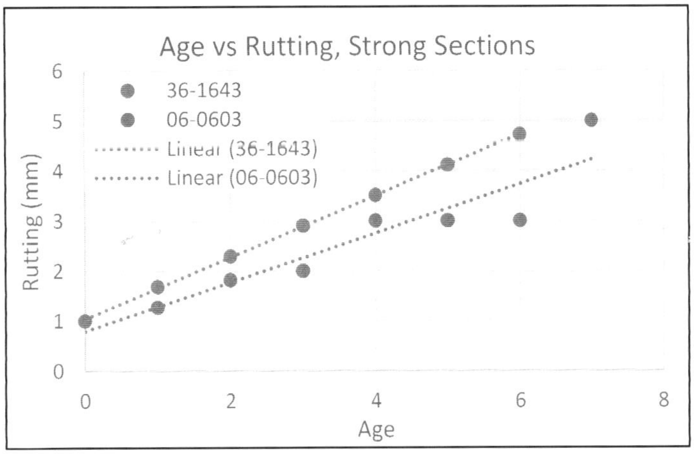

Figure 12: Rutting Progression on Strong Sections

6 Modeling And Deterioration Rates

The authors performed multiple regression analysis to develop deterioration rates for the four test sections. The regression models were developed using pavement service life (age), Traffic load (AADTT) and climate impact (Freezing Index) as independent variables. The historical climate data used to calculate the freezing index was downloaded from www.weatherunderground.com. The Freezing Index (FI) was calculated using the Formula 1 (PavementInteractive 2017) (Rutherford 1985).

[1] $FI = \sum(\bar{T} - 32°\mathrm{F})$

Where $\bar{T}$ is the mean daily temperature.

The regression analysis is reported in this paper along with the models' R-squared values. The authors acceptable R-squared values between 0.8 to 1.0 as an indication of the development of a representable and reliable model. Rutting models were developed using two independent variables (Pavement Age and AADTT). The transverse cracking models were developed using three independent variables (Pavement Age, AADTT and Freezing Index). Table 1 presents deterioration models developed using linear regression.

Table 1: Regression Analysis for Deterioration Modeling

State	Section	Distress	Developed Model	R-Squared
CA	06-0603	Rutting	y =.397*age+.000124*AADTT-1.98	0.89
CA	06-0606	Rutting	y=0.411*age+0.0014*AADTT-8.57	0.92
CA	06-0603	Transverse Crack Count	y=4.606*age+.0511*AADTT+40.09*FI-140.11	0.82
CA	06-0606	Transverse Crack Count	y=4.38*age+0.046*AADTT-24.67*FI-102.681	0.93
NY	36-1643	Rutting	y=0.623*age-0.0000742*AADTT+1.439	0.93
NY	36-A350	Rutting	y=-1.041*age+0.00349*AADTT-7.562	0.84
NY	36-1643	Transverse Crack Count	y=2.894*age-0.00995*AADTT+0.406*FI+50.5632	0.82
NY	36-A350	Transverse Crack Count	y=20.8*age+0.0055*AADTT-0.9066*FI-8.362	0.91

7 Conclusion

This paper presented a comprehensive evaluation of four LTPP test sections located in California and New York State. Two sections were selected from each State. The sections selection was performed to create a controlled comparison between sections with identical traffic spectrum and others with comparable structural capacity. The analysis in this paper was performed on two levels: a) In-State analysis and b) Cross State analysis. The comparison of the two test sections located in California (06-0603 and 06-0606) concluded that structural strength has limited impact on transverse cracking. However, the reduction of 1.5" of AC layer's thickness resulted in a significant increase in rutting depth and an accelerated rutting progression in section 06-0606.

The in-state analysis performed on sections located in New York State concluded that 0.8" chip seal layer installed on section 36-A350 had the worst resistance to transverse cracking and rutting compared to the thin overlay applied on section 36-1643. The section treated by thin overlay was subjected to 0.7" rutting on the 1st in-service year. However, the deterioration rate of the thin HMA overlay is lower than the deterioration of the chip seal layer installed on 36-A350.

The cross-state analysis was performed by analyzing the sections with comparable structural capacity and located in two different states. The weak pavement sections in California and New York State were 06-0606 and 36-A350 respectively. The strong pavement sections in California and New York State were 06-0603 and 36-1643 respectively. The analysis of weak sections concluded that chip seal layer had low resistance to thermal cracking even if subjected to low traffic loads. The ability of chip seal layer to resist rutting is limited even in wet, freeze climate zone. Section 06-0603 located in California lower rutting depth although it is located in warmer climate than section 36-1643. The analysis of strong pavement sections concluded that sections in California and New York exhibited comparable rate of transverse crack counts. The transverse cracks in California were slightly lower than that in New York. This result was concluded even though the traffic on the section in California is subjected to 150% additional truck loads (AADTT) compared to the section in New York. The rutting resistance of both sections was comparable and insignificant difference was concluded using T-test.

The authors developed deterioration models for transverse cracking and rutting for all four LTPP sections. The deterioration models were developed using multiple regression analysis. The independent variable included pavement age, AADTT and Freezing Index.

References

Elkins, G.E., Thompson, T., Simpson, A., and Ostrom, B. 2012. Long-Term Pavement Performance Information Management System Pavement Performance Database User Reference Guide. Washington, DC : Turner Fairbank Highway Research Center, FHWA-RD-03-088.

Federal Highway Administration (FHWA). 2011. Results of Long-Term Pavement Performance SPS-3 Analysis: Preventive Maintenance of Flexible Pavements. Washington, DC : Turner Fairbank Highway Research Center, FHWA-HRT-11-049.

Federal Highway Administration (FHWA). 2009. Long Term Pavement Performance Program: Accomplishments and Benefits 1989 - 2009. McLean, Turner Fairbank Highway Research Center. FHWA-HRT-10-071.

Federal Highway Administration (FHWA). 1992. SPS-6: Rehabilitation of Jointed Portland Cement Concrete Pavements: Construction Report. Turner Fairbank Highway Research Center, FHWA-AZ92-379-11. FHWA-AZ92-379-11.

Hall, K.T., Correa, C.E., and Simpson, A.L. 2002. LTPP Data Analysis: Effectivness of Maintenance and Rehabilitation Options. Washington, DC : National Cooperative Highway Research Program, NCHRP Web Document 47 (Project 20-50(3/4)).

Interactive, Pavement. [Online] [Cited: July 31, 2017.] http://www.pavementinteractive.org/calculation-of-frost-depth/.

Kandhal, P., and Cooley, A. Jr. 2003. Accelerated Laboratory Rutting Tests: Evaluation of the Asphalt Pavement Analyzer. Washington, DC : Turner Fairbank Highway Research Center, NCHRP Report 508.

Morian, D.A., Gibson, S.D., and EPPS, J.A. 1998. Maintaining Flexible Pavements - The Long-Term Pavement Performance Experiment SPS-3 5-Year Data Analysis. Washington, DC : Turner Fairbank Highway Research Center, FHWA-RD-97-102.

Program, Long-Term Pavement Performance (LTPP). [Online] [Cited: July 31, 2017.] https://infopave.fhwa.dot.gov/.

Rutherford, M.S., Mahoney, J.P., Hicks, R.G. 1987. Guidelines for Spring Highway Use Restrictions. Washington State Department of Transportation, Planning, Research and Public Transportation Division, Washington, DC : Turner Fairbank Highway Research Center, FHWA-TS-87-209.

Building Tomorrow's Society
Bâtir la Société de Demain

Fredericton, Canada
June 13 to 16, 2018 / *13 au 16 juin, 2018*

DESIGN OF CYCLE TRACKS AT BUS STOPS: CASE STUDY AND GUIDELINES

Kamaras-Garland, Joshua[1,4], Baig, Mohammed[1], Rafsanjani, Khan[1], Bertrand, Jeremy[1], Di Stefano, Nicolas[1], Marsden, Graham[1], Bouchard, Christina[2], Hollingworth, Brian[2], and Easa, Said[3]
[1] Civil Engineering Student, Ryerson University
[2] Industry Clients, IBI Group, Toronto, Ontario
[3] Technical Advisor, Department of Civil Engineering, Ryerson University
[4] jkamaras@ryerson.ca

Abstract: With the approval of a new ten year Cycling Network Plan**,** the cycling infrastructure in the City of Toronto is currently experiencing tremendous growth and investment. This paper identifies and suggests enhancements to existing and proposed bus stops that interact with cycle tracks. In order to accomodate for cycling infrastructure growth in the City of Toronto two designs were developed to be implemented as a standard for both new construction and reconstruction projects. Vehicles are no longer the sole transportation mode of choice; cycling and public transit continue to exceed expectations, soaring in popularity. As a result, city streets must adapt and adjust to these changes in human transportation behaviour: bus-stops being one such example. This report outlines two alternative designs to existing bus-stops on streets with cycle tracks which enhance safety and improve operations while maintaining low costs: A pedestrian safety island; and a sidewalk cutaway design. Reducing conflict zones between cyclists, pedestrians, passengers, public transit vehicles and motorists is of utmost priority, and the creation of a safe urban environment that values all modes of transportation proportionately is critically important. A qualitative and quantitative evaluation was used to determine the best alternatives based on the four preliminary designs. This will allow for a standardized approach which the City of Toronto can use when adding new bus-stops or upgrading existing facilities.

1 INTRODUCTION

Cycle tracks are becoming a popular addition to urban road design as cycling continues to gain traction in growing cities like Toronto. The implementation of cycle tracks create additional conflict zones, safety issues, and design flaws at locations where they meet with public transit stops. Bus stops require a large area for level boarding, and typically, cycle lanes run alongside the curb, forcing the bus to stop within the cycle lane. The goal of this project is to create a standardized bus stop that the City of Toronto can adhere to while creating and expanding their network of cycle tracks in the future. This new design standard must prioritize safety of all stakeholders and be an efficient and effective use of space since urban centres are typically very dense. If a creative and dynamic solution can be reached, cycling as a primary mode of transportation may be encouraged in urban centres. Moreover, shifting the focus from vehicular based transportation designs to more sustainable and environmentally conscious cycling and pedestrian designs is an important step in modernizing our urban communities.

The activities of the project conducted by the authors are shown in Fig. 1. As a component the report, the authors have completed a literature review to better understand the interaction between protected bicycle lanes and transit. Specifically, members studied Vision Zero, Toronto's road safety plan from 2017 to 2021 (City of Toronto n.d.A), as well as a study conducted by the City of Toronto on collisions involving both bicycle and motor-vehicles (City of Toronto, 2003). A member focused specifically on cycling in the City of Toronto, this consisted of the city's Ten Year Cycling Network Plan as well as a cycling study from 2001 to 2011 (City of Toronto n.d.B, Toronto 2011). Members also examined case studies in European locations which are known as global leaders in the area of sustainable transportation: Denmark, specifically Copenhagen (City of Copenhagen 2015); and The Netherlands, specifically Amsterdam (Pucher and Buehler 2007A & B). Case studies regarding the interaction between transit and cyclists were also considered, a member consulted Reports from the City of Winnipeg, State of Massachusetts and the Australian Road Authority (Suderman and Redmond n.d., Massachusetts 2011, Ker,Yapp and ,Moore,2005). Existing conditions were observed at a standard bus stop on Toronto near Sherbourne St. and Wellesley St. (D'Andrea 2016). Finally, a member studied the methodology behind protected bike lanes, this involved a thesis written in 2014 at Portland State University on models predicting Bicyclist Comfort in Protected Bike Lanes (Foster 2014, Burchfield 2009). Following these reviews a proposal illustrating multiple designs of a bus stop involving a protected cycling lane in the City of Toronto was drafted.

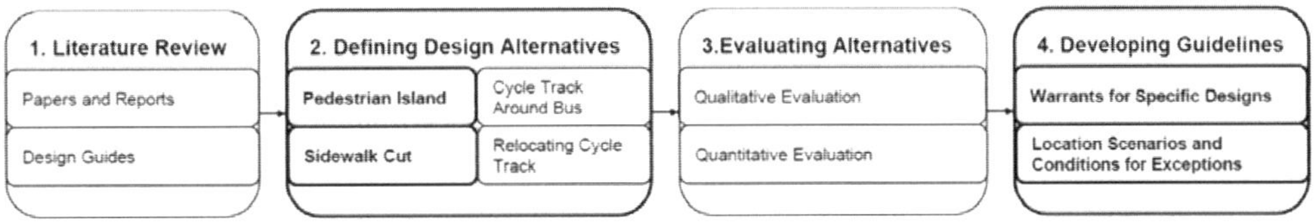

Figure 1: Activities of the project

2 DESIGN GUIDELINES

In order to design an effective bus stop, minimum standards across varying agencies must be met. These regulatory values are predetermined for comfort, safety, and accessibility. Some of these regulations fall under the discretion of the City of Toronto, Ministry of Transportation of Ontario (MTO), and City of Toronto Accessibility Guidelines (Toronto 2014 - 2017, MTO 2000, MTO 2013, Toronto Transit Commission 2017, Toronto 2001). Each of the alternative designs outlined in the report will follow these standards and meet each minimum requirement in terms of lane width, cycle track width, grade, and elevation.

Table 1: Design guidelines for bus stops based on critical values from varying standards

Facility	Element		Minimum Design Guideline
Cycle Track	Width	-	1.2 m (channelized) ; 1.5m (typical) ; 2m (desired)
	Taper [a]	-	≥ 3:1 (length to width ratio)
Bus Island [b]	Width	-	2.7 m (2.4 m free of any obstruction)
	Length	-	12 m (free of obstruction) ; 19m (for 18m buses)
	Railing	-	0.9 –1 m (For clear sightlines) ; 0.15-0.3m from curb
	Elevation	-	0.15 m (to match existing curb heights)
Lanes	Width	-	3.3 m (bus lanes) ; 3.1 m (streetcar lanes)

[a] Tactile material to be used along edge of cycle track and sidewalk.

[b] Tactile Strips must be used at each edge for crossings and boarding. Crossings width should be 1.85 m minimum.

3 DESIGN ALTERNATIVES

Four alternatives were developed through studying existing cycling infrastructure around the world and adapting these facilities for Toronto and the prescribed mandates. Note that in all of the design alternatives outlined below, it is assumed that the cycle track is raised to sidewalk elevation through typical road sections without access points or bus stops. All boarding and alighting areas will be marked by a standard accessible tactile strip and the cycle track and curb edge would be lined with a tactile surface for pedestrians who are visually impaired. Where snow removal is difficult heated concrete floors may be considered, such as the loading and waiting area for bus passengers. In considering the pavement structure, special consideration should be taken as buses are heavy slow moving vehicles and rutting can occur at bus stops. It is recommended to consider rigid pavement at the bus stop.

3.1 Alternative 1: Pedestrian Island

The first design alternative, a pedestrian island within the right of way between the cycle track and the right traffic lane, focuses on optimizing pedestrian and cyclist safety by minimizing conflict areas. The focal point of this design is the safety island that keeps waiting pedestrians separate from the cycle traffic, while keeping the cycle track and sidewalk elevations separate from the road. The cycle track detours around the pedestrian island so that cyclists remain largely unimpeded by public transit stops. The design also features street furniture which serve as a physical barrier between the cycle lane and waiting pedestrians. Furthermore, a level boarding experience can be achieved for public transit passengers without having to interfere with cyclist traffic. Pavement markings will further delineate the space, making it clear for all stakeholders which portion of the space is reserved for their use. This will be done using a combination of coloured paint and patterns.

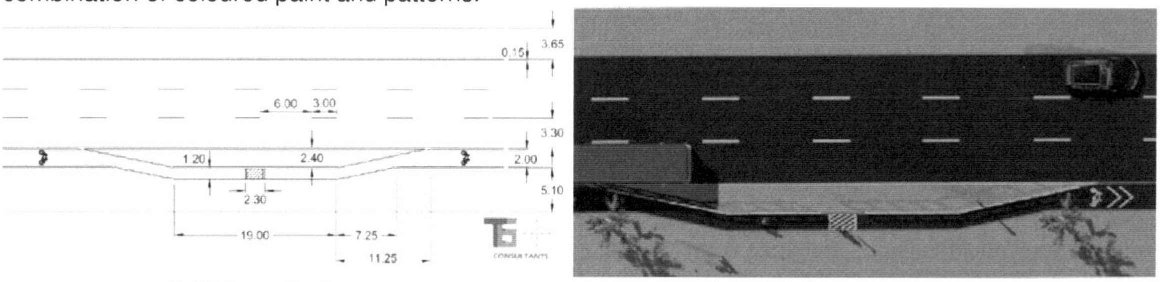

(a) Schematic diagram (b) Aerial view of rendered model

Figure 2: Design alternative 1

3.2 Alternative 2: Cycle Track Around Bus

The second design alternative, altering the path of the cycle track to pass the stopped bus on the left side, focuses on minimizing delays for all road users. This design is based on observed behaviour of current cyclists interacting with transit vehicles at bus stops. The cycle track will ramp down to street level as it approaches the location of the stopped bus at a public transit stop. Painted lines would delineate a tapered path for cyclists that would direct them around the bus loading zone and realign back on the other side, where the track would ramp back up to sidewalk level. Segments of dashed lines will instruct bus operators where to merge or diverge through the track. During the tapering of the cycle path, the track would also narrow to enforce a more cautious approach from cyclists. Additionally, warning signs and coloured zones would be implemented to direct all users to proceed with caution.

3.3 Alternative 3: Sidewalk Cut

The third design alternative, putting the bus stop entirely within the cycle track, builds upon the most common operations on existing cycle tracks such as Richmond/Adelaide. This design attempts to address the safety and operational issues that were found in the current bus stop configurations while reducing the change to existing services and surrounding facilities. On the sidewalk the furniture is reconfigured to open up the full sidewalk width in order to reallocate the right of way. Where the total cycling facility (including buffers) is not wide enough to accommodate a stopped bus a sidewalk or boulevard cut is made to ensure that when the bus is stopped it is clear of the adjacent traffic lane. A yield line and accompanying signage will be placed at the entrance to the bus stop, as well as coloured zones indicating any conflicts as an additional warning. Following the yield line the cycle track will lower down to street level, and a pavement marking will highlight the grade change. Diamonds or equivalent pavement markings will indicate that the lane is only for cyclists and buses.

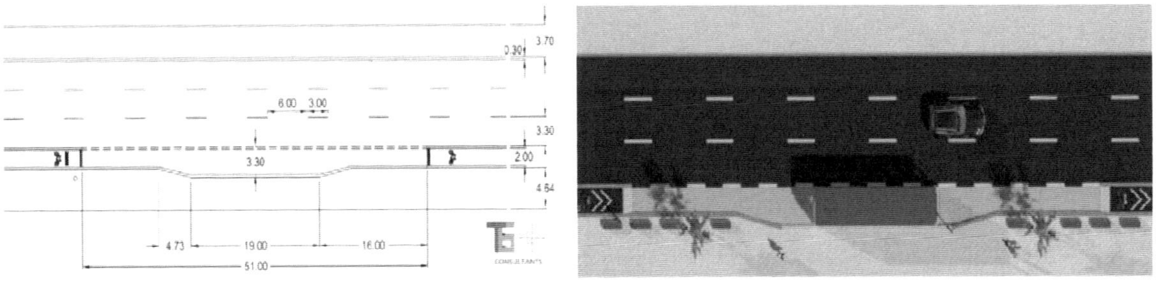

(a) Schematic diagram (b) Aerial view of rendered model

Figure 3: Design alternative 3

3.4 Alternative 4: Relocating the Cycle Track

Our fourth and final design, relocating the cycle track to the left side of the street, completely eliminates the bus-bicycle conflict at bus stops. This design is only applicable on one way streets with cycle tracks, such as the Richmond/Adelaide corridor. It would be recommended that sidewalk furnishings and the concrete sidewalk be upgraded to meet the Accessibility for Ontarians Act as well as various City of Toronto and TTC standards (TTC 2005). A partial shelter and benches would greatly improve the waiting conditions for passengers, and a furniture reorganization would allow for clear access and egress to and from the buses. Planters or a separated curb, similar to Sherbourne St. and Wellesley St. W., could be utilized here in place of the elevated track. Right turn boxes could be implemented where necessary and most cyclist left turns will be uninhibited by this design. Bus operations and motor operations would remain the same, however an advanced left may be required to maintain sufficient LOS levels.

4 EVALUATION OF ALTERNATIVES

To gain a comprehensive understanding of all the alternatives, a qualitative and quantitative assessment was conducted deriving from the triple-bottom line: society; environment; and economy. As such, design parameters of safety, design utility, cost, and the environmental impacts were examined. The quantitative approach enabled the evaluation of criteria through the assignment of weighting factors, while the qualitative approach addressed the various impacts of an alternative on stakeholders with respect to the status quo conditions. These stakeholders include transit users, cyclists, pedestrians, motorists and the City of Toronto. One recurring assumption used for both evaluations was that all stakeholders act rationally, which was done to simplify the process and generate accurate ratings. Due to limitations in available software the design models can not be modelled, and our evaluation will rely on the previously mentioned methods. These methods enabled us to better serve our clients by allowing us to develop solutions that are both pragmatic and inclusive.

4.1 Quantitative Methodology

The classical quantitative method for scoring alternatives was conducted by using a set of 19 evaluation criteria based on safety, design utility, cost, and environmental factors, as seen in Table 1. Weightings were strategically assigned to each of the 4 factors in order to create an optimal design tailored for our stakeholders from a value from 1-5 which was utilized in Equation 1.

[1] $S_i = W_j R_{ij}$

Where: S_i = score of alternative i ranging from 1-10 with 1 being the lowest and 10 the highest; W_j = weight of factor j; and R_{ij} = rating of alternative i with respect to factor j.

Scores were assigned based on their importance, since safety has the potential to be a hazard and create risk is a top priority for engineering alternatives, so this criteria was adjusted by a factor of five, whereas design utility criteria was adjusted by four since it can create significant problems for a governing agency due to items such as snow and ice control. Although cost factors were not emphasized for this project, a multiplier of two was assigned. The environmental factors were taken as a singular unit given that the nature of the project carried minimal environmental implications with respect to the status quo. For efficiency, all members scored the selected criteria as a team and discussions were kept as objective as possible. The scores for the quantitative analysis can be seen in Table 2.

Table 2: Results of quantitative evaluation

		Alternatives						Alternatives			
		1	2	3	4			1	2	3	4
	Traffic Impacts	7	4	6	8		Perceived Safety	8	3	7	9
	Transit Integration	9	3	7	9		Cycle-Ped Conflict	8	10	10	10
	User Concern	9	3	5	9	Safety	Cycle-Bus Conflict	10	5	8	10
	Attractiveness	8	4	9	7		Cycle-Vehicle Conflict	10	9	10	10
Design Utility	Accessibility	9	6	9	9		Bus-Vehicle Conflict	9	8	7	9
	Street Furniture	9	6	7	8		Total	225	175	210	240
	Operations & Maintenance	8	8	9	9		Material & Construction	3	7	7	4
	Constructability	4	9	6	9		External Infra. Upgrades	8	4	4	7
	Total	252	172	232	272	Cost	Constructability	4	5	6	4
	Construction Waste	6	9	8	9		Operations & Maintenance	8	8	9	9
Environment	Green Space Potential	7	6	7	8		Total	46	48	52	48
	Total	13	15	15	17		Net Score	536	410	509	577

The first theme evaluated was safety which prioritized vulnerable road users. The highest scores were given to alternatives 1 and 4, followed by 3 and finally 2. Conflicts were distinguished between modes, with each alternative scoring similarly, less alternative 2, due to its inherently conflicting nature. In regard to Design Utility, specifically traffic impacts, all categories scored well except alternative 2 due to motorists' flow being disturbed by moving cyclists. Transit integration showed to be polarizing as alternatives 1 and 4 scored very well, while alternatives 2 and 3, received lower scores due to the need for transit to merge through traffic. Accessibility showed exemplary results in all alternatives except 2. Operations and maintenance was a strong suit of all alternatives since the current maintenance practices were adopted and improved. Finally, constructability received scores ranging from 4 to 9. Alternatives 2 and 4 received the highest, due to their simplicity, while alternative 1 being more complicated scored the lowest. Cost was distinguished between materials and construction cost, as well as the need for external infrastructure upgrades. Initial costs ranged significantly, with Alternatives 1 and 4 scoring lowest, due to their complexity and need for concrete work. Alternatives 2 and 3 scored significantly higher as much of their work involved primarily line painting. External infrastructure upgrades followed this trend, with alternatives 1 and 4 being more expensive. The alternative with the most waste generated was alternative 1, while the other alternatives scored relatively low due to their simplicity. Green space potential scored similarly across the board, as all alternatives make good use of street furniture and available space.

4.2 Qualitative Methodology

The advantage of the qualitative method stems from an emphasis on the stakeholders at hand. Stakeholders are identified by C, BD, BP, P, Ci and M; Cyclists, Bus Drivers, Bus Passengers, Pedestrians, the City/Municipality and Motorists. Using the previously defined evaluation criteria for the quantitative evaluation, this method compares the effectiveness of each alternative to the status quo, with respect to the stakeholders. As seen in Table 2, arbitrary values of 1,0,-1 were assigned corresponding with a colour scheme of green, yellow, and red respectively for ease of visual interpretation and analysis. If a factor belonging to a criteria improved the user experience for a given stakeholder, a value of 1 was assigned. A value of -1 was assigned if a user's experience was worsened and a value of 0 was assigned if there was no change. The culmination of all scores for each stakeholder will aid in identifying the best design. Like the quantitative method, the scores for this method were generated holistically to avoid bias.

Table 3: Results of qualitative evaluation

		Alternative 1						Alternative 2						Alternative 3						Alternative 4					
		C	BD	BP	P	Ci	M	C	BD	BP	P	Ci	M	C	BD	BP	P	Ci	M	C	BD	BP	P	Ci	M
Safety	Perceived Safety	1	0	-1	0		0	-1	-1	0	0		-1	1	0	0	0		0	1	1	0	0		1
	Cycle-Passenger Conflict	-1			-1			0			0			0			0			0			0		
	Cycle-Bus Conflict	1	1					-1	-1					1	1					1	1				
	Cycle-Vehicle Conflict	1					1	-1					-1	0					0	1					1
	Bus-Vehicle Conflict		1				1		0				0		0				0		1				1
Design Utility	Traffic Impacts	1	1		1		-1	-1	-1		0		-1	0	1		0		0	1	1		1		-1
	Transit Integration		1	1					-1	1					1	1					1	1			
	User Comfort	1	1	-1	1		-1	-1	-1	0	0		-1	1	1	0	0		0	1	1	0	0		-1
	Attractiveness	1	1	1	1		1	0	0	0	0		0	0	0	0	0		0	0	0	0	0		0
	Accesibility	1		1	1			-1		1	1			1		1	1			1		1	1		
	Street Furniture Utility	1		1	1			0		1	1			0		1	1			1		1	1		
	Operation and Maintenance		1			0			-1			0			1			-1			1			0	
	Constructability					-1						-1						-1						0	
Cost	Materials and Const. Cost					-1						-1						-1						0	
	External Infra. Upgrades					0						0						-1						0	
	Constructability					-1						-1						-1						-1	
	Operation and Maintenance					0						0						-1						0	
Environment	Construction Waste					-1						-1						-1						0	
	Green Space: Planters					1						1						1						1	
	Score (%)	78	88	33	57	-38	17	-67	-75	50	29	-38	-67	44	63	50	29	-75	0	78	88	50	43	0	17

4.3 Alternative 1: Pedestrian Island

The first alternative, the bus stop island, was evaluated and showed a positive result overall. The culminating colour was green which shows that in comparison to the status quo, the bus stop island showed an overall improvement. Further analyzing the design, safety scored a net benefit, clearly improving the safety of most stakeholders due to the effective use of channelization and the mitigation of bus/vehicle interactions with cyclists. Cyclist-passenger conflicts showed a net decrease as passengers are now required to cross a live cycle track. This conflict is less critical however than conflicts between vehicles and pedestrians/cyclists which have a higher level of KSI collisions. Overall, most conflicts are eliminated by implementing this design, and safety is inherently improved. The analysis of the design utility shows a net benefit result as it proves to be more accessible, and visually appealing to all stakeholders across the board. Constructability also shows a net decrease due to the increased complexity of the design. Traffic impacts are minimal but likely to improve for most stakeholders. The costs of the design is larger due to the requirement for more materials, space, and labour to construct. As a result most categories show a net decrease in this regard. Fortunately for maintenance it is believed that the the costs should be similar to the status quo. The environment is relatively unaffected but the addition of "green" street furniture will certainly benefit the design.

The benefits are clear based on the safety and design utilities of this particular model. The negative effect on bus passengers is relatively small, and the overall improvements outweigh the negative factors which were discussed. As a result of this qualitative assessment this design is to be the frontrunner since design alternative 4 will be excluded for the purposes of this study.

4.4 Alternative 2: Cycle Track Around Bus

As discussed, this alternative requires the bus to merge at two points, it is largely for this reason that it scored poorly in the evaluation. While this design served well in both the accessibility and street furniture portions of design utility it scored poorly in nearly every other category. Perceived safety is sorely lacking for all stakeholders and negative scores were assigned. While there is no cycle-passenger conflict, or an impact on bus vehicle conflicts, there is a potential for cycle-bus and cycle-vehicle conflicts, which are some of the most severe.

As the bus has to merge twice the scores for design utility and user comfort are negative. Attractiveness, accessibility and street furniture utility are strengths of this design as they showed positive or neutral scores. Operation and maintenance as well as constructability scored poorly, due to its intricacy. Cost has scored poorly as the additional elements of safety are more expensive than the status quo however, this alternative does not require any external infrastructure upgrades. Environmental scores for this alternative vary depending on the category. While the potential for green space is improved over the status quo, the construction waste is increased. This alternative clearly scores the worst among the four analyzed in this report, largely due to safety as there are more potential conflict points when compared with other alternatives. For this reason, it will not be recommended further.

4.5 Alternative 3: Sidewalk Cut

In Alternative 3, the sidewalk is cut back to allow adequate space for a bus to safely pull out of the lane of traffic. This alternative scored well in the quantitative evaluation as it is the status quo but with additional safety measures. Safety was enhanced by the additional signage, coloured zones, and grade separation clearly indicating the right-of-way. The design utility of this concept scored higher than the status quo due to the improved use of space, additional greenery, safer delineation, and improved conflict resolution. This alternative improves the existing traffic for motorists by allowing the bus to fully merge out of the lane of traffic. The transit integration is improved due to the less ambiguous right of way between cyclists and buses. The accessibility is improved for cyclists of all ages and abilities and more space is implemented in the design to comply with the Accessibility for Ontarians with Disabilities Act (AODA).

This option is easy to maintain with the use of plows and sweepers. Overall the alternative is a low cost option but requires some additional expenses such as the added pavement markings. Maintenance fees may also be lowered as the existing facility has a high cost to replace damaged bollards and faded pavement markings. The sidewalk cutaway design allows for a properly planned urban landscape and had a positive score in the category of street furniture. The design can be easily integrated into many reconstruction projects and requires minimal disruption of existing traffic networks. It is recommended to be further analyzed for future use where inadequate right of ways make it difficult to implement Alternative 1.

4.6 Alternative 4: Relocating the Cycle Track

The final alternative, relocating the cycle track into the left lane, appeared to show the best result of all the alternatives. In comparison to the status quo it is either better or equivalent in almost all aspects, except for motorist traffic impacts and comfort, and the city's constructability concerns. With regards to safety it scored positively in most categories as the bus-bicycle conflict is completely eliminated for all bus stops, as well as bus-vehicle merging conflicts. The actual conflicts between cyclists and motorists are the same as the status quo, the only difference being a conflict with right turns as opposed to left turns. As for design utility, this design essentially takes the existing design and improves it. This design offers the chance to reorganize sections of roadways and more appropriately place boarding areas, drop-offs, and bus stops. Further traffic analysis will be required to determine if re-timing signals or if protected left phases are warranted. The cost observed no comparable difference under the assumption that any new cycle track will also be raised but may require the installation of new signal heads and signage for the protected left turns.

Overall this alternative appears to be superior. It eliminates the majority of existing conflicts, does not create any additional conflicts between motorists and cyclists, is easy to implement, and the operations and maintenance can be completed using existing equipment and methodologies. However, it is not recommended to pursue this alternative further for this project. The relocated cycle track is only feasible on the Richmond-Adelaide corridor, and largely cannot be used elsewhere in the cycle track network. The goal of this project is to create a standardized bus stop that the City of Toronto can adhere to while expanding or enhancing their cycle track network. We strongly recommend that this alternative be investigated further and potentially a new standard be created for the Richmond-Adelaide Corridor between Bathurst Street and Parliament Street.

5 WARRANTS

This section aids in simplifying the selection process for designing bus stops at cycle tracks. The designs in question are the sidewalk cutaway, and pedestrian island. The Sidewalk Cutaway design is advantageous in cases where there is a restricted ROW and the island is beneficial in areas of high cycle/bus ratio. A low bus volume may suggest the simple sidewalk cutaway be advantageous. Additional warrants can be found below assuming a cycle track is warranted by surpassing 2000 cyclists/day as follows: If two or more of the following warrants are true, the Sidewalk Cutaway should be used, otherwise the Pedestrian Island is to be used. All warrants should be considered in the peak hour for all transportation modes (# of cyclists, pedestrians, motorists, and busses):

1. The roadway operates at 60% capacity or greater.
2. The volume of Cyclist/Bus ratio is equal to or greater than 2.
3. The volume of Cyclist/Pedestrian ratio is equal to 3 or less.

Special provisions should occur for areas where there is a high number of conflicts between cyclists and buses, cyclists and pedestrians or cyclists and vehicles or the location has a history of collisions.

6 CASE STUDY

A case study was taken at TTC stop #11887 to showcase how the newly developed standards can be implemented during a road reconstruction of Adelaide St. West. This street has an average daily cyclist count of 2564 during an 8hr period for 2016 (City of Toronto 2018) and approximately 62 buses/day at this location between Monday to Friday (Toronto Transit Commission 2018). This is a midblock stop with access points on either side of the stop, one being a private access and the other being Peter St. This creates a unique longitudinal space restriction that may impact the design of the pedestrian island. The large right-of-way has provided an adequate width to implement either alternative 1 or 3, the bus island or sidewalk cutaway, as they were the top scoring. Stop #11887 is host to street furniture that blocks the boarding and alighting of public transit customers. This is another characteristic that will showcase the two design's abilities to adapt to a variety of conditions.. The accessibility of the stop and overall appeal of the area will be improved with either design choice. This has been applied through the use of tactile strips, street furniture and coloured conflict zones. Grading and drainage are major considerations that pose a concern in any design. The extensive underground utilities in this region create a challenge in handling drainage and special attention has been provided to this matter. The island itself will feature a heated concrete pad to minimize maintenance operations whereas the sidewalk cutaway will not require this attribute due to its linear design. Asphalt concrete will be utilized for the stop as there is no concern of rutting due to the low bus volume. A bus shelter is optional for this location due to the minimal rider usage but benches have been provided in designated areas. Right-of-ways have been clearly established with pavement markings, lines, and grade separations. A sample model of the pedestrian island has been provided in Figure 4 below:

Figure 4: Model of the pedestrian island implemented at TTC Stop #11887

7 CONCLUSIONS

Four design alternatives have been created in this project as standardized solutions for transit stops that interact with cycle tracks. This report hosted a series of challenges such as, no software being able to analyze the impacts of each alternative and there being no current accepted references for these facilities in Ontario. After undergoing a detailed qualitative and quantitative evaluation of all four alternatives, some conclusions can be drawn about which alternatives show the most promise, and which alternatives to continue forward with. The leading design in both the qualitative and quantitative evaluations was design alternative #4, relocating the cycle track to the opposite side of the road on one-way streets. This design alternative is the simplest, and most effective, but due to the fact that its use is limited to one-way streets, selecting this as a standardized option is not feasible and this option will be intentionally overlooked. This being said, it is recommended that an additional project be opened that focuses on this alternative as the primary option for one-way streets. Design alternatives #1 and #3, the bus-stop island and the sidewalk cutaway respectively, both scored relatively well on the qualitative and quantitative evaluations. Although the bus-stop island scored slightly higher than the sidewalk cutaway, the space limitations that exist on many streets make this space-intensive design difficult to standardize on its own. As a solution, the recommended approach is to combine both the bus-stop island and the sidewalk cutaway alternatives into a single city standard. The bus-stop island will always be the first choice, but on streets where space is limited, the sidewalk cutaway will be implemented as a second option that is almost as effective. The trade-off in efficiency and effectiveness is minimal as shown by the qualitative and quantitative assessments which makes this dual standard approach not only possible, but preferred.

Acknowledgements

The authors are grateful to Dr. B. Farooq, M. Navarra, and M. Eskandar for their help in various aspects of this project.

References

Burchfield, Robert. "Cycle Tracks: Lessons Learned." National Association of City Transportation Officials.
February 4, 2009. https://nacto.org/wp-content/uploads/2011/03/Cycle-Track-Lessons-Learned.pdf.
City of Toronto. "Open Data Catalogue" City of Toronto. 2018 https://www.toronto.ca/city-government/data-research-maps/open-data/open-data-catalogue/
"Construction Drawings for Roadworks." Toronto. 2014-2017.
https://www1.toronto.ca/wps/portal/contentonly?vgnextoid=d0d73d3085131410

VgnVCM10000071d60f89RCRD&vgnextchannel=9deeabbf06721410VgnVCM 10000071d60f89RCRD.
"Copenhagen City of Cyclists The Bicycle Account 2014."City of Copenhagen, & Cycling Embassy of Denmark. May 2015. http://www.cycling-embassy.dk/wp-content/uploads/2015/05/Copenhagens-Biycle-Account-2014.pdf.
"Cycling Network 10 Year Plan." City of Toronto. n.d. A https://www.toronto.ca/services-payments/streets-parking-transportation/cycling-in-toronto/cycle-track-projects/cycling-network-10-year-plan/.
D'Andrea, Michael. "Sherbourne Street and Wellesley Street Cycle Track Review and Design Improvements." City of Toronto. November 3, 2016. https://www.toronto.ca/legdocs/mmis/2016/pw/bgrd/backgroundfile-98101.pdf.
Foster, Nicholas Mark-Andrew. "Predicting Bicyclist Comfort in Protected Bike Lanes." PDXScholar. May 8, 2014. https://pdxscholar.library.pdx.edu/cgi/viewcontent.cgi?article=2969&context=open_access_etds.
Ker, Ian, Steve Yap, and Philip Moore. "Bus-Bike Interaction within the Road Network." Washington State Department of Transportation. 2005. http://www.wsdot.wa.gov/NR/rdonlyres/8458A9D7-0654-4C93-A545-ADA8F35856D5/0/05BusBike.pdf.
"Lane Widths Guideline Version 2.0." Toronto. June 2017 2017.https://www.toronto.ca/wp-content/uploads/2017/11/921b-ecs-specs-roaddg-Lane_Widths_Guideline_Version_2.0_Jun2017.pdf.
"MassDOT Separated Bike Lane Planning & Design Guide: Chapter 5, Curbside Activity Design." Massachusetts Department of Transportation. 2011 http://www.massdot.state.ma.us/Portals/8/docs/SBLG/Chapter5_Curbside.pdf
" Ontario Traffic Manual - Book 11 - Pavement, Hazard and Delineation Markings." Ministry of Transportation of Ontario. March 2000. https://www.library.mto.gov.on.ca/SydneyPLUS/Sydney/ViewRecord.aspx?template=Books&record=7fede97e-4361-4435-85dc-4c94d04c6131&lang=en-US.
"Ontario Traffic Manual - Book 18 - Cycling Facilities." Ministry of Transportation of Ontario. December 2013. http://www.raqsb.mto.gov.on.ca/techpubs/eps.nsf/0/825810eb3ddd203385257da0063d934/$FILE/Ontario%20Traffic%20Manual%20-%20Book%2018.pdf.
Pucher, John, and Ralph Buehler. "At the Frontiers of Cycling: Policy Innovations in the Netherlands, Denmark, and Germany." Rutgers University. November 18, 2017. http://www.fietsberaad.nl/library/repository/bestanden/ATFC%20Nov%2021.pdf.
Suderman, Scott, and Judy Redmond. "Pembina Highway Buffered Bike Lane Project - A Solution to Managing Cyclists, Buses and Pedestrians at Transit Stops." Transportation Association of Canada. 2013. http://conf.tac-atc.ca/english/annualconference/tac2013/session19/suderman.pdf.
"Toronto Bicycle/Motor-Vehicle Collision Study (2003)." City of Toronto. 2003. http://council.london.ca/CouncilArchives/Agendas/Environment%20and%20Transportation%20Committee%20Agendas/ETC%20Agendas%202004/2004-02-23%20Agenda/Item%206.pdf.
Toronto Transit Commission. "TTC Bus Routes" Toronto Transit Commission. 2018
"Vision Zero: Toronto's Road Safety Plan." City of Toronto, n.d.B https://www.toronto.ca/wp-content/uploads/2017/11/990f-2017-Vision-Zero-Road-Safety-PlanJune1.pdf.

Building Tomorrow's Society
Bâtir la Société de Demain

Fredericton, Canada
June 13 to 16, 2018 / *13 au 16 juin, 2018*

FREEWAY WORK ZONE: CURRENT TRENDS AND FUTURE NEEDS

Pilanavithana, Udila[1,4], Qu, Xiaobo,[2] and Easa, Said[3]
[1] Ryerson University, Canada
[2] Chalmers University of Technology, Gothenburg, Sweden
[3] Ryerson University, Canada
[4] udilashalitha.pilana@ryerson.ca

Abstract: Work zones are recognized as a significant source of freeway incidents as they involve frequent lane closures and lane merges that last for different periods of time. Work zones generate road user delays, traffic incidents, safety hazards, and vehicle emissions. This article presents a comprehensive literature review of the mobility, safety, driver behaviour, and cost and planning aspects of state of the art work zones. Over 100 publications, including research papers, master thesis, and dissertations were reviewed in order to understand the spread of work zone applications. Mobility and safety have become important measures that are used to optimize the performance of work zones by minimizing delay and traffic collisions. Work zone configuration plays a major role in all aspects of work zones while capacity reduction and speed limits are critical issues for mobility. The review revealed that safety is mostly modeled and predicted using collision data as the main factor. The impact of nighttime safety attracted less research attention. Cutting-edge smart work zone concepts have emerged as a key component in most of the phases of work zone studies. Driver behaviour is explored as one of the main considerations that affects the entire work zone aspects. In this article, the important elements of each publication are tabulated to give the reader quick access to the relevant references. Apart from the existing approaches, the design variables, limitations, and future research needs are listed separately for prospective users. It is our hope that this paper will serve as a quick freeway work zone reference and a guide for potential researchers and practitioners.

1 Introduction

Highway work zones have been an area of major concern for decision makers and researchers when conducting multiple activities on road networks, including construction and maintenance. Many work zone studies have been published. These studies examine different work zone aspects such as mobility, safety, driver behaviour, and planning. Work zones are nationally recognized as a significant source of highway incidents as they involve frequent lane closures and lane merges that last for different periods of time according to the type of construction or maintenance and their intensity. These scenarios generate road user delays, traffic incidents, safety hazards and vehicle emissions. The Highway Capacity Manual (HCM 2010) defines a work zone as "a segment of highway in which maintenance and construction operations reduce the number of lanes available to traffic or affect the operational characteristics of the traffic flowing through the segment".

The past few decades have seen major improvements in work zone analysis with the adoption of cutting-edge science and technology. A review of the work zone literature revealed that the majority of the research contributions have been in the areas of mobility and safety. Driver behaviour, planning and cost represented only 8% of the research contributions to work zone analysis. The publications reviewed included 70 journals, 14 MASc theses and 16 PhD dissertations. Many of these papers were published in highly reputable international journals. Despite the large number of work zone studies, the exposure is not comprehensive and only gives a picture of work zone application site. The references are summarized in a tabular form under each of the area sub-headings with a synopsis for readers. The limitations of the existing approaches and future needs are also briefly mentioned for prospective users.

2 Analysis of Work Zone Applications

The following subsections review different work zone applications. All of these applications have been classified into four major areas: (a) Mobility (b) Safety (c) Driver Behaviour and (d) Cost and Planning. Each section describes one of these major areas and contains tables that list all of the reviewed studies published in this area.

2.1 Mobility Applications

Son (1999) incorporated fluid approximations to establish a queueing model for the average delay of lane closures on two-lane highways. Fluid approximations and stochastic models were originally developed by Newell in 1969. Turley (2002) presented a Generalized Linear Model (GLM) for speed by evaluating the Arrow-Panel Caution Displays. An online algorithm for variable speed limit control in work zones was introduced by Kang et al. (2004). The proposed model and algorithm optimize the speed-limit based on the evolution of dynamic traffic conditions and macroscopic traffic characteristics. Chin (2004) developed a moving slot model and framework for operational strategies for slot/lane assignment rules. This model is used to manage highway space for interactions between the entrance and exit processes in order to maximize capacity while accounting for vehicle maneuvers. Kang (2006) developed an advanced Dynamic Lane Merge (DLM) control model and operational algorithm as well as a dynamic Variable Speed Limit (VSL) model and operational algorithm. The optimization took into account the interactions between the speed, flow and work zone capacity.

Kang and Chang (2006) proposed a speed control strategy called the Time-of-Day Speed Limit (TOD SL). This strategy enables researchers and designers to establish optimal real-time speed limits by maximizing the use of available historical data. Kim (2008) investigated the upstream traffic characteristics of a work zone and developed a mechanistic model of capacity based on merging behaviour. This model estimates work zone capacity according to the lane closure configuration and traffic conditions such as heavy vehicle percentage, merging patterns, merging time and the queue discharge flow rate. Radwan et al. (2011) evaluated the operational effectiveness of the Simplified Dynamic Lane Merge Systems (SDLMS) i.e. Early Merge and Late Merge in the presence of VSL. The simulation results revealed that a late merge, with or without VSL, produced higher mean throughputs for all adherence levels (compliance rates) of motorists. Kurker (2013) evaluated a Dynamic Traffic Assignment (DTA) system equipped with Early Merge, Late Merge and Fixed Cycle Work Zone Signal Merge Control (FCWZSMC) using VISSIM simulations. He also developed guidelines for the selection of merge concepts according to the traffic flow. FCWZSMC was recommended for multiple lane closures and high traffic flows.

Du and Chien (2014) developed guidelines for the use of the road shoulder to increase the capacity in work zones. Patil (2015) evaluated the use of Portable Traffic Signal (PTS) systems in long rural two-lane highway work zones. This study compared three different conditions for the control of one-lane traffic involving the deployment of flagger and PTS system combinations in conjunction with pilot car operations. A PTS unit without flagger operation was recommended to control one-lane traffic for long rural two-lane highway work zones. Yeom (2015) developed capacity and Free Flow Speed (FFS) models which were calibrated using three sources (field, sensor and literature achieves). The introduced capacity model may be amalgamated into the next issue of the Highway Capacity Manual (HCM). Abdelmohsen (2016) conducted studies involving a large number of parameters to obtain more realistic scenarios. The primary focus of the model was minimizing traffic delay, however, the model also made contributions to the area of cost and safety. Table 1 presents the references involving work zone mobility.

Table 1: Mobility references on the mobility aspect

No	Sample References	Type	Country*	Analytical Tools Used
1	Turley (2002)	Journal	USA	Regression Analysis, Statistical Analysis
2	Liu et al. (2017)	Journal	China, USA	Statistical Analysis
3	Bie et al. (2016)	Journal	Canada	Regression Analysis, Statistical Analysis
4	Chung (2011)	Journal	Korea	Statistical Analysis
5	Radwan et al. (2011)	Journal	USA	Statistical Analysis
6	Zheng et al. (2011)	Journal	SWZ, NLD	Fuzzy Logic Artificial Neural Networks
7	Zhu (2015)	MASc	USA	Regression Analysis, Sensitivity Analysis
8	Abdelmohsen (2016)	PhD	USA	Non-Dominated Sorting Genetic Algorithm
9	Patil (2015)	MASc	USA	Statistical Analysis
10	Yeom (2015)	PhD	USA	Regression Analysis, Statistical Analysis
11	Kim (2008)	PhD	USA	Probability Analysis
12	Kang (2006)	PhD	USA	Extended Kalman Filtering (EKF) Algorithm
13	Qu (2014)	PhD	USA	Genetic Algorithm, Statistical Analysis
14	Li (2008)	MASc	USA	Statistical Analysis, Cost-Benefit Analysis
15	Chin (2004)	PhD	USA	Probability Analysis
16	Zhang (2016)	MASc	USA	Agent-Based Dynamic Traffic Assignment
17	Du and Chien (2014)	Journal	USA, China	Statistical Analysis, Sensitivity Analysis
18	Kang and Chang (2006)	Journal	USA	Min-Max Optimization Criterion
19	Ahmed et al. (2016)	Journal	UAE, USA	Statistical Analysis
20	Ng (2012)	Journal	USA	Integer Linear Programs
21	Kang et al. (2004)	Journal	USA	Lindo-API Linear Optimization Program
22	Isaiah (2014)	MASc	USA	Qualitative Analysis, Citable Test Analysis
23	Son (1999)	Journal	Korea	Statistical Analysis
24	Kurker et al (2014)	Report	USA	Statistical Analysis, Sensitivity Analysis
25	Kurker (2013)	MASc	USA	Statistical Analysis
26	Ding et al. (2013)	Journal	China, USA	Regression Analysis
27	Weng and Meng (2011a)	Journal	Singapore	Decision Tree-Based Approach

* USA – United State of America UAE – United Arab Emirates SWZ- Switzerland NLD – Netherland

2.2 Safety Applications

Chambless et al. (2002) determined the important characteristics that were over-represented in work zone crashes. These included severity, driver characteristics, location and misjudgement. This study involved a large crash dataset from Alabama, Michigan and Tennessee. Schrock and Ullman (2003) conducted research on the spacing of law enforcement pullout areas in highway work zones. They concluded that the spacing for law enforcement pullout areas should be 2 to 3 miles and no longer than 3 miles for alternative work zones with closed emergency shoulders. They also identified the speed reduction potential of fluorescent orange sheeting, innovative message signs and changeable message signs with radar. Li (2007) proposed 4 Crash Severity Index (CSI) models including 2 Driver-Independent

(DI-CSI) models and 2 Driver-Dependent (DD-CSI) models. The DI-CSI models estimate driving risk levels in work zones by incorporating travel conditions without human factors. The DD-CSI models estimate driving risk levels in work zones by assessing driver factors such as demographic characteristics and driver error.

Li and Bai (2008) conducted a study to identify the deficiencies of current Temporary Traffic Control (TTC) measures. Chen (2008) developed a crash prediction model using regression analysis. The results revealed that the construction type was a significant factor in work zone crashes. Chen (2008) also developed a VISSIM model and did a conflict analysis based on the simulation results. Table 2 presents the references involving work zone safety.

Table 2: Safety references on the safety aspect

No	Sample References	Type	Country	Analytical Tools Used
1	Qi and Zhao (2017)	Journal	USA	Statistical Analysis
2	Li and Bai 2008(a)	Journal	USA	Regression Analysis, Statistical Analysis
3	Mathes (2012)	MASc	USA	Regression Analysis, Statistical Analysis
4	Abdelmohsen (2016)	PhD	USA	Optimization - NSGA2, Statistical Analysis
5	Chen (2008)	PhD	USA	Regression Analysis, Statistical Analysis
6	Chambless et al. (2002)	Journal	USA	Graphical & Statistical Analysis
7	Elghamrawy (2011)	PhD	USA	Optimization, Statistical & Sensitivity Analysis
8	Nannapaneni (2011), Miatudila (2016)	PhD	USA	Statistical Analysis
9	Li (2007)	PhD	USA	Statistical Analysis, Regression Analysis
10	Wong (2009)	MASc	USA	Cost-Benefit & Regression Analysis
11	Bai et al. (2010)	Journal	USA	Statistical Analysis
12	Li and Bai (2009)	Journal	USA	Statistical Analysis, Regression Analysis
13	Phanomchoeng et al. (2010)	Journal	USA	Sound Wave Mechanics, Graphical Analysis
14	Schrock and Ullman (2003)	Journal	USA	Delphi Method, Graphical Analysis
15	Kurker et al. (2014)	Report	USA	Statistical Analysis, Sensitivity Analysis
16	Ding et al. (2013)	Journal	China	Regression Analysis
17	Meng et al. (2010)	Journal	Singapore	Quantitative Risk Assessment Model

Wong (2009) introduced a Risk Index (RI) to measure the level of injury risk in work zone. Wong conducted extensive studies of the safety benefits of ArmorGuard™ barriers and concluded that these barriers could be used to assure worker safety in work zones. Elghamrawy (2011) conducted field experiments to evaluate the effectiveness of temporary rumble strips prior to and at the edge of work zones by analyzing the sound levels generated by the tires. Nannapaneni (2011) used questionnaires and field tests to evaluate the effect of Steady Burn Lights (SBLs) on channelizing drums vs crash occurrence at night. The results revealed that SBLs on drums increased the occurrence of risky driver behaviour and higher speeds. Mathes (2012) introduced the job site line of communication model to assure the safety of workers.

Ding et al. (2013) assessed the safety of work zones by looking at the length of the transition area and the speed limit values. Kurker et al. (2014) used his work zone VISSIM model vehicle trajectories in FHWA's Surrogate Safety Assessment Model (SSAM) to conduct the traffic conflict assessment to quantify the safety. Abdelmohsen (2016) developed a model to minimize the occurrence of work zone crashes using multi-objective optimization techniques. Miatudila (2016) introduced multinomial logit models to analyze crash severity and countermeasures to mitigate crash occurrence. Qi and Zhao (2017) presented an interesting signalized lane control strategy and cycle lengths to minimize conflicts related to lane merge and rear-end collisions on freeways.

2.3 Driver Behaviour Applications

Whitmire (2007) investigated the effectiveness of in-vehicle information technologies. The results indicated that audio information requires less reaction time than visual information. Heaslip et al. (2011) evaluated the effectiveness of selected work zone guidelines published by FHWA for older drivers and pedestrians. Adeli (2014) evaluated driver speed variations according to speed limits and road work signs. The results revealed that drivers generally complied with the speed limit. The results also revealed that age, road familiarity and experience had a noteworthy impact on speed limit compliance. Zehtabi (2014) conducted a study on driver perception of surroundings in work zones. The prospect-theory and Genetic Algorithm (GA) were used to develop the longitudinal acceleration model. The variables taken in to account were the work zone length, TTC barrier types, intensity and speed. Moradpour et al. (2015) compared the Conventional Lane Merge (CLM) configurations used in MUTCD and the Missouri Department of Transportation (MoDOT) configurations according to TTC and demographical influence. Table 3 presents the references involving driver behaviour in work zones.

Table 3: Driver behaviour references on the driver behaviour aspect

No	Sample References	Type	Country*	Analytical Tools Used
1	Hamdar et al. (2016)	Journal	USA, Lebanon	Optimization – Genetic Algorithm, Statistical Analysis
2	Heaslip et al. (2011)	Journal	USA	Statistical Analysis
3	Shakouri et al. (2016)	Journal	USA	NASA-TLX & Motion Sickness Assessment
4	Adeli (2014)	MASc	USA	Statistical Analysis
5	Nadathur and Narayanan (2016)	MASc	USA	Optimization – Solver
6	Zehtabi (2014)	MASc	USA	Optimization – Genetic Algorithm, Statistical Analysis
7	Moradpour et al. (2015)	Journal	USA	Statistical Analysis
8	Whitmire (2007)	PhD	USA	NASA-TLX & Motion Sickness, Statistical Analysis
9	Weng and Meng (2011b)	Journal	Singapore	Traffic Flow Fundamental Diagram

Shakouri et al. (2016) used questionnaires and driving simulator studies to examine driver merging behaviour in work zone configurations. A motion sickness assessment was also done for all of the participants in order to ensure confidence when taking the test. The findings supported previous research on the effects of work zone configurations and traffic density on performance variables and the subjective work load of drivers.

2.4 Planning Applications

Schonfeld et al. (2002) used a multi-objective optimization technique to develop a model to minimize total cost by optimizing work zone length and cycle times. Sukumaran et al. (2006) presented a model for the stochastic analysis of factors affecting work zone schedules using Monte-Carlo simulations and the Analytic Hierarchy Process (AHP). Tang and Chien (2008) also developed a model to schedule work zones by considering a discrete time-cost relation using a genetic algorithm optimization process. Kaewmoracharoen (2009) developed a work zone visualization model using 3-D models linked with scheduling data. He also examined the industrial influence on work zone visualization using a questionnaire. The results indicate that work zone visualization is a low cost communication technique that can also increase mobility. Table 4 presents the references involving work zone cost and planning.

Table 4: Planning references on planning aspect

No	Sample References	Type	Country	Analytical Tools Used
1	Abdelmohsen (2016)	PhD	USA	Optimization-NSGA2, Statistical Analysis
2	Schonfeld et al. (2002)	Journal	USA	Optimization , Sensitivity Analysis
3	Elghamrawy (2011), Qu (2014)	PhD	USA	Optimization – GA, Statistical & Sensitivity Analysis
4	Du and Chien (2014)	Journal	USA, China	Statistical Analysis, Sensitivity Analysis,
5	Tang and Chien (2008)	Journal	USA	Optimization – GA, Elitist Selection, Sensitivity Analysis
6	Sukumaran et al. (2006)(a)	Journal	USA	Monte Carlo Simulation, Statistical Analysis
7	Sukumaran et al. (2006)(b)	Journal	USA	Monte Carlo Simulation, Analytic Hierarchy Process (AHP)
8	Schonfeld et al. (2002)	Journal	USA	Optimization – Iteration, Sensitivity Analysis
9	Kaewmoracharoen (2009)	PhD	USA	Statistical Analysis
10	Yang (2010)	PhD	USA	Simulated Annealing, Statistical Analysis
11	Chou (2010)	PhD	USA	Simulation, Regression Analysis, Sensitivity Analysis
12	Qu (2014)	PhD	USA	Optimization – Genetic Algorithm, Sensitivity Analysis
13	Wiegand (2007)	MASc	USA	Statistical Analysis
14	Hajdin and Lindenmann (2007)	Journal	Switzerland	Optimization
15	Sukumaran et al. (2006)	Journal	USA	Monte Carlo Simulation, Statistical Analysis

Yang (2010) presented a framework and systematic methodology for optimizing critical work zone decisions by considering cost-effectiveness. Yang also developed analytical and simulation models using Two-Stage Modified Population-Based Simulated Annealing (2PBSA) to search for near-optimal solutions. Chou (2010) presented the Simulation Based Secondary Incident Filtering (SBSIF) methodology to evaluate proposed Traffic Incident Management (TIM) programs. Elghamrawy (2011) worked on developing an optimization model to minimize work zone costs involving construction costs, user delay costs and accident costs. Qu (2014) developed a work zone schedule model that considers work zone cost variables and incorporates Travel Time Reliability (TTR) measures. Du and Chien (2014) presented a model to optimize work zone length by considering the dynamic traffic volume and road capacity. They also created guidelines for the use of the road shoulder for various work zone scenarios. Abdelmohsen (2016) developed a multi-objective optimization model for work zone layouts to minimize construction costs. The study evaluated TTC measures and other work zone layout parameters including flaggers, spotters and TTC devices.

3 Design Variables and Limitations

The literature review revealed that a large variety of decision variables were taken into account when studying all of the aspects related to work zones. The most common decision variables for mobility studies included work zone configurations, AADT, speed, design speed, delay, volume, lane capacity, head-way distance, head-way times, directional split, vehicle length, weather, lighting, vehicle classifications, heavy vehicle percentage, VSL, road class, position of closed lane and length of ramp. Aside from traffic flow characteristics, fatal and crash data, crash characteristics, TTC types, VMSs, Intelligent Transportation Systems (ITS) technologies, regulations and policies, work zone layout details, work zone types, traffic control type, road surface, gender, age, time, day, type of collision, accident location, light and weather were the most common decision variables for safety studies. TTC barrier types, work zone intensity, age, education level, income, car ownership, driving experience, driving stress and human related factors, acceleration, braking and lane position were the most common decision variables in driver behaviour studies. Agency cost (maintenance, traffic mitigation, equipment/labour idling, routine maintenance) and user cost (delay, vehicle operating, expected accident cost) were the most significant decision variables cost and planning studies.

Beyond the remarkable applications reviewed in this paper, there were also limitations. Most of the researchers listed the limitations of their studies. Real time high precision data acquisition, data screening, accessibility to emergency responders, oversaturated flow conditions, heavy truck impact, access & egress methods and lane merge position (left to right or right to left) were some of the highlighted limitations in the mobility studies. The unavailability of crash data, less information in crash inventories, validation of the proposed models, influence of different channelizing devices, driver cognitive process (gap acceptance, environment), nighttime luminance (brightness) requirements and the effect of the property damage only crash type were some of the highlighted limitations in the safety studies. Fewer participants and a lack of proper demographical representation for driving simulators, the adoption of advanced data collection technologies, practices in autonomous vehicles, and improvements to the car-following theory were some of the key limitations pertaining to driver behaviour studies. The enhancement of the robustness of the developed models, emission and fuel consumption aspects through simulation results, test the convergence of optimization and relationship between maintenance cost and project duration cost were some of the noteworthy limitations in the planning and cost studies.

4 Future Research

Mathematical theories such as the prospect theory can be used to approximate the human decision making process for the development of car following and lane merge models in work zones. The effect of horizontal and vertical alignment can be studied using microsimulation and the development of equations to estimate traffic flow parameters such as delay, capacity and average speed. It is critical to develop guidelines for the establishment of work zone speed limits and how speed reduction distribution regulate through up stream to merging area considering design speed of road, traffic flow data and work zone layout parameters. It would also be valuable to develop matrices to measure emissions due to work zones and minimization strategies involving the optimization of mobility. The application of multi-objective optimization concepts considering two or more parameters would be useful in order to generate the optimal trade-offs between them. The development of in-vehicle advanced warning systems using sophisticated ITS technologies which warn drivers of upcoming work zones is also inevitable. These technologies should provide advice on which lane to chose and which lane is closed in order to prevent aggressive maneuvering queue propagation. It is noteworthy that there were very few studies conducted on the joint merge concept. It would be valuable to analyze this concept and its applicability to selected highways in order to enhance mobility and safety in work zones.

Although there have been studies on the use of barriers at work zones, it is necessary to develop guidelines and recommendations for barrier types according to their benefits and applicability considering mobility and safety. As many studies have shown, the development of a work zone related standard crash database is paramount in order for researchers and policy makers to save more lives with more reliable countermeasures. There is also a need for more nighttime safety related studies involving things such as luminosity requirements for high visibility traffic control devices, signs and workers. The development of guidelines and criteria for the use of positive protection devices in work zones is also necessary in order to minimize safety hazards due to errant vehicles. Driving simulator studies with an increase in participants and more realistic conditions will lead to higher accuracy in the prediction of human behaviour in work zones. Guidelines can also be developed to plan and schedule work zones using optimization techniques. A review of the work zone literature also revealed a lack of research focused on Canadian work zones, especially regarding safety.

With the rapid advances in science and technology, the topic of “Autonomous Vehicles” is becoming more significant with each passing day. With the collaboration of intelligent transportation systems and cutting-edge mechatronic platforms, it will be necessary to study the behaviour of autonomous vehicles with the presence of smart work zones and the concepts already proposed. Microsimulation tools and driving simulators can be incorporated for research purposes. On the other hand, the development of concepts and guidelines for smart work zones that include sensor networks, real time data acquisition, data warehousing, real time signalization and merge concepts (early merge and late merge), and the application of positive protection devices with high accuracy will be important for future work zone research.

5 Conclusions

This article reviewed different aspects of work zone applications such as mobility, safety, driver behaviour, cost and planning. Work zones are a significant source of freeway accidents, generating road user delays, traffic incidents, safety hazards and vehicle emissions. Work zone mobility and safety were the mostly conducted research studies among the literature reviewed which were penetrated in to the subareas of capacity optimization, delay minimization and safety assurance. Work zone configuration played a major role in all aspect and cutting-edge smart work zone concepts were applied in many phases of work zone studies. Demographical and psychological phenomenon were thoroughly examined whereas modern technology mostly involves driver behavioural measures. Finally, optimal planning and cost minimization strategies were proposed to incorporate in reliable scheduling and refined policies in states and agencies to present more effective and user ready work zone environment. It is important to mention that a limited number of studies and a brief summary of variables, limitations and future research needs were described in this paper due to space constraints. In this review, the important elements of each publication were tabulated to give the reader quick access to the relevant references. Apart from the existing approaches, the design variables, limitations, and future research needs are listed separately for prospective users. It is our hope that this paper will serve as a quick freeway work zone reference and a guide for potential researches and practitioners.

References

Abdelmohsen, A. Z. 2016. Optimizing the construction planning of highway work zones. ProQuest Dissertations Publishing.

Adeli, A. 2014. Work zone speed analysis using driving simulator data. ProQuest Dissertations Publishing.

Ahmed, K., Al-Zoubi, K., Siddiqui, M. A. and Anas, M. 2016;2015. Evaluation of the effectiveness of portable variable message signs in work zones in united arab emirates. *IET Intelligent Transport Systems* 10 (2): 114-21.

Bai, Y., Finger, K. and Li, Y. 2010. Analyzing motorists' responses to temporary signage in highway work zones. *Safety Science* 48 (2): 215-21.

Bie, Y., Xu W., Lei, N. and Tony, Z. Q. 2017. Effect of speed limits at speed transition zones. *Canadian Journal of Civil Engineering* 44 (1): 10-7.

Chambless, J., Ghadiali A. M., Lindly J. K. and McFadden, J. 2002. Multistate work-zone crash characteristics. *Institute of Transportation Engineers. ITE Journal* 72 (5): 46.

Chen, Y. 2008. A highway work zone design and traffic management decision system. ProQuest Dissertations Publishing.

Chin, C. 2004. A slot model for highway flow optimization through entry, exit, and flow control. ProQuest Dissertations Publishing.

Chou, C. 2010. Understanding the impact of incidents and incident management programs on freeway mobility and safety. ProQuest Dissertations Publishing.

Chung, Y. 2011. Assessment of non-recurrent traffic congestion caused by freeway work zones and its statistical analysis with unobserved heterogeneity. *Transport Policy* 18 (4): 587-94.

Ding, M., Zhang, C., Yan, J. and Tsai, J. 2013. Key factors analysis of work zone safety and mobility based on micro-simulation. Procedia - Social and Behavioral Sciences 96: 582-90.

Du, B. and Chien, S. I. 2014. Feasibility of shoulder use for highway work zone optimization. Journal of Traffic and Transportation Engineering (English Edition) 1 (4): 235-46.

Elghamrawy, T. M. 2011. Optimizing work zone practices for highway construction projects. ProQuest Dissertations Publishing.

Hajdin, R. and Lindenmann, H. 2007. Algorithm for the planning of optimum highway work zones. *Journal of Infrastructure Systems* 13 (3): 202-14.

Hamdar, S. H., Hiam, K. and Zehtabi, S. 2016. A simulator-based approach for modeling longitudinal driving behavior in construction work zones: Exploration and assessment. *Simulation* 92 (6): 579-94.

Heaslip, K., Collura, J. and Knodler, M. 2011. Evaluation of work zone design features to aid older drivers. *Institute of Transportation Engineers. ITE Journal* 81 (3): 36.
Isaiah, S. E. 2014. Testing and evaluation of methods for automated speed identification on highway work zones. ProQuest Dissertations Publishing.
Kaewmoracharoen, M. 2009. Feasibility of visualization and simulation applications to improve work zone safety and mobility. ProQuest Dissertations Publishing.
Kang, K. 2006. Development of optimal control strategies for freeway work zone operations. ProQuest Dissertations Publishing.
Kang, K. and Chang, G. 2006. A robust model for optimal time-of-day speed control at highway work zones. *IEEE Transactions on Intelligent Transportation Systems* 7 (1): 115-23.
Kang, K., Chang, G. and Zou, N. 2004. Optimal dynamic speed-limit control for highway work zone operations. Transportation Research Record: Journal of the Transportation Research Board 1877: 77-84.
Kim, C. 2008. A mechanistic model of work zone capacity. ProQuest Dissertations Publishing.
Kurker, M. G. 2013. Evaluation of freeway work zone merge concepts. ProQuest Dissertations Publishing.
Li, X. 2008. Evaluate the effectiveness of the speed monitoring display for work zones in las vegas. ProQuest Dissertations Publishing.
Li, Y. 2007. Analyzing highway work zone crashes and traffic control effectiveness. ProQuest Dissertations Publishing.
Li, Y. and Bai, Y. 2009. Highway work zone risk factors and their impact on crash severity. *Journal of Transportation Engineering* 135 (10): 694-701.
Li, Y. and Yong, B. 2009. Effectiveness of temporary traffic control measures in highway work zones. Safety Science 47 (3): 453-8.
Liu, P., Zhang, J., Qu, J., Lu, J., Cheng, Y. and Tan, H. 2017. Evaluation of mobility impact on urban work zones using statistical models. Journal of Central South University 24 (6): 1513-21.
Mathes, J. A. 2012. Integrated risk management for improving internal traffic control, work zone safety, and mobility during major construction. ProQuest Dissertations Publishing.
Meng, Q., Weng, J., and Qu, X., 2010. A probabilistic quantitative risk assessment model for the long-term work zone crashes. Accident Analysis and Prevention, 42(6), 1866-1877.
Miatudila, A. S. Sr. 2016. Modeling the crash injury severity in work zone areas on freeways. ProQuest Dissertations Publishing.
Moradpour, S., Wu, S., Long, S., Ming, C. L., Dincer, K. and Qin, R. 2015. Use of traffic simulators to determine driver response to work zone configurations. *Proceedings of the International Annual Conference of the American Society for Engineering Management*: 1.
Nadathur, V. R. and Narayanan, H. 2016. Data analysis for driving pattern identification and driver's behavior modeling in a freeway work zone. ProQuest Dissertations Publishing.
Nannapaneni, P. L. V. 2011. Evaluating the use of steady burn warning lights on drums for work zone safety. ProQuest Dissertations Publishing.
Ng, M. 2012. Traffic flow Theory–Based stochastic optimization model for work zones on two-lane highways. *Journal of Transportation Engineering* 138 (10): 1269-73.
Patil, S. S. 2015. Evaluation of portable traffic signals in conjunction with pilot car operations at two-lane, two-way temporary rural work zones in kansas. ProQuest Dissertations Publishing.
Phanomchoeng, G., Rajamani, R. and Hourdos, J. 2010. Directional sound for long-distance auditory warnings from a highway construction work zone. IEEE Transactions on Vehicular Technology 59 (5): 2266-76.
Qi, Y. and Zhao, Q. 2017. Safety impacts of signalized lane merge control at highway work zones. *Transportation Planning and Technology* 40 (5): 577.
Qu, T. 2014. Travel time reliability based work zone scheduling. ProQuest Dissertations Publishing.
Radwan, E., Zaidi, Z. and Harb, R. 2011. Operational evaluation of dynamic lane merging in work zones with variable speed limits. *Procedia - Social and Behavioral Sciences*16: 460-9.
Schonfeld, P., Steven C. and Yimin T. 2002. Optimizing work zones for two-lane highway maintenance projects. *Journal of Transportation Engineering* 128 (2): 145-55.
Schrock, S. and Ullman, G. 2003. Spacing of law enforcement pullout areas in highway work zones. *Transportation Research Record: Journal of the Transportation Research Board* 1824: 37-43.
Shakouri, M., Laura, H. I., Fereydoun, A. and Sherif, I. 2016. Drivers' merging behavior data in highway work zones. *Data in Brief* 6: 829-32.

Son, Y. T. 1999. Queueing delay models for two-lane highway work zones. *Transportation Research Part B* 33 (7): 459-71.
Sukumaran, P., Bayraktar, M. E., Hong, T. and Hastak, M. 2006. Model for analysis of factors affecting construction schedule in highway work zones. *Journal of Transportation Engineering* 132 (6): 508-17.
Sukumaran, P., Hong, T., Bayraktar, M. E. and Hastak, M. 2006. Validation of a model for predicting schedule changes in highway work Zones—Case studies. Journal of Transportation Engineering 132 (8): 638-48.
Tang, Y. and Chien, S. 2008. Scheduling work zones for highway maintenance projects: Considering a discrete time-cost relation. *Transportation Research Record: Journal of the Transportation Research Board* 2055: 21-30.
Turley, B. M. 2002. Daniel B. fambro student paper award: Dancing diamonds in highway work zones: A evaluation of arrow panel caution displays. Institute of Transportation Engineers. ITE Journal 72 (11): 34.
Weng, J., and Meng, Q., 2011a. Decision tree-based model for estimation of work zone capacity. Transportation Research Record, 2257, 40-50.
Weng, J., and Meng, Q., 2011b. Modelling speed-flow relationship and merging behavior in work zone merging areas. Transportation Research Part C, 19(6), 985-996.
Whitmire, J. II. 2007. The effect of in -vehicle warning systems on driver response in work zones. ProQuest Dissertations Publishing.
Wiegand, J. D. 2007. State transportation agencies' utilization of work zone congestion mitigation strategies. ProQuest Dissertations Publishing.
Wong, J. M. 2009. Analysis of ArmorGuard™ work zone protection barrier system. ProQuest Dissertations Publishing.
Yang, N. 2010. Optimization of highway work zone decisions considering short-term and long-term impacts. ProQuest Dissertations Publishing.
Yeom, C. H. 2015. Statistical and simulation models of freeway work zones. ProQuest Dissertations Publishing.
Zech, W. C., Mohan, S. and Dmochowski, J. 2005. Evaluation of rumble strips and police presence as speed control measures in highway work zones. Practice Periodical on Structural Design and Construction 10 (4): 267-75.
Zehtabi, S. 2014. Work zone characteristics in microscopic acceleration modeling: Calibration and numerical analysis. ProQuest Dissertations Publishing.
Zhang, K. 2016. Traffic impact analysis of several dynamic lane management strategies for congestion mitigation based on DTA model. ProQuest Dissertations Publishing.
Zheng, N., Andreas, H., Serge H., Henk, V. Z. and Peters, D. 2011. A comparison of freeway work zone capacity prediction models. Procedia - Social and Behavioral Sciences 16: 419-29.
Zhu, W. 2015. Two-lane highway work zone capacity model and control analysis. ProQuest Dissertations Publishing.

Building Tomorrow's Society

Bâtir la Société de Demain

Fredericton, Canada

June 13 to 16, 2018 / *13 au 16 juin, 2018*

FUZZY PROBABILISTIC APPROACH FOR RISK ASSESSMENT OF BOT TOLL ROADS IN INDIAN CONTEXT

Patel, Tirth[1,4], Bhatt, Tejas[2], and Bhatt, Naimish[3]
[1] U.V.Patel College of Engineering, Ganpat University, India
[2] U.V.Patel College of Engineering, Ganpat University, India
[3] Pandit Deendayal Petroleum University
[4] tdpatel1610@gmail.com

Abstract: The BOT toll model has been a common approach for highway construction in India .Due to the large amount of investment, many stakeholders and long concession period, BOT toll roads become susceptible to different risks. Risk assessment is one of the essential steps of risk management that enables professionals to quantify and analyze the risks that may hamper the project performance in terms of cost, quality, safety, and time. The risk assessment is one of the difficult task because BOT toll roads have complex structure. In this study, total seventy-one risks have been identified and categorized under fourteen risk factors on a basis of case-studies of Indian toll roads and literature review. A structured questionnaire was prepared then filled by contractors, government officers, academicians, project managers and consultants. For risk assessment, a systematic quantitative based fuzzy probabilistic model is proposed with the help of LAB-VIEW, as a risk assessment technique to simulate the impreciseness of human judgment and to improve the assessment accuracy. This comprehensive risk assessor model can be used in BOT toll roads in Indian scenario for prioritization of the critical risk. It is important for the public and private sectors to establish an effective risk assessor model for public-private partnership (PPP) projects in order to achieve win-win condition for both.

1 INTRODUCTION

BOT (Built-transfer and operate) is the most adopted model of PPP model, in which the concessionaire is responsible to design, build, operate (for concession period) and transfer back the infrastructure to the government or public body in return of that concessionaire will collect the revenue from road user (worldbank, 2018). From last decade, the BOT toll model has played important role in the infrastructure sector in India. Unfortunately, BOT toll roads have suffered various risks that are responsible for cancellations and/or remarkable renegotiations in result of that time cost overrun and cost overrun of the projects happens .In the last few years, India has faced a challenge of private investment in infrastructure and the growth of gross fixed capital formation (GFCF) of India has been disappointing due to unfavorable market condition (Department of Economic Affairs, India, 2015).Recently, Indian government emphasized the need for increasing public investment in infrastructure as well as a review for the mechanism of PPP model. The arrangements complexity and conflicting nature of the contract of PPP projects have led to increased risk exposure for all the stakeholders. (Woodward 1995). The most difficult task in risk management is the risk assessment because it contains the evaluation of the probability of occurrence of risks and their impacts on a project's objectives (Thomas, 2006). It is hard to tackle all potential risks in a construction project. There is essential to focus on the critical risks, because to identify all the risks is time consuming and baleful (El-Sayegh, 2008).It is required to identify the most critical risks and control them (Barkley, 2004)

Due to long lead times, the requirement of the large capital expenditure and frequent change in regulations BOT projects are highly susceptible of risk .Risk Management can play significant role in the successful delivery of the project, as it provides a chance for stakeholders to understand the risks better and assess the potential problems then mitigate them (Zhang, 2007).

This Study considered different risks of BOT toll roads in Indian context and the assessment of different risks by fuzzy probabilistic model.

2 BACKGROUND OF STUDY

India has the second largest road network in the world, having a road network of over 6,603,293 kilometre kilometer (NHAI, 2018). Due to Economic growth, Increased the thrust of government, inclination towards the road in freight traffic, eruption of private participation and outburst in passenger traffic and vehicle density are pivotal growth drivers in road sector investments. In the next five years, investment via PPPs are predicted to be in the India of USD31 billion for National highways. (IBEF, 2018).So, PPP models are rapidly becoming the preferred approach to provide infrastructure India because PPPs have been seen as a solution to cope up with inefficient governmental funds for infrastructure development in India The twelfth Plan estimated an investment of USD 833.725 billion (at current prices) in infrastructure during the Plan. The following figure 1 illustrates the awarded road projects under BOT model and also shows declination trend of awarding the road project has been clearly remarked. (IBEF, 2018)

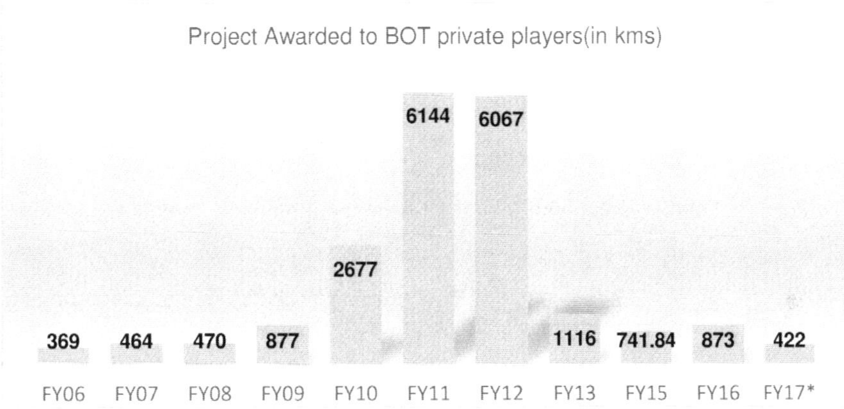

Figure 1: Awarded project year wise (Source: IBEF, 2018)

Risk management is considered crucial and integral tool of project management (Tang, 2007). For BOT toll roads, risk management should be done effectively by identifying the risks associated with each activity of the project. These risks can be assessed or measured in terms of probability and impact.

The objective of the study is to identify different risk factors and its categorization by literature review and expert interviews and to do risk ranking by the fuzzy probabilistic approach by lab-view.

3 RISK MANAGEMENT

Risk management is the field that identifies various risks in condition, evaluates their significant impact on important performance measures, enact measures to control or prevent the underlying risks in project and allocate the risks to external parties (Smith, 2006).

Risk can be understood as an evolution of the probability, severity and exposure to all hazards for an activity (Almishari, 2003).

Risk is an intricate event that has physical, monetary, and social feature (Loosemore, 2008).The risk identification is the very first step of the risk assessment. A structured questionnaire was prepared after reviewing literatures and interacting field experts which includes seventy-one risks .These risks have been categorized under fourteen risk factors. The questions were related to the probability and impact of each risk on a five-point Likert scale.

Fore response planning, risk priority can be used for response planning. The risk priority can be used to prioritize risks for further quantitative assessment or response planning (PMI, 2008).

4 IDENTIFIED RISKS

Following risks have been identified after reviewing different literature and interacting with field experts.

Table 1: Identified risks of BOT toll roads

Risk Factor	Sub Risk
Project Management Risk	Use of appropriate planning tools and techniques by project manager
	Lack of induction and training of human resources
	Negative attitude of project manager
	Lack of coordinating ability and rapport of project manager with other contractors at site
	Lack of effective monitoring and feedback by project manager
	Absence of safety management
Construction Risk	Change in the scope of work
	Construction cost overrun
	Construction time overrun
	Poor quality workmanship
	Frequent change of subcontractor (Disputes between contractors and Subcontractors)
	Insolvency/default of sub-contractors or supplier
	Poor site management and supervision by contractor
	Improper construction methods/quality variations
Contractual Risk	Contract clause being ambiguous/one-sided
	Conflict in Contract documents
	Poorly Tailored Contract forms
	Original contract duration is rigid and has no scope to accommodate any changes
Consultant Related Risk	Insufficient data collection and methodology for traffic forecasting
	Delay in performing inspection and testing by consultant
	Design deficiency
	Inflexibility of consultant
	Non- use of advanced engineering design software
	Unclear and inadequate details in drawings
	Inadequate experience of consultant with regard to type of project

Political Risk	Legislation change
	Politically motivated Resistance
	Not good relation with Government Relations
	Delay due to government Policy restriction and land transfer
	Termination of agreement by government.
	Attitude of Government toward foreign investor/ investor
Development Risk	Defect in request for proposal
	Delays in approval submittals
	Bid risk
	Delays in Preparation submittals
Financial Risk	Change in Currency rate
	Change in Interest Rate
	Equity Risk
	Inflation
	Liquidity
	Improper choice/methodology of raising capital
	Not able to achieve the reasonable profit targets
Legal risk	Change in Law
	Change in regulation
	Industrial regulatory change
	Labor regulation change
Natural Risk	Force majeure
	Geotechnical condition
	Weather
	Environment
Operational Risk	Operation cost overrun
	Operational revenues below expectation
	Low operating productivity
	Maintenance costs higher than expected
	Maintenance more frequent than expected
Regulatory Risk	Environmental clearances
	Social Impact Assessment
	Bribes
Social risk	Lack of tradition of private provision of public services
	Public opposition to project

	Land Acquisition
Resource Risk	Labor shortage/strike
	Delay in materials delivery
	Changes in material types and specifications during construction
	Improper selection of equipment
	Equipment breakdowns
	Shortage of equipment
	Lack of planning with respect to material scheduling and assigning priorities
	Unqualified workforce
Traffic Risk	Traffic shortfalls
	Impossibility to charge user at the projected rates

5 RESEARCH METHODOLOGY

For evaluation and analyzing the data of identified risks, a rule based fuzzy approximation model has been designed. The study covers two antecedents and one consequent for modeling of a fuzzy inference engine. The antecedents are probability and impact, one consequent is risk.

5.1 Collection of data through questionnaire

Questionnaires are undoubtedly the quickest and most efficient approach of learning the views of all project stakeholders and permitting these views to be scrutinized and compared. A structured questionnaire was prepared on the basis of the literature and case-study review. Then, Face-to-face Interviews have been carried out. A questionnaire included a five-point Likert- scale, facilitating statistical evaluation of the risk. The questionnaire included two parameters, probability and impact of the risk. Total forty-five questionnaires were presented academia and industry experts representing various organizations: contractor, government officers, academicians, project managers and consultants followed by interviews with each participant.to the various experts of the field. Out of that forty experts Author got thirty one responses filled by the various field experts.

5.2 Fuzzification

From the questionnaire survey data of probability and impact have derived, which are crisp inputs. By Fuzzification process, the author has converted crisp inputs into fuzzy inputs. In this process, membership function is applied to the measurements and the degree of truth in each premise is determined. Determination of all system variables requires for the measurement purpose and then membership value is given.

5.2.1 System Variables for analysis

The system contains two input variables and one output variable with different predicates. Probability has five predicates very-low, low, moderate, high and very-high; Impact has five predicates very-low, low, and moderate, high and very-high. Accordingly, output variable risk can be determined. Risk has five predicates very-low, low, and moderate, high and very-high. The Value of two inputs can be obtained from the arithmetic mean of that particular risk which has been enumerated by questionnaire survey. Following figure 2 shows the fuzzy-logic procedure

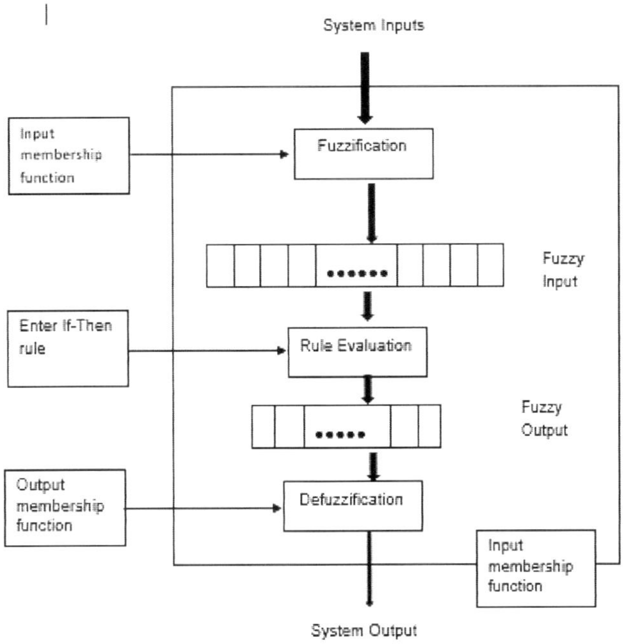

Figure 2: Flow chart of fuzzy logic

5.2.2 Fuzzy inference engine

An inference engine concludes the fuzzy control actions by emulating the decision-making in depicting and implementing knowledge about how best to control the process. It is formed by a set of IF-THEN rules and membership functions (Yen & Langeri, 2005). The fuzzy rule is:

If <antecedent 1> Θ <antecedent 2>... Θ <antecedent N> then <consequent 1> Θ <consequent 2>... Θ <consequent N>

Θ represents the logical operator connected by AND logical operator. The author has designed total 25 rules for risk evaluation, risk evaluation can be determined by executing rules. Some of the fuzzy rules for a risk evaluation are as follows:

- If (Probability is Very Low) and (Impact is Very low) then (Risk is Very Low risk)
- If (Probability is Medium) and (Impact is Medium) then (Risk is Intermediate risk)
- If (Probability is Very high) and (Impact is Very high) then (Risk is Very high risk)

5.2.3 Membership Function

In a fuzzy set theory, the main objective of the membership function is the translation of real world values to fuzzy values. A trapezoid membership function is used here and its syntax is:

[1] Trapmf(x, [a b c d]) $= 0 \quad x \le a$

$= (x - a) / (b - a) \quad a \le x \le b$

$= 1 \quad b \le x \le c$

$= (d - x) / (d - c) \quad c \le x \le d$

$= 0 \quad d \le x$

This represents, trapezoid curve as function of vector x and depends on four scalar parameters a, b, c and d. The parameters a, d represent the feet of a trapezoid and the parameters b, c represent the shoulder of trapezoid (Yen & Langeri, 2005).

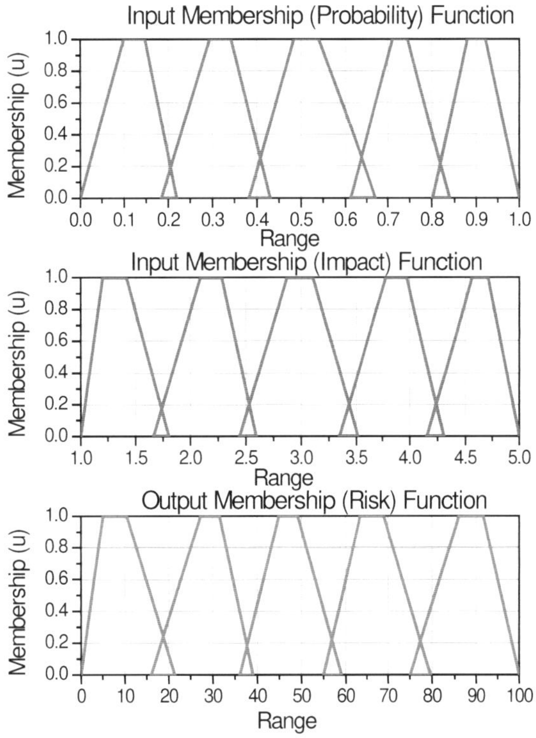

Probability Membership Function
Very Low [0 0.0989 0.147 0.219]
Low [0.184 0.294 0.342 0.430]
Medium [0.382 0.484 0.540 0.671]
High [0.612 0.711 0.745 0.839]
Very high [0.802 0.879 0.922 1]
Impact Membership Function
Very low [1 1.203 1.417 1.802]
Low [1.663 2.0909 2.283 2.593]
Medium [2.443 2.871 3.106 3.513]
High [3.342 3.780 3.973 4.304]
Very high [4.155 4.561 4.711 5]
Output Variable: Risk Score
Very low risk [0 5.080 10.427 21.390]
Low risk [16.043 27.272 31.550 39.037]
Intermediate risk [36.096 44.919 49.197 59]
High risk [55 63.369 68.984 79.679]
Very high risk [75.133 86 91.711 100]

Figure 3 : Membership function of inputs and output

5.2.4 Defuzzification

Defuzzification converts the conclusions of the inference engine into crisp output. All the fuzzy outputs values effectively modify their respective output membership functions. Author used one of the most commonly used techniques in defuzzification is center of the area or centroid method.

6 DATA ANALYSIS WITH RISK ASSESSOR MODEL

The fuzzy simulation output displays a road map of the whole fuzzy inference process. Here, author has put the different values of probability and impact into risk assessor model. As the result of simulation, different rules has triggered and value of risk generated. Figure 4 represents a rule viewer as a part of simulation result and land risk has been assessed with the risk assessor model. Author got the risk score of land risk by entering the value of probability and impact.

To see the entire output of risk surface of the system, surface viewer illustrates the entire span of the output set based on the entire span of the input set of probability and Impact. Fig. 5 represents the surface viewer for the risk with respect to probability and impact.

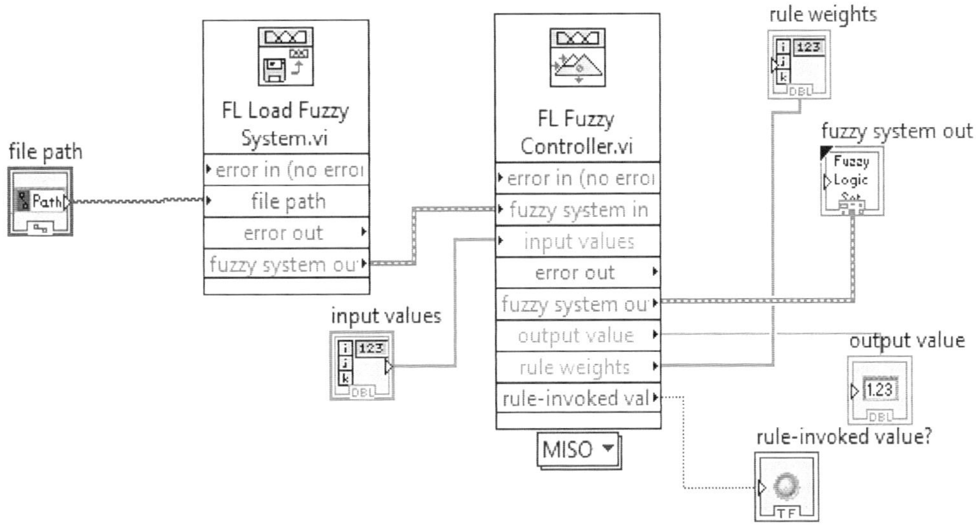

Figure 4: Front Panel of Simulated Risk Assessor model

Here, Figure 6 shows the behaviour of risk with probability and Figure 7 illustrates the behaviour of risk with Impact. By putting the values of land acquisition risk probability value 0.91 and impact 4.52 into the input variables, author got the 88.048 percentage output variable.

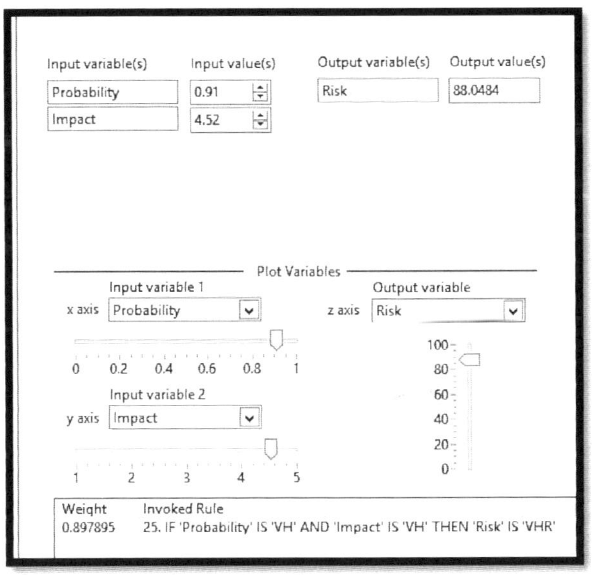

Figure 5: Rule Viewer of land risk

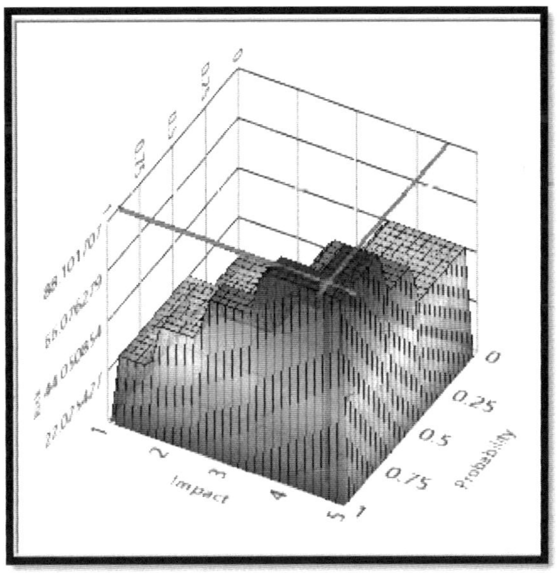

Figure 6: Surface viewer of land risk

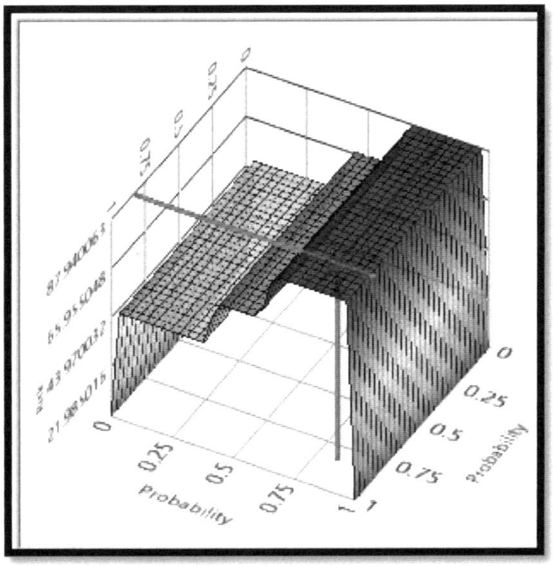

Figure 7: Surface viewer of traffic risk-Probability

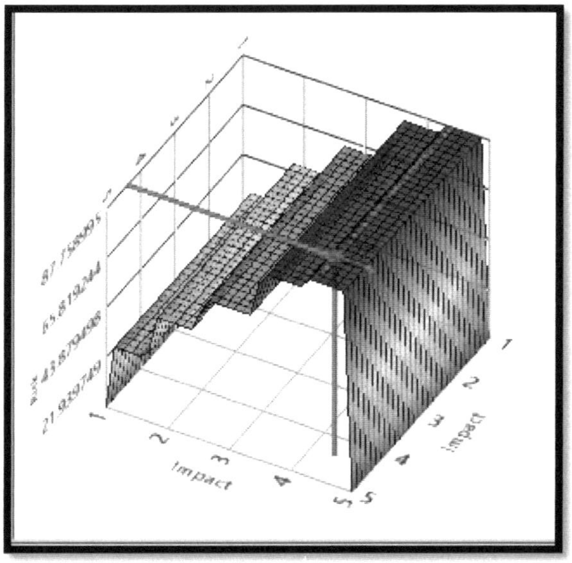

Figure 8: Surface viewer of traffic risk-Impact

Figure 7 shows the behaviour of risk with probability and figure 8 shows the behaviour of risk with impact. From this simulation result, we have identified the critical risk of different the risk factor. Here some of critical risk have mentioned with risk score.

Table 2: Most Critical Risk of BOT toll roads in Indian scenario

Risk Factor	Type Of Risk	Risk Score
Land Acquisition	Social risk/macro risk	88.04
Construction time overrun	Construction risk	87.82
Traffic shortfalls	Traffic related risk	87.76
Construction cost overrun	Construction risk	78.14
Change in Law	Political Risk	74.85
Original contract duration is rigid and has no scope to accommodate any changes	Contract related risk	71.04
Operation cost overrun	Operational risk	67.23
Insufficient data collection and survey before design	Consultant related risk	67.21
Inflation	Financial Risk	67.2
Legislation change	Political Risk	67.18
Lack of effective monitoring and feedback by project manager	Project Management Risk	67.04
Social Impact Assessment	Social Risk	67.03
Not able to achieve the reasonable profit targets	Traffic related risk	67.01
Delay due to government Policy restriction and land transfer	Political Risk	67

Environmental clearances	Social Risk	66.99

7 CONCLUSION

BOT projects are suffering loss due to lack risk assessor model. The result shows that BOT toll road are facing many risk factors. The most critical risk of BOT toll roads in Indian context is Land acquisition risk. So, the government should make such arrangement by that 80-100% land should be acquired at the time of award of the road to the concessionaire. The government should start an awareness program for Indian citizens to understand that BOT toll roads are crucial to ensure high-quality roads. A project completion risk guarantee provision that should cover the developers by providing a guarantee to BOT toll roads against risk factors such as land acquisition risks, delay in obtaining environmental and forest clearances, local protests, delay in provision of right-of-way etc. should be fixed. By Hybrid PPP model and OPRC (operational and performance based Road Contracts) can be the solution of these kind of risks. The delays in approval submittals is identified as most crucial risk, so government also take step ahead for make single window clearance. Many government officers are not enough skilled for monitor the BOT toll project. So, skills strengthening program should be done. Contract document should be made such that proper risk allocation can be done and win-win condition can be happened between government and concessionaire. Contractors often claims that operational revenues are below than expectation. The government should implement some yardstick for the consultant work and consultant has to improve the techniques. India's BOT toll road project experienced the change in scope for the project. But, due to Original contract duration is rigid and has no scope to accommodate any changes, disputes have been happened lot time. Risk carrying tool like insurance should be developed. Here, concessionaire can assess risk properly by the risk assessor model, risk response strategy can be made accordingly. So, concessionaire can get expected profit from the project.

References

Ahmed, O. J. (2003). Risk assessment in construction. *Journal of Construction Engineering and Management, 129(5)*, 492-500.

Almishari, O. A. (2003). Risk Assessment in Construction. *Journal of Construction Engineering and Management ASCE, 129*, 492-500.

Barkley, B. (2004). *Project risk management.* New York: McGraw-Hill.

Department of Economic Affairs, India. (2015). *COMMITTEE ON REVISITING AND REVITALISING PUBLIC PRIVATE PARTNERSHIP MODEL OF INFRASTRUCTURE.* Delhi: Department of Economic Affairs,Ministry of Finance, India.

El-Sayegh, S. a. (2008). Risk assessment and allocation in the UAE construction industry. *International Journal of Project Management, 26(4)*, 431-438.

IBEF. (2018). *ROADS.* IBEF.

Loosemore, M. a. (2008). Perceptions of contractual risk allocation in construction supply chains. *134(1)*, 95–105.

PMI. (2008). *A Guide to the Project Management Body of Knowledge.* Pennsylvania: Project Management Institute.

Smith, N. J. (2006). *Managing risk in construction projects.* Malden, MA: 2nd Ed., Blackwell Science.

Tang, W. Q. (2007). Risk management in the Chinese construction industry. *Journal of construction Engineering Management, 133(12)*, 944-954.

Thomas, A. K. (2006). Modelling and assessment of critical risks in BOT road projects. *Construction Management Economics, 24(4)*, 407-424.

worldbank. (2018, March 20). Retrieved from https://ppp.worldbank.org/ppp/agreements/concessions-bots-dbos: https://ppp.worldbank.org/ppp/agreements/concessions-bots-dbos

Yen, J., & Langeri, R. (2005). *Fuzzy Logic- Intelligence, Control, and Information* (3rd Edition ed.). Pearson Education.

Zhang, G. a. (2007). Fuzzy analytical hierarchy process risk assessment approach for joint venture construction projects in. *Journal of Construction Engineering and management, 133(10)*, 771–778.

Building Tomorrow's Society
Bâtir la Société de Demain

Fredericton, Canada
June 13 to 16, 2018 / *13 au 16 juin, 2018*

GEOGRID REINFORCEMENT IN FLEXIBLE PAVED ROADS

Ahmed, Mai[1,2], Abdelhamid, Mennatallah [1], Hussain, Sarah [1], Khedr, Safwan[1], Breakah, Tamer [1], Saudy, Maram [1], Elkadi, Omar [1], Abou-Zeid, Mohamed [1]
[1] The Department of Construction Engineering, The American University In Cairo, Egypt
[2]Mai_ibrahim@aucegypt.edu

Abstract: The scope of this study is to explore and evaluate the effect of reinforcing the interface between the base and asphalt concrete layers in a flexible pavement section with bi-axial geogrids; using both large scale experimental tests and numerical modeling using three-dimensional nonlinear finite elements analysis. A laboratory characterization for each layer of the pavement section according to a laboratory programs that included soil classification, plate loading test, determination of asphalt concrete dynamic modulus for and CBR ratio for different elements in pavement section. A laboratory model of the pavement with and without the geogrid reinforcement, located at the top of the base layer was constructed. A dynamic loading scheme was applied on the experimental pavement model. Concurrently, a numerical model was developed to simulate the effect of such type of reinforcement on the pavement section in terms of the road's performance enhancement and reduction in rutting. The numerical model was proved to verify simulation of the pavement experimental model when comparing the overall trend of stress distribution throughout the pavement structure.

1 INTRODUCTION

Geogrids are one type of geosynthetic reinforcement systems. They are made of polymers that are connected in parallel sets of tensile ribs with openings in between them (Abu-Farsakh, 2016). According to literature, geogrids can effectively decrease rutting in the asphalt concrete layer in flexible pavement (Gu, 2016). This may be attributed to the mechanism of the geogrids interaction with the pavement layers (Moghaddas-Nejad, 1996). When the asphalt concrete layer is compacted over geogrids, the aggregate particles penetrate through the openings of the grid resulting in the creation of a strong interlock and therefore the lateral movements of the unbound base material or mix particles are reduced drastically. This interlock acts as a resistor for rutting development such that stresses are transferred by tensile forces and thus the performance of the road is enhanced (Abu-Farsakh, 2011).

Geogrid reinforcement is utilized as a part of flexible paved roadways in two noteworthy application zones: base reinforcement and subgrade stabilization Limited research has been conducted on base reinforcement applications, so the scope of this research was to explore and test the effect of reinforcing the interface between the base and the asphalt concrete layers in a flexible pavement section using bi-axial geogrids.

The methodology of this research was divided into two phases; experimental work and numerical modeling. The experimental work consisted of lab characterization of the different materials used in the different layers of flexible pavement; namely, untreated base, subgrade soil, geogrid and asphalt concrete. A full scale pavement laboratory model was subjected to dynamic loading, simulating traffic loads on a road. The pavement section structure was decided through proper conventional design analysis. Iterations and trials on the finite element modeling software "MIDAS" were used in order to simulate the effect of geogrids on the paved road section using a 3D model and verified by the full scale dynamic testing.

2 EXPERIMENTAL PROGRAM

2.1 Lab Materials Characterization

Different tests were conducted on the various layers of the pavement section in order to measure and determine the parameters to be used to characterize the layers as input to the numerical model. The results are shown in Table 1. Properties that were not tested were reasonably assumed based on previous experiences as well as from the literature.

Table 1: Small Scale Tests

Layer	Test Name	ASTM
Subgrade	Gradation	C136/C136M - 14
	Water Content	D2216-98
	Compaction	D698-12e2
	CBR	D1883-16
	Specific Gravity	D854-14
	Direct Shear	D3080
Base	Gradation	C136/C136M - 14
	Water Content	D2216-10
	Modified Proctor Compaction	D1557-12
	CBR	D1883-16
	Specific Gravity	C127-15
Asphalt Concrete	Flow and Stability	D6927-15
	Extraction	D2172/D2172M-17
	Theoretical Maximum Specific Gravity (G_{mm})	D2041/D2041M-11
	Dynamic Modulus (E^*)	D3497
Geogrids (Tenax LBO 330)	Tensile Strength	D6637

2.1.1 Materials Properties

Soil, base material and asphalt mix constituents were obtained from a construction site. Tables 2 to 4 shows the measured characteristics of the materials used as subgrade, base and asphalt concrete for the tested model pavement section.

Table 2: Subgrade Layer Properties

Property	Result
Specific Gravity	2.70
Optimum Water Content (%)	7.6
$\gamma_{dry\ max}$ (kN/m^3)	17.76
CBR (%)	26
Modulus of Elasticity (E), (MPa)	15.8
Cohesion (C)	0
Angle of Internal Friction (ϕ^o)	30
Unit Weight in Full Scale Exp Model, (kN/m^3)	18
Poisson's Ratio (ν) (assumed)	0.3

Table 3: Base Layer Properties

Property	Result
AASHTO Classification	A
Specific Gravity	2.56
CBR (%)	142.7
Modulus of Elasticity (E) (MPa)	22
Angle of Internal Friction (ϕ^o)	40
Unit Weight in Full Scale Exp Model, (kN/m^3)	20
Poisson's Ratio (ν) (assumed)	0.35

Table 4: Asphalt Concrete Layer Properties

Property	Result
Flow (2.5mm)	10.8
Stability (kN)	1.8
Optimum Asphalt Content (%)	5.66
Maximum Specific Gravity G_{mb}	2.1
Theoretical Maximum Specific Gravity (G_{mm})	2.41
Air Voids (Core Sample) (%)	13.7
Dynamic Modulus (E^*) (MPa)	161.5
Unit Weight in Full Scale Exp Model, (kN/m^3)	22
Poisson's Ratio (ν) (assumed)	0.45

Table 5: Geogrids Layer Properties

Property	Result
Tensile Strength (kN/m)	40.9
Modulus of Elasticity (MPa)	161.5
Unit Weight (KN/m^3)	1
Poisson's Ratio (ν) (assumed)	0.15

2.2 Full Scale Test

2.2.1 Test Design

Two full-scale pavement structures were tested; a control section and a geogrid reinforced section. The geogrids were placed in between the asphalt concrete layer and the base layer. The lab model had dimensions of 2x2x1 meters. The model boundaries did not significantly interfere with the pavement stress distribution as checked by the numerical model (MIDAS). The layers' thicknesses were selected in accordance to the Egyptian code of design, which suggested typical section with 10 cm of asphalt concrete and 25 cm of untreated base. This left subgrade soil layer of 65 cm in the lab model.

To achieve the desired degree of compaction, the weights according to maximum unit weight were prepared and compacted till pre-marked depth. Compaction was done using a plate compactor. The subgrade layer was divided and compacted on 3 layers: 0.25, 0.25, and 0.15 m. The degree of compaction for the subgrade using sand cone test was 102%. The base layer was divided and compacted on 2 layers, 12.5 cm each. The degree of compaction for the base layer was 100.23%. Asphalt concrete layer was placed and compacted on 2 layers, 5 cm thick each.

2.2.2 Test Setup

As shown in figure 1, the test setup comprises an actuator, external LVDTs, a loading plate, a data logger, and the full-scale tank.

a. Actuator and LVDT

b. Complete Test Setup

Figure 1: Full Scale Test Setup

2.2.3 Dynamic Load Pattern

The load pattern chosen is shown in Figure 2. This loading simulated the load of vehicles at intersections, toll stations and slow speed roads. It was selected because rutting increases at these locations of the roads and the asphalt concrete becomes more vulnerable to the applied loads of the trucks than in the locations of higher speeds. Each loading cycle lasted for one second in which:

0.2 second of constant load of 0.5 ton
0.3 second of loading from 0.5 ton to 10 tons
0.3 second of unloading from 10 tons to 0.5 ton
0.2 second of constant load of 0.5 ton

The minimum load which is the 0.5 ton was used instead of zero load to avoid the impact due to load application.

The load pattern was assigned through the control system of the actuator. Five thousand loading cycles were applied and stopped for the hydraulic pump to cool down and then the next 5000 cycles were applied and so on.

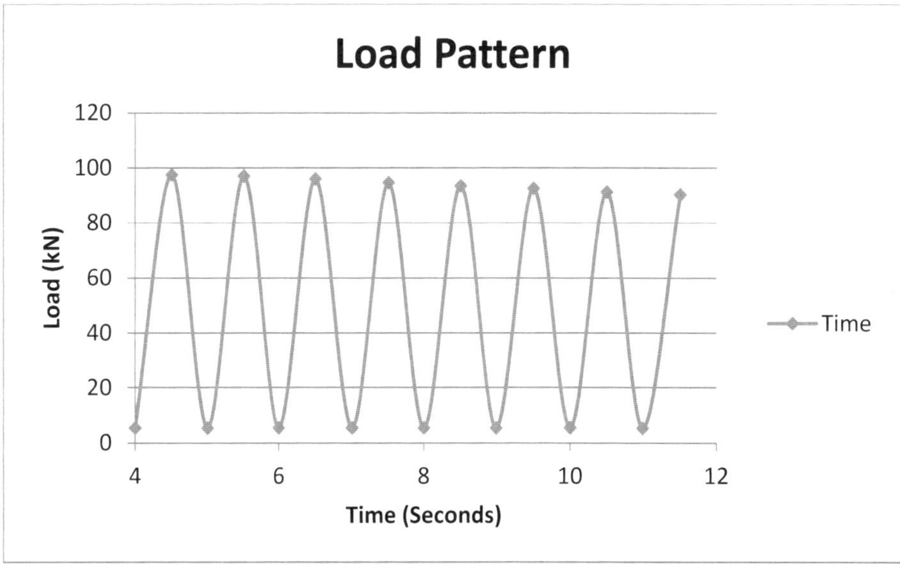

Figure 2: Load Pattern

2.2.4 Number Of Cycles

The applied maximum load of 10 tons (98 kN) was converted to ESALs using the ESAL equivalency table presented in the Egyptian Code of Design and AASHTO 1995. The total load on an axle was 200kN and thus each loading cycle was equivalent to 39 ESALs. The test was continued up to 30,000 cycle, which means that the maximum ESALs applied throughout the test was 1,170,000 for each of the test setups.

2.2.5 Tire Contact Area

According to previous research, a loading area size of 0.31 x 0.20 m was used to simulate the loading imprint.

2.2.6 Pressure Gauges Setup

To observe and study the stresses distribution in the different layers of the road, ten pressure gauges were placed inside the full scale model. The pressure gauges were placed at different positions as shown in Figure 3. The pressure gauges were labelled from P1 to P10. The distribution is clarified in the figure (P1 and P8 were placed laterally in order to measure vertical pressures).

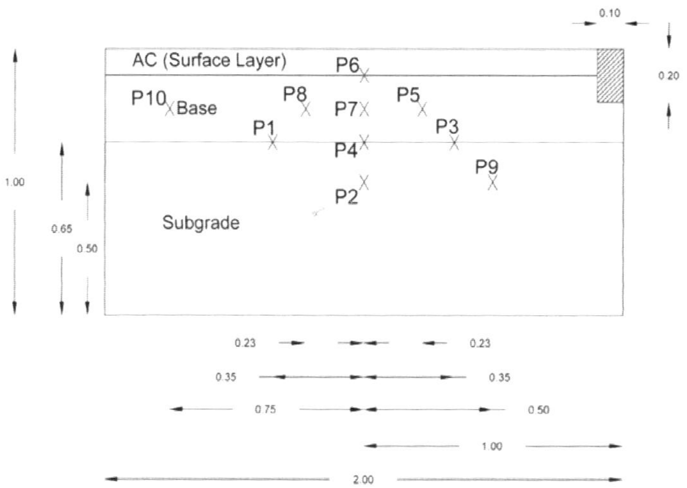

Figure 3: Pressure Gauges Locations (Elevation)

2.2.7 Full Scale Model Testing Results

The rutting results were acquired from mounted LVDT's. The readings were taken each 50 milliseconds, and the readings included time and load, which was measured by a load cell and deformation. The residual deformation for each 5000 cycles was subtracted from the values obtained for the next 5000 cycles to get the plastic deformation caused by the load. This modification was made to the 30,000 cycles to be able to get the plastic deformation at each load throughout the 30,000 cycles.

The modified deformation versus cycle curve is shown in Figure 4.

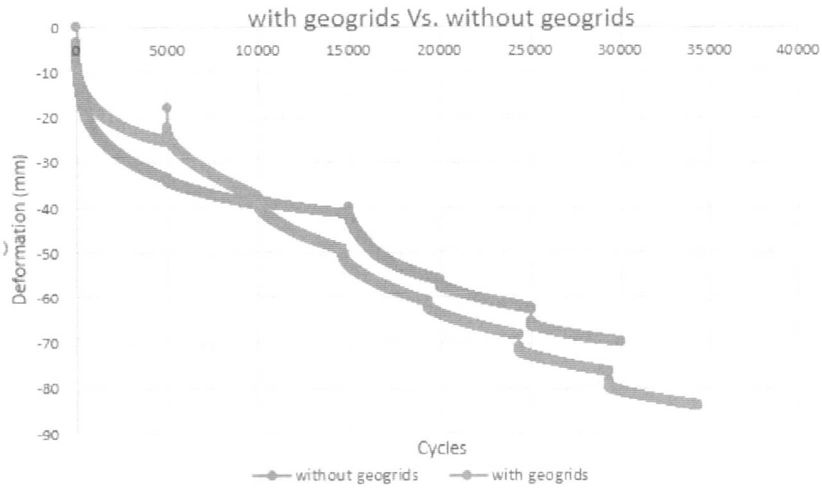

Figure 4: Modified Deformation versus Cycles Graph

When comparing the results of the two models, it was observed that the sample with geogrids had less deformation until cycle number 10,000. The area before the cycle 10,000 is the area that should be analyzed because the rutting appeared in the sample without geogrids in the cycle 1000, which is equivalent to 39,000 ESALs.

By visual inspection to the model without geogrids, after the asphalt layer was removed, an apparent deformation in the base layer was found. This can be explained that the stresses were carried by the base layer rather than the asphalt layer, which led to lower deformation as after the cycle number 10,000, and this shows the stresses dissipation through the asphalt concrete layer.

For the sample with geogrids, the asphalt layer and the geogrids carried all the stresses, which was proved when the asphalt layer was removed and no deformation in the base layer was found.

In the sample without geogrids, after 1000 cycles, the rutting had already appeared which means that the asphalt already reached failure condition. While in the sample with geogrids, the rutting started to appear in the asphalt layer after nearly 3000 cycles, which is equivalent to 117,000 ESALs. This means that geogrids tripled the lifetime of the asphalt.

The deformation of the LVDT was used to plot the deformation against time for loading cycles to show the resilient deformation that happened within the cycle as the load was added and then removed and the residual deformation that is accumulated through the loading cycles as shown in Figure 5.

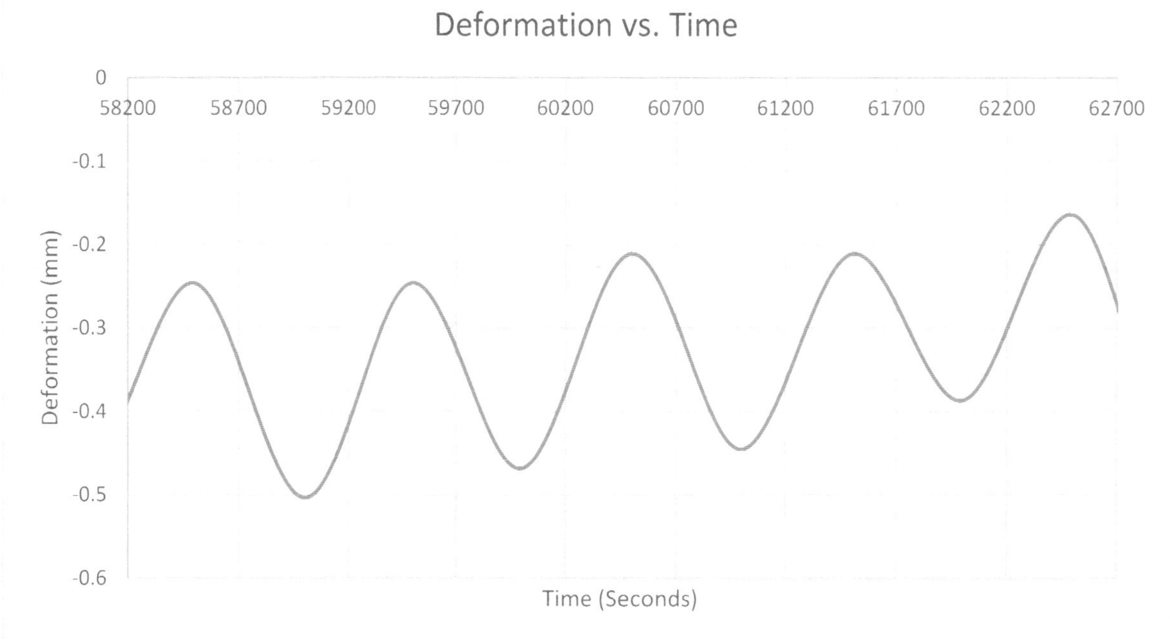

Figure 5: Deformation versus Time Graph

2.2.8 Rutting

The total rutting deformation for the control section was 7.3cm and for the reinforced section was 5.2cm. The shape of the models after the end of the test is shown in Figure 6.

a) Without Geogrids b) With Geogrids

Figure 6: Rutting Appearance

3 NUMERICAL MODELING

3.1 Preleminary Model

A preliminary model with assumed material properties was firstly developed to decide on the boundary conditions of the tank that were needed to be fabricated for the full-scale test.
A table for the constitutive models used and the mesh sizes used for the model is as follows:

Table 6: Constitution Models and Mesh Sizes Table

Layer	Mesh Size (m)	Constitutive Model
Subgrade	0.2	Mohr-Coulomb
Base	0.15	Mohr-Coulomb
Asphalt Concrete	0.12	Elastic

Several iterations were done to reach the model dimensions. The selected dimensions were 2 x 2 x 1 m, results showed that the boundary conditions did not significantly affect the stresses. It also showed that the deformation is negligible at the sides of the tank, which means the 2 x 2 m dimension of the tank is sufficient and that the deformation is not affected by the height of the tank, which is 1 meter high. The vertical deformation of the cross section as calculated in the numerical model is shown in Figure 7.

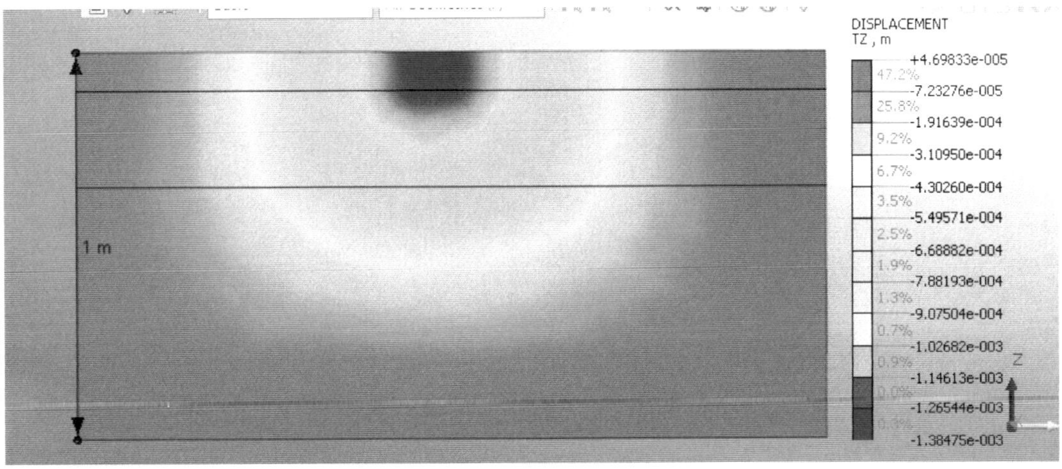

Figure 7: Cross-section for the vertical deformation

3.2 Final Model

After the full-scale test, another two models were developed for both the lab model with geogrids and the sample without geogrids. The model was based on an equivalent static load analysis, where the dynamic modulus of the asphalt was used.

3.2.1 Without Geogrids

The results showed that the stresses are highest at the center of loading and decreases towards the sides, thus; the deformation is maximum at the center and decreases towards the sides. The model also showed the points where plastic points had occurred which means that there is plastic deformation at these points and they will not return back to their positions. The results of the model are shown in Figure 8.

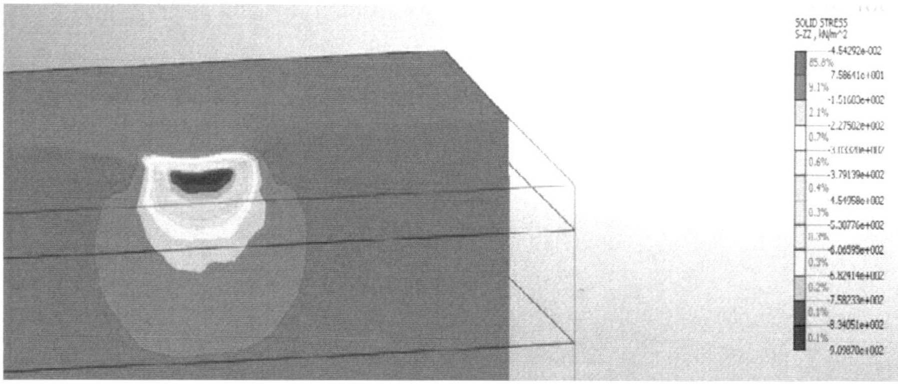

Figure 8: Cross-section for the vertical stresses

3.2.2 With Geogrids

The model with geogrids showed a better stress distribution than that of the sample with no geogrids as the geogrids helped in distribution of forces through tensioning the geogrids which also caused interlocking of particles in the asphalt layer, thus, resulting in a better distribution of stresses in the base layer. The results of the model are shown in Figure 9.

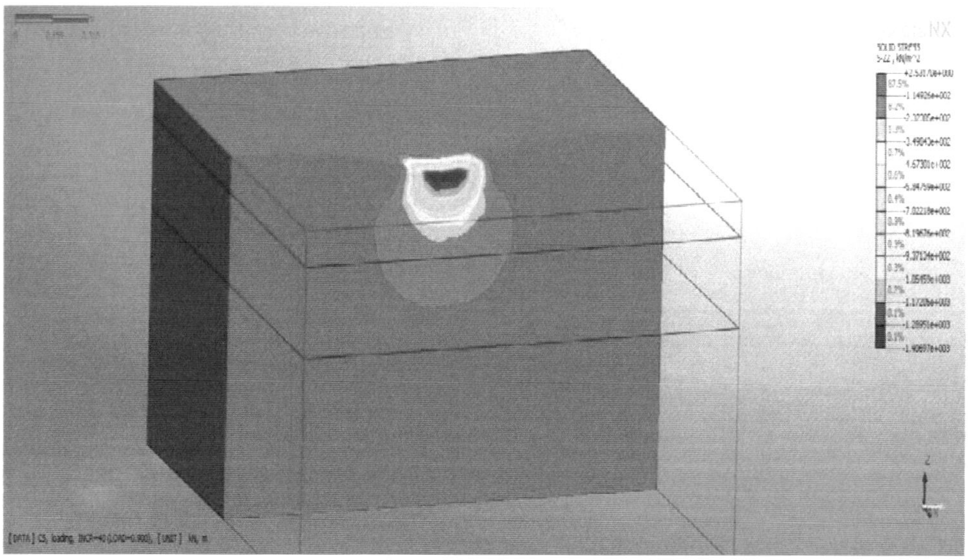

Figure 9: Cross-section for the Vertical Stresses

The model with geogrids also illustrated the tension that occurred in the geogrids layer and how the forces are distributed in the xx direction. The tension in the geogrids caused an interlocking of particles of the asphalt that lead to better distribution of stresses.

4 ANALYSIS OF RESULTS

The interpretation of the full-scale test results and visual inspection showed occurrence of residual deformation in the base layer with the absence of geogrids. Adding geogrids decreased this phenomenon and consequently reduced pavement rutting significantly (almost 30% less). The experimental model was modeled afterwards on MIDAS Software with and without geogrids. The model without geogrids showed a numerical failure at 60% of the load, which means that the model failed to find convergence due to instability; thus, the asphalt failed. However, the numerical failure in the model with geogrids was at 90% of the load. The numerical failure difference indicates that the geogrids caused 30% enhancement in terms of the pavement load carrying capacity.

A better distribution of stresses in the model with geogrids was observed, since the geogrids led to the interlock of particles of the asphalt. The results of the model could not be numerically compared to that of the full-scale test because the numerical model was based on an equivalent static load analysis and the full-scale test was done based on dynamic loading. However, the overall trend of stress distribution was the same in both the full-scale test and the model.

5 CONCLUSION

From the results and analysis stated above, it was concluded that the geogrids had a major effect on the stresses distribution inside the section of the road tested, as the stresses did not cause a deformation in the base layer in the case of presence of the geogrids. Additionally, from the full-scale test it was concluded that the presence of geogrids increases the lifetime of the asphalt concrete by nearly three times.

6 RECOMMENDATIONS

For further research, it is recommended to:

1. Evaluate the Fatigue Cracking caused by the cyclic loading
2. Try using Geogrids in different positions, which may enhance the results of the stresses
3. Test the dynamic modulus at different temperatures instead of testing it at the only one temperature, to account for the change in the properties of the asphalt concrete due to climate changes
4. Test the exact shear properties of the Subgrade and Base layer rather than using reasonable assumptions
5. Modeling the full-scale test using dynamic loading instead of equivalent static loading on MIDAS

References

Abu-Farsakh, Murad, Shadi Hanandeh, Louay Mohammad, and Qiming Chen. "Performance of geosynthetic reinforced/stabilized paved roads built over soft soil under cyclic plate loads." Geotextiles and Geomembranes 44, no. 6 (2016): 845-53. doi:10.1016/j.geotexmem.2016.06.009.

Abu-Farsakh, Murad Y., and Qiming Chen. "Evaluation of geogrid base reinforcement in flexible pavement using cyclic plate load testing." International Journal of Pavement Engineering 12, no. 3 (2011): 275-88. doi:10.1080/10298436.2010.549565.

American Association of State Highway and Transportation Officials. 1995. AASHTO provisional standards. Washington, D.C.: American Association of State Highway and Transportation Officials.

Breakah, Tamer M., and R. Christopher Williams. "Stochastic finite element analysis of moisture damage in hot mix asphalt." Materials and Structures 48, no. 1-2 (2013): 93-106. doi:10.1617/s11527-013-0170-x.

Gu, Fan, Xue Luo, Rong Luo, Robert L. Lytton, Elie Y. Hajj, and Raj V. Siddharthan. "Numerical modeling of geogrid-reinforced flexible pavement and corresponding validation using large-scale tank test." Construction and Building Materials 122 (2016): 214-30. doi:10.1016/j.conbuildmat.2016.06.081.

Moghaddas-Nejad, Fereidoon, and John C. Small. "Closure to “Effect of Geogrid Reinforcement in Model Track Tests on Pavements” by Fereidoon Moghaddas-Nejad and John C. Small." Journal of Transportation Engineering 125, no. 3 (1999): 267. doi:10.1061/(asce)0733-947x(1999)125:3(267)

Building Tomorrow's Society
Bâtir la Société de Demain

Fredericton, Canada
June 13 to 16, 2018 / *13 au 16 juin, 2018*

IDENTIFYING THE FACTORS THAT INCREASE THE SEVERITY OF PEDESTRIANS' INJURIES WHEN STRUCK BY VEHICLES

Bridgwater, Spencer[1], Dabbour, Essam[2, 3] and Haider, Murtaza[1]
[1] Ryerson University, Canada
[2] Abu Dhabi University, United Arab Emirates
[3] essam.dabbour@adu.ac.ae

Abstract: This research focuses on identifying the factors that correlate with the increase in the severity of pedestrian injuries when struck by motorized vehicles. Using a dataset of all collisions involving pedestrians from Washington State, the objective of this research is to identify demographic attributes of at-risk drivers who are more likely to be involved in collisions resulting in death or serious injuries to pedestrians. Other factors being investigated include environmental conditions, visibility, urban/rural location, and the type of the roadway. A multinomial logit model was estimated to identify the most significant factors associated with death or serious injuries for pedestrians. This paper finds young, male, and impaired drivers are more likely to be involved in collisions resulting in death of or serious injuries to pedestrians. Poor visibility and overcast conditions are also associated with higher odds for pedestrian injury or death. Similarly, rural locations and interstate facilities also found to be associated with higher odds. A key limitation of this paper is the lack of demographic attributes of pedestrians.

1 INTRODUCTION

In 2015, the National Highway Traffic Safety Administration reported 5,376 pedestrian deaths from traffic crashes in the U.S., a 9.5% increase from pedestrian fatalities in 2014, and is the highest number of pedestrians killed annually since 1996 (National Highway Traffic Safety Administration [NHTSA], 2017). Although overall traffic fatalities in USA have largely declined over the past two decades, the percentage of total fatalities involving pedestrians has risen from 11% in 2006 to 15% in 2015, with no single year in decline (NHTSA, 2017).

Vehicle-pedestrian collisions typically have a high degree of severity. In 2015, 15% of all traffic fatalities and nearly 3% of injuries in traffic crashes were pedestrians (NHTSA, 2017). This suggests that vehicle-pedestrian collisions are generally more severe for injury severity than single-vehicle or multivehicle collisions. Injury severity is related to both collision speed and vehicle type. For high-speed collisions, pedestrian death is by far the most likely outcome. Further, at low speeds, pedestrians struck by a large sports-utility vehicle (SUV) or a pick-up truck are more likely to have higher odds of injury severity and death compared to those struck by a smaller passenger car (Ballesteros et al. 2004). Pedestrians struck by vehicles other than passenger cars at low speeds were twice as likely to endure traumatic brain, thoracic, and abdominal injuries, although at higher speeds there is a significantly higher chance of death overall (Ballesteros et al. 2004).

A collision involves first a potential situation, and if a conflict exists, then either an evasive action could be taken, which may or may not ultimately prevent a collision (Allen et al. 1978). In two-party situations, it is often the unpredictability of one of the parties that contributes to a collision, and a greater unpredictability typically correlates with a lower window of reaction time available to each party to evade (Allen et al.,

1978). Further, the failure of any party to notice a potential hazard or delayed responses to hazards in general can result in traffic collisions (Ewing & Dumbaugh, 2009).

Results from population regression models across 21 countries show that a reduction in urban speed limits from 60 km/h to 50 km/h could reduce pedestrian fatalities by 25% (Fieldwick & Brown, 1987). Other studies have demonstrated that since the stopping distance for a vehicle is proportional to the square of the initial speed, even small reductions in the speed can greatly lower the severity of injury and probability of death (Anderson, et al., 1997).

In 2015, 74% of all pedestrian fatalities in traffic collisions occurred at night, with the highest total percentage (26%) occurring from 6pm to 9pm, and the next highest total percentage (23%) occurring from 9pm to 12am (NHTSA, 2017). It was also found that providing continuous roadway lighting reduces the risk of collisions and severity of injuries more than partial lighting, and similarly partial lighting more than no lighting (Dewar, 2007).

When drinking was a contributing factor, the probability of a severe injury greatly increased overall, with fatal injury probability increasing by 73%, and the probability of a complain of pain or no injury decreasing by 37% (Kim, et al., 2013). The authors also note that because drinking is correlated with risk-taking behaviour, and that driver intoxication may be endogenous (Kim, et al., 2013).

Infrastructure design that incorporates effective safety engineering has been addressed in the literature. At the neighbourhood level, road and zoning layouts such as cul-de-sac conventional networks, grid networks, and offset networks provide different advantages and disadvantages for both drivers and pedestrians (Wei & Lovegrove, 2012). In particular, 3-way offset layouts, a Dutch initiative, encouraged mode split (i.e. increased walking, bicycling, and transit) and improved road safety, at the cost of a small reduction in vehicular accessibility (Wei & Lovegrove, 2012).

The objective of this paper is to identify the significant factors contributing to the injury severity of pedestrian when struck by vehicles. Several explanatory variables are investigated, including those related to the driver (such as driver's gender, age, and state of intoxication) and those related to the environment (such as lighting and weather conditions).

2 DATA COLLECTION

All pedestrian-related collisions were extracted from aggregate datasets on traffic collisions maintained in the Highway Safety Information System (HSIS) data files for Washington State between 2003 and 2013 (Nujjetty & Mohamedshah; HSIS, 2014). For every analysis year, accident files and vehicle files were merged together to obtain the relevant variables for each incident. The recorded incident number, which is unique to each collision, was used to link the two files together for each year.

Between 2003 and 2013 there were 492,676 total traffic collisions reported in Washington State. The accident reporting threshold at the time of data collection was a minimum of $750 in property damage or personal injury. After eliminating incomplete records, there were 6,797 traffic collisions involving pedestrians. The data were then further filtered to remove collisions in which a pedestrian was not struck by a vehicle (2,115), injury severity was unstated (41), driver's gender was unstated (1,005), driver's age was unstated (110), driver's intoxication condition was unstated (651), lighting conditions were unstated (21), weather conditions were unstated (12) and road surface condition was unstated (5). The final dataset used for analysis comprised 2,837 records of collisions where vehicles collided with pedestrians. We estimated multinomial logit models using R software.

Within the Accident sub-file of the HSIS data, the severity variable is of primary interest for the scope of this research. For injury severity, the HSIS files denote a range of categorical values including unstated, no injury, possible injury, non-disabling injury, disabling injury, died at hospital, dead on arrival, and dead at scene. A limitation of this dataset is the HSIS definition of the severity variable, which it explains as the injury of greatest severity in a crash. While pedestrians involved in vehicle collisions are much more likely to sustain serious injuries compared to the driver, it is possible for scenarios to exist in which the driver is more seriously injured than the pedestrian. For example, a significantly older driver may incur more serious injuries than a younger pedestrian during low-speed collisions as will be discussed later in a subsequent section of this paper. However, these cases likely represent a small minority of cases overall. Indeed, for the adjusted dataset which includes only collisions involving pedestrians, there are 5 instances of driver death, all of which were reported to have the same severity as the involved pedestrian. It is therefore assumed that severity in general refers to pedestrian injuries.

3 MODEL

Statistical models for injury severity in motor-vehicle crashes are usually conditioned on both observable and unobservable factors that affect the probability of a particular injury severity category (J-K. Kim, Ulfarsson, Shankar, & S. Kim, 2008). The literature on injury severity mentions several empirical methods for modelling injury severity. These methods include logistic regression, multinomial logit models, and ordered probit models (Ballesteros, et al., 2004; Ulfarsson and Mannering, 2004; Kockelman and Kweon, 2002). Since HSIS categorises *injury severity* into mutually exclusive categories, we use multinomial logit models to determine the determinants of injury severity. Logit modeling allows for the estimation of the *k*-1 logit equations for nominal dependent variables with *k* categories, and follows a logit distribution. The approach designates one response category as a reference and reports the log-odds of the linear function generated by the predictors as per the following formula (Rodriguez, 2007):

$$[1]\ logit(Y_i) \;=\; \eta_{ij} \;=\; \log\left(\frac{\pi_{ij}}{\pi_{iJ}}\right) \;=\; \alpha_j + x_i'\beta_j$$

Where α_j is a constant and β_j is a vector of regression coefficients, for $j = 1, 2, \ldots, J-1$. In this case, we also have:

$$[2]\ \pi_{ij} \;=\; \Pr\{Y_i = j\}$$

The log-odds generated by the multinomial logit model may also be written in terms of their original probabilities, π_{ij}, such that:

$$[3]\ \pi_{ij} \;=\; \frac{\exp\{\eta_{ij}\}}{\sum_{k=1}^{J} \exp\{\eta_{ik}\}}$$

for $j = 1, \ldots, J$. Equation 3 will yield probabilities that sum to 1 for each *i* (adapted from Rodriguez, 2007).

4 EMPIRICAL RESULTS

As mentioned earlier, our dependent variable is *injury severity*, captured as five categories ranging from no injry to death. Injury severity tabulations are presented in Table 1. Of the 2,837 collisions in the data set, approximately 7% collisions reported no injury, approximately 7.9% of the collisions resulted in death, and another 85.1% of the collisions involved some form of injury. This research estimated multinomial logistic regression models using the *multinom* function from the *nnet* package in R (Ripley and Venables, 2016). Results of the multinomial logit regression are summarized in Table 2. It should be noted that driver's age was not analyzed as a continuous variable. This is due to the high crash rates at both young age (from lack of experience) and old age (due to physiological reasons) as found by McCartt and Teoh (2015). Therefore, driver's age was analyzed as categorial variable, as shown in Table 2. The coefficient values represent the log-odds for comparing a baseline value of a predictor (e.g. for driver gender, the baseline is male) to another categorical value for the same predictor (for driver gender, this would be female), which is then further compared to the dependent categorical variable severity (adjusted) instances, as they relate to the reference of "No Injury". For comparison purposes, all coefficients were found to be statistically significant at the 0.01 confidence level except where stated.

Estimates presented in Table 3 show that male drivers are 1.70 times more likely to cause a pedestrian death from collision than female drivers. However, compared to male drivers, female drivers are 1.20 times as likely to cause disabling injury to a pedestrian from a collision, 1.47 times as likely to cause non-disabling injury, and 1.51 times as likely to cause possible injury. These results are statistically significant at the 1% confidence level. Further, drivers under 25 were significantly more likely to be involved in a collision causing severe injury or death of pedestrians when compared to drivers aged 25-74. However, drivers aged 75 and over were most likely to be involved in crashes causing fatal injury compared to all other drivers, with approximately a 1.13 odds ratio compared to drivers under 25. This held also for disabling injury (1.47 times as likely), non-disabling injury (1.74 times as likely), and possible injury (1.62 times as likely). For non-disabling and possible injury severities, there was no other statistically significant finding for age groups, the closest being possible injury for drivers 65-74, at the 10% confidence level.

Impaired driving was found to have increased odds ratios for both fatal injury (1.60 times) and disabling injury (1.26 times), and decreased odds ratios for non-disabling injury (0.87 times) and possible injury (0.68 times).

For fatal injury, the most significant odds ratio compared to daylight was the scenario with the lowest likelihood of adequate visual acuity. Drivers who collided with pedestrians in the dark (with no streetlights) were 10.21 times as likely to cause fatal injury than compared to collisions in the daytime, all other factors remaining constant. Increased odds compared to daytime driving were also found for disabling injury. No statistical significance was found for non-disabling injury due to dark driving conditions (without streetlights) compared to daytime driving. Further, dark driving conditions with streetlights (both on and off) also saw increased odds ratios for fatal injury, with lights on at 3.67 times more likely, and lights off at 3.19 times more likely. Compared to daylight lighting conditions, both dusk and dawn saw increases in all odds ratios for all severities except for fatal injury at dawn. For fatal injury, daylight saw an increased odds ratio of 1.59 compared to dawn. Dusk was the only lighting condition that was found to have increased odds ratios for all severities compared to daylight, with fatal injury specifically being 2.05 times more likely. All weather types compared to reference (clear/partly cloudy), except for overcast were found to have decreased odds ratios for fatal injury. Snowy weather was found to have the largest reduction in odds ratio for fatal injury severity compared to clear weather, at 0.53 times as likely. However, increases in odds ratios from snowy weather for disabling injury (1.71 times) and possible injury (1.14 times) were found to be statistically significant. Although rainy weather showed a small (but statistically significant) decrease in odds for fatal injury compared to clear weather (0.94 times as likely), there were statistically significant increases in odds for disabling injury (1.26 times), non-disabling injury (1.25 times), and possible injury (1.57 times). Overcast weather was found to have increased odds ratios for all severity types compared to clear weather, but only fatal injury (1.24 times) and possible injury (1.27 times) were significant.

Table 1: Adjusted summary counts for *severity* categorical variable

Injury Severity	Count (*N*)	% of TOTAL
Death from injury	223	7.9%
Disabling injury	528	18.6%
Non-disabling injury	1,005	35.4%
Possible injury	883	31.1%
No injury	198	7.0%
TOTAL	6,752	100%

Table 2: Pedestrian injury severity model estimation results

Independent Variable		Dependent Variable			
		Death From Injury	Disabling Injury	Non-Disabling Injury	Possible Injury
Driver Age (Ref: Under 25)	25-54	-0.322*** (0.079)	-0.257*** (0.067)	-0.013 (0.061)	-0.052 (0.063)
	55-64	-0.103** (0.047)	-0.442*** (0.080)	-0.058 (0.075)	-0.029 (0.075)

	65-74	-0.183*** (0.009)	0.045 (0.058)	0.060 (0.080)	0.145* (0.075)
	75+	0.118*** (0.004)	0.386*** (0.016)	0.553*** (0.068)	0.484*** (0.060)
Driver Gender (Ref: Male)	Female	-0.530*** (0.012)	0.188*** (0.067)	0.386*** (0.058)	0.411*** (0.060)
Lighting Condition (Ref: Daylight)	Dawn	-0.464*** (0.0003)	0.300*** (0.003)	0.170*** (0.003)	0.091*** (0.003)
	Dusk	0.717*** (0.001)	0.605*** (0.005)	0.373*** (0.006)	0.472*** (0.006)
	Dark, street lights on	1.299*** (0.062)	0.747*** (0.070)	0.292*** (0.065)	0.096 (0.068)
	Dark, street lights off	1.160*** (0.001)	-0.925*** (0.001)	-0.608*** (0.002)	-0.793*** (0.002)
	Dark, no street lights	2.324*** (0.051)	0.905*** (0.044)	0.050 (0.044)	-0.228*** (0.027)
Weather (Ref: Clear)	Overcast	0.211*** (0.008)	0.044 (0.086)	0.095 (0.073)	0.235*** (0.075)
	Raining	-0.061*** (0.014)	0.229*** (0.082)	0.219*** (0.073)	0.448*** (0.075)
	Snowing	-0.635*** (0.001)	0.539*** (0.003)	0.135*** (0.002)	0.692*** (0.003)
	Other	-0.116*** (0.002)	-0.052*** (0.004)	-0.012** (0.005)	-1.153*** (0.002)
Day of Week (Ref: Weekday)	Weekend	0.145*** (0.012)	-0.171** (0.079)	-0.167** (0.071)	-0.360*** (0.075)
Driver Intoxication (Ref: Not Impaired)	Impaired	0.471*** (0.005)	0.229*** (0.007)	-0.145*** (0.007)	-0.385*** (0.004)
Environment Location (Ref: Rural)	Urban	-1.125*** (0.033)	-0.974*** (0.071)	-0.486*** (0.084)	0.025 (0.065)
Road Type (Ref: Interstate)	Arterial	-1.048*** (0.018)	0.107 (0.071)	0.428*** (0.095)	0.549*** (0.066)
	Collector	-1.057*** (0.005)	0.128*** (0.023)	0.747*** (0.023)	1.424*** (0.027)
Vehicle Year (Continuous)	-	-0.020*** (0.0001)	-0.012*** (0.0001)	-0.017*** (0.0001)	-0.007*** (0.0001)
Constant	-	42.225*** (0.0003)	25.304*** (0.0001)	35.254*** (0.0001)	14.196*** (0.0001)

*p<0.1; **p<0.05; ***p<0.01, McFadden R^2 = 0.052, Log-likelihood = -3846.0, "Ref" = Reference value for categorical variables, Standard errors are in parentheses.

Weekends (Saturdays and Sundays) were found to have a significant impact on increased odds for fatal injury compared to weekdays, at 1.16 times more likely, at the 1% confidence level. However, weekends also were found to have decreased odds ratios for disabling, non-disabling, and possible injury severities, at 0.84, 0.85, and 0.70 times those of weekdays, respectively. The confidence intervals are at the 5% level for disabling and non-disabling injury, and 1% for possible injury.

Compared to rural areas, urban areas were found to have statistically significant reductions in odds ratios for fatal injury (0.33 times), disabling injury (0.38 times), and non-disabling injury (0.62 times). A small increase in odds ratio was found for urban areas compared to rural areas for possible injury (1.03 times), but it was not significant. Compared to interstate roads, both arterial and collector roads were found to have statistically significant decreases in odds ratios for fatal injury, at 0.35 times and 0.34 times, respectively. It is noted that in general, pedestrians are not permitted on interstate roads, which suggest that other factors exist (such as intoxication levels). However, both arterial and collector roads were found to have increases in odds ratios for all other severity types, with statistical significance at the 1% level for all cases except for arterial roads and disabling injury, which was not significant at the $P < 0.1$ level. In particular, collector roads had the highest odd ratio increases for disabling injury (1.14 times), non-disabling injury (2.11 times), and possible injury (4.12 times), compared to interstate roads.

Vehicle year (a continuous variable) representing the age of the vehicle was found to have a small but statistically significant impact on injury severity. A one unit (year) increase in vehicle year was found to correlate with a small decrease in odds ratio (0.997 times) for fatal injury and non-disabling injury (0.994 times). Small odds ratio increases were found for disabling injury and possible injury (both less than a 1% increase). The complete odds ratios are given in *Table 3*.

5 DISCUSSION

Injury severity is found in the literature to be most affected by pedestrians' physical condition (age, alcohol use), vehicle speed, driver and pedestrian perception and reaction (weather, lighting), and area of impact (vehicle type) (Lee & Abdel-Aty, 2005). We found that an increase in injury severity to fatal injury is likely correlated with driver age, driver gender, light condition, weather condition, temporal conditions, driver intoxication, road functional class and setting, and vehicle year. Using the model presented, it was found that the log-odds ratios for fatal injury increased for younger driver age groups (under 25), as reported in Table 2. These results are similar to findings by McCartt and Teoh (2015) and Kim et al. (2008), where young drivers (under 25) were found to have significantly higher rates of fatal injury overall compared to other age groups (McCartt & Teoh, 2015). The model outlined in this paper suggests that elderly drivers (75+) are more likely to inflict severe or fatal injury on pedestrians compared to all other driver ages. Conversely, McCartt & Teoh (2015) report a model in which elderly drivers were found to have a lower incidence rate of fatal injury and incapacitation injury to pedestrians, despite a higher rate of crash incidence overall (McCartt & Teoh, 2015; Kim et al., 2008). Furthermore, the model in this paper suggests that male drivers are more likely to fatally injure pedestrians than females. Conversely, female drivers were found to be more likely to injure pedestrians in a non-fatal fashion compared to males, all other factors remaining constant. This could be due to a number of contributing factors, such as risk aversion proportion of driver gender, but similar findings were reported by McCartt & Teoh (McCartt & Teoh, 2015). This skew towards males as contributing to more severe injuries is also found on the pedestrian side of the equation within the literature.

While males in general contribute to a larger share of traffic collisions overall behind the wheel compared to females, they also make up a significant portion of the fatalities as pedestrians involved in collisions. In 2015, 70% of the pedestrians killed in traffic crashes were males (NHTSA, 2017). This equates to a pedestrian fatality rate per 100,000 people of 2.37 for males, more than double for females (NHTSA, 2017). This statistic has remained relatively consistent even since the early 1990s. In 1994, 68% of all U.S. pedestrian fatalities were male (LaScala, Gerber, and Gruenewald, 2000).

Table 3: Summarized odds ratios for injury severity by the model

Independent Variable		Dependent Variable			
		Death From Injury	Disabling Injury	Non-Disabling Injury	Possible Injury
Driver Age (Ref: Under 25)	25-54	0.724	0.773	0.987	0.950
	55-64	0.902	0.643	0.943	0.971
	65-74	0.832	1.046	1.062	1.156
	75+	1.125	1.472	1.738	1.622
Driver Gender	Female	0.589	1.206	1.471	1.508

(Ref: Male)					
Lighting Condition (Ref: Daylight)	Dawn	0.629	1.350	1.185	1.096
	Dusk	2.047	1.832	1.452	1.602
	Dark, street lights on	3.667	2.111	1.339	1.100
	Dark, street lights off	3.190	0.397	0.544	0.452
	Dark, no street lights	10.213	2.472	1.051	0.796
Weather (Ref: Clear)	Overcast	1.235	1.045	1.099	1.265
	Raining	0.940	1.257	1.245	1.565
	Snowing	0.530	1.715	1.144	1.997
	Other	0.890	0.949	0.989	0.316
Day of Week (Ref: Weekday)	Weekend	1.157	0.843	0.846	0.698
Driver Intoxication (Ref: Not Impaired)	Impaired	1.602	1.257	0.865	0.681
Environment Location (Ref: Rural)	Urban	0.325	0.378	0.615	1.026
Road Type (Ref: Interstate)	Arterial	0.351	1.113	1.534	1.732
	Collector	0.348	1.137	2.110	4.155
Vehicle Year (Continuous)	-	0.980	0.988	0.983	0.993

In terms of visual acuity, daylight provides a greater capacity for both driver and pedestrian awareness than compared to a lack of natural lighting during night hours. Fatal injury was found to be the most likely outcome for all non-daylight/dawn scenarios. It is theorized that drivers have a greatly reduced visual acuity and therefore reduced reaction time during low-light settings compared to highly visible settings.

6 DATA LIMITATIONS

Vehicle type could not be analyzed due to accuracy-related issues at the data-collecting-level (i.e. the Washington State HSIS manual did not state vehicle subtypes in their explanatory pages, as the staff believed the codes to be inaccurate. Analogous manuals from other states could not be used as there was no consistency between reported codes for vehicle type). Furthermore, vehicle speed was not available within the data and therefore its impact on injury severity cannot be reported on here. It is hypothesized that greater speed will result in higher injury severity, due to speed's involvement in both momentum and force. However, speed in general likely has an overarching effect on the overall model, causing the effect of other significant variables to appear small in comparison. Other predictors that would ideally be tested in this model are pedestrian characteristics, such as age, gender, and intoxication, as it is hypothesized that physiological aspects of the victim play a contributing role in injury severity. Hour of the day as a temporal parameter may also be included in an ideal model, though was also not available in the data. However, the presented model parameters for light condition act as a moderate proxy toward time of day, as they contain indices of natural lighting (or lack thereof), though there are other conditions as well.

7 CONCLUSIONS

Modelling pedestrian injury severity in vehicle collisions requires consideration of numerous driver characteristics and environmental factors. We estimated Multinomial logistic regression models to analyze pedestrians' injury severity in collisions with automobiles. It was found that an increase in fatal injury was associated with younger and elderly drivers, male drivers to a greater extent than female drivers, dark lighting conditions compared to daylight, overcast weather conditions, specific days of the week, driver intoxication, larger/busier interstate roads compared to smaller arterial roads, rural environments more than urban environments, and older vehicles. The presented model does not address vehicle speed, or pedestrian age, gender, or intoxication. It would be beneficial to include these factors in future models in order to assess their impact.

Future work could also look at the impact of vehicle type, which would help address the nature of the impact itself. Further, injury severity here is presented on a categorical basis overall, but does not address specific injury type, recovery rate and costs, as well as intangible costs due to fatality and severe injury (including mental health injury). The results presented here could be used to formulate public policy and promote awareness about the risks faced by pedestrians. Educating younger drivers about safety and advising pedestrians about being vigilant at times when visibility is poor could help improve safety for pedestrians.

References

Allen, B. L. Shin, T. B., and Cooper, P. 1978. Analysis of traffic conflicts and collisions. Transportation Research Record. 67-74.

Anderson, R. W., McLean, A. J., Farmer, M. J., Lee, B. H., and Brooks, C. G. 1997. Vehicle travel speeds and the incidence of fatal pedestrian crashes. *Accident Analysis and Prevention,* **29**(5), 667-674.

Ballesteros, M. F., Dischinger, P. C., and Langenberg, P. 2004. Pedestrian injuries and vehicle type in Maryland, 1995–1999. *Accident Analysis and Prevention,* **36**(1), 73-81. doi:10.1016/s0001-4575(02)00129-x

Dewar, R. 2007. The human factors in traffic safety, 2nd edition. Tucson, AZ: Lawyers and Judges Publishing Company, Inc.

Ewing, R., and Dumbaugh, E. 2009. The built environment and traffic safety. *Journal of Planning Literature*, **23**(4), 347-367.

Fieldwick, R., and Brown, R. J. 1987. The effect of speed limits on road casualties. *Traffic Engineering and Control,* **28**, 635-640.

Nujjetty, A. p., and Mohamedshah, Y. M. 2014. *State Data Files: WASHINGTON* [Data file and guidebook]. Washington: Highway Safety Information System.

Kim, J., Ulfarsson, G. F., Shankar, V. N., and Kim, S. 2008. Age and pedestrian injury severity in motor-vehicle crashes: A heteroskedastic logit analysis. *Accident Analysis and Prevention,* **40**(5), 1695-1702. doi:10.1016/j.aap.2008.06.005

Kockelman, K.M., Kweon, Y., 2002. Driver injury severity: an application of ordered probit models *Accident Analysis and Prevention*, **34**(3), 313–321.

LaScala, E. A., Gerber, D., and Gruenewald, P. J. 2000. Demographic and environmental correlates of pedestrian injury collisions: a spatial analysis. *Accident Analysis and Prevention,* **32**(5), 651-658. doi:10.1016/s0001-4575(99)00100-1

Lee, C., and Abdel-Aty, M. 2005. Comprehensive analysis of vehicle–pedestrian crashes at intersections in Florida. *Accident Analysis and Prevention,* **37**(4), 775-786. doi:10.1016/j.aap.2005.03.019

Mccartt, A. T., and Teoh, E. R. 2015. Tracking progress in teenage driver crash risk in the United States since the advent of graduated driver licensing programs. *Journal of Safety Research,* **53**, 1-9. doi:10.1016/j.jsr.2015.01.001

National Highway Traffic Safety Administration. 2017. Pedestrians: 2015 data. (Traffic Safety Facts. Report No. DOT HS 812 375). Washington, DC: National Highway Traffic Safety Administration. Retrieved from https://crashstats.nhtsa.dot.gov/Api/Public/Publication/812375

Riply, B., and Venables, W. 2016. Software for feed-forward neural networks with a single hidden layer, and for multinomial log-linear models. https://cran.r-project.org/web/packages/nnet/index.html

Rodriguez, G. 2007. WWS 509: *Multinomial Response Models* [Online Lecture Notes]. Retrieved from http://data.princeton.edu/wws509/notes/c6.pdf

Ulfarsson, G.F., and Mannering, F.L. 2004. Difference in male and female injury severities in sport-utility vehicle, minivan, pickup and passenger car accidents. *Accident Analysis and Prevention*, *36*(2), 135–147.

Wei, V. F., and Lovegrove, G. 2012. Sustainable road safety: A new (?) neighbourhood road pattern that saves VRU lives. *Accident Analysis and Prevention,* **44**(1), 140-148. doi:10.1016/j.aap.2010.12.005

Building Tomorrow's Society
Bâtir la Société de Demain

Fredericton, Canada
June 13 to 16, 2018 / *13 au 16 juin, 2018*

INCORPORATING ASSET VALUE IN LIFE CYCLE COST ANALYSIS FOR TRANSPORTATION INFRASTRUCTURE ASSETS

Alyami, Zaid[1,3], and Tighe, Susan L.[2]
[1] Project Manager, Jacoobs
[2] Deputy Provost and Associate Vice-President, Integrated Planning and Budgeting
Norman W. McLeod Professor in Sustainable Pavement Engineering
President 2017 – 2018 Canadian Society for Civil Engineering
[3] zalyami@uwaterloo.ca

Abstract: Asset valuation is an essential component of effective asset management. It is an important method to demonstrate proper management of public assets and effective utilization of tax payers' money. Asset valuation is used in standard reporting, depreciation schedules, auditor requirements and condition assessments. As such, it allows agencies to demonstrate funding needs for asset preservation. On the other hand, Life Cycle Cost Analysis (LCCA) is used in asset management to evaluate different maintenance and rehabilitation strategies. In addition to maintenance and rehabilitation costs, agencies started to include other factors such as salvage value and user costs and environmental costs in the LCCA. However, to date, the impact of asset value on LCCA is not considered. Incorporating asset value in LCCA as a means of evaluating the return on investment is imperative. The paper presents a methodology to incorporate asset value as a component of LCCA. The proposed method is illustrated through an example using data from the Ministry of Transportation (MTO) Pavement Management System (PMS2).

1 Background

Asset management has gained movement over the last two decades. Asset Management in basic terms is a systematic business process that employs strategic, engineering and economical means to provide holistic approach to manage infrastructure assets to meet specified performance measures' Level of Services (LOS). There are many definitions of Asset Management in the literature; however, a widely used definition is that of the Federal Highway Administration (FHWA) US department of Transportation also adopted by Transportation Association Canada (TAC) (FHWA 1999; TAC 2013):

"Asset management is a systematic process of maintaining, upgrading and operating physical assets cost-effectively. It combines sound business practices and economic theory, and it provides tools to facilitate a more organized logical approach to decision making. Thus, asset management provides a framework for handling both short- and long-range planning."

From the definition, it can be noted that implementing asset management necessitates implementing sound business practices and economic theory. It is a decision making tool or a framework that spans over an extended time horizon (FHWA 1999). In other words, implementing asset management is the development of business plans and programs of maintaining, upgrading and operating infrastructure assets over a specified time horizon. Therefore, as in any business practice, it is important to understand the value of the assets being managed.

On the other hand, with the implementation of asset management, several government regulatory bodies mandate agencies to report their Tangible Capital Assets' (TCA) values within their annual statement. For example, the Canadian Public Sector Accounting Board (PSAB), the Governmental Accounting Standard Board (GASB) and the New Zealand International Financial Reporting Standards (NZ IFRS) to name a few (Alyami and Tighe 2016).

Asset valuation is an essential component of effective asset management (TAC 2013). It is an important method to demonstrate proper management of public assets and effective utilization of tax payers' money. Asset valuation is used in standard reporting, depreciation schedules, auditor requirements and condition assessments (Byrne 1994). As such, it allows agencies to demonstrate funding needs for asset preservation (Lugg 2005).

Life Cycle Cost Analysis (LCCA) is used in asset management to evaluate different maintenance and rehabilitation strategies. In addition to maintenance and rehabilitation costs, agencies started to include other factors such as salvage value and user costs and environmental costs in the LCCA (Bryce et al. 2014; Mallela et al. 2011; Ozbay et al. 2004; Smith and Fung 2006; TAC 2013; Torres-Machí et al. 2015). However, to date, the impact of asset value on LCCA is not considered. Incorporating asset value in LCCA as a means of evaluating the return on investment is imperative.

2 Scope and Objective

This paper presents a methodology to incorporate asset value in LCCA for transportation infrastructure assets. The proposed method is illustrated through a pavement example based on data obtained from MTO PMS2. However, the method can be applied to other infrastructure assets.

3 Asset Valuation

In transportation infrastructure context, asset valuation- or asset management in general, are implemented to fixed and unfixed tangible assets within or out of the right of way (ROW) (TAC 2001, 2013). Example of fixed assets within the ROW include: pavements, bridges, signs, signals, etc. Fixed and unfixed assets out of the ROW include: maintenance depots (Ex. salt sheds, fuel tanks, etc.), material stockpiles, laboratories, communication equipment, computer hardware, etc. In addition, Haas and Raymond identified other non-tangible assets such as intellectual property, land, etc. (Haas and Raymond 1999).

The American Institute of Real Estate Appraisers (AIREA) defines asset valuation as the process of estimating the value of a specific asset at a given date, and it measures the relative value or wealth of asset over time (AIREA 1987). Marston et al. defined asset valuation in the context of engineering as "the art of estimating the fair monetary measure of the desirability of ownership of specific properties for specific purpose...engineering valuation is the art of estimating the value of specific properties where professional engineering knowledge and judgment are essential. ... based fundamentally upon [the asset's] ability to produce some kind of useful service during its expected future life in service...." (Marston et al. 1963).

There are various valuation methods that can be utilized to estimate infrastructure asset values such as book value, replacement cost, and written down replacement cost. Table 1 presents examples of asset valuation methods and basic definitions. It is recognized that there is no universally accepted method by the international community. However, it is noted that the book value, the replacement cost and the written down replacement cost methods are commonly used in highway infrastructure valuation (Cowe Falls et al. 2004; Dewan and Smith 2005; McNeil et al. 2000; OECD 2000).

Table 1: Asset valuation methods and basic definition (Adapted from (Amekudzi et al. 2002; TAC 2001)

Asset Valuation Method	Overview
Book Value	Present value based on historical costs depreciated to the present (commonly used for financial accounting purposes)
Replacement Cost	Present value based on cost of replacing/rebuilding the asset

Written Down Replacement Costs	Present value based on current replacement cost depreciated to asset current condition (commonly used for management accounting purposes)
Equivalent Present Worth in Place	The worth "as is". The book Value adjusted for inflation, depreciation, depletion and wear; i.e., the (accounts for changes in prices and usage; applicable to comparing with other investments)
Productivity Realized Value	The value in use. Net present value of benefit stream for remaining service life (provides a reflection of relative importance of the asset)
Market Value	Price buyer is willing to pay
Net Salvage Value	Cost to replace the facility less the cost of returning it to 'new condition' Cost of materials
Salvage Value	Present worth of the amount obtainable from disposing or recycling Facility
Option Value	Value of asset in specific circumstances (Used by private sector)

4 Life Cycle Cost Analysis

Life Cycle Cost Analysis (LCCA) refers to the direct financial costs associated with a project (TAC 2013). LCCA evaluates competing alternatives by evaluating costs incurred along the project life cycle including initial construction costs, maintenance and rehabilitation costs to maintain functional condition along the service life. This process is widely applied because it can evaluate differences between design options such as pavement type and various feasible design cross sections. Agencies of all levels have used LCCA to evaluate new technologies, develop alternatives, and to provide defensible decisions for alternative financing and procurement of projects (Smith and Fung 2006). LCCA is used by agencies to assist with long-term planning asset management plans and budget estimates.

In LCCA, it is important to account for the change of the time value of money (FHWA 2002; Markow 2012; TAC 2013). In other words, costs at different times must be converted to their value at a common point in time using a discount rate. To evaluate competing alternatives, infrastructure investments are converted to a single variable, the Net Present Cost (NPC). The NPC allows for comparing the total costs of the alternatives in today's dollars. The NPW is calculated as follows:

[1] NPC = $IC + \sum_{j=1}^{k}(M\&R_j \ x \ (\frac{1}{1+i})^{n_j})$

Where

NPC= Net Present Cost ($)

IC= Initial Cost ($)

K = Number of future maintenance, preservation and rehabilitation

$M\&R_j$= Cost of jth future maintenance, preservation and rehabilitation activity ($)

i= Discount rate

n_j= Number of years from the present to the jth future maintenance, preservation and rehabilitation treatment

5 Proposed Methodology

5.1 Asset Value Loss: An Integration Mechanism

In the context of asset management, it is paramount to establish a value of an asset and be able to manage it, maintain and enhance its value. The following desired characteristics are identified in incorporating asset value in asset management decision making: 1) The asset valuation method selected should be readily and easily calculated. 2) The valuation method directly relates to the asset condition, reflecting the needs and returns on investments for assets' preservation. 3) Capturing the value to the various stakeholders. 4) Addressing the challenges in predicting future asset values due to the instability of economic forces and the difficulty to predict future unit prices. In other words, due to the change in unit prices due to market forces, asset values may increase or decrease regardless of any asset management stewardship.

To address the challenges identified above to provide a stable measure that can be used in asset management decision making, the Asset Value Loss (AV_L) is introduced as a ratio of the depreciated asset value to that of a new asset value, expressed as follows:

[2] Asset Value Loss (AV_L) = $\frac{RC - WDRC}{RC}$

Where; AVL is the Asset Value Loss ratio, WDRC is Written Down Replacement Cost, and 17 RC is the Replacement Cost

The RC and WDRC methods are straightforward methods and easily calculated, understood and communicated. Using the ratio eliminates the impact of changes to unit prices and inflation or discount rate, as the percentage loss will remain constant regardless of any changes to unit prices.

To illustrate, consider a pavement section with 15 years analysis period based on MTO PMS2 data presented in Table 2. The section received rehabilitation at year 13. The section attributes, condition, replacement unit cost, and interest rate are provided in Table 2.

Table 1: Asset value loss illustration example

Year	PCI	RC (1)	WDRC (2)	Value Loss ($) (1-2)	VL (%) (1-2)/(1)
1	97	$ 4,140,281	$ 4,006,964	$ 133,317	3%
2	96	$ 3,755,357	$ 3,617,160	$ 138,197	4%
3	95	$ 3,406,219	$ 3,236,930	$ 169,289	5%
4	94	$ 3,089,541	$ 2,901,388	$ 188,153	6%
5	88	$ 2,802,305	$ 2,463,506	$ 338,799	12%
6	87	$ 2,541,773	$ 2,200,921	$ 340,852	13%
7	86	$ 2,305,463	$ 1,974,168	$ 331,295	14%
8	80	$ 2,091,123	$ 1,664,743	$ 426,380	20%
9	79	$ 1,896,710	$ 1,499,919	$ 396,792	21%
10	76	$ 1,720,372	$ 1,313,676	$ 406,696	24%
11	73	$ 1,560,428	$ 1,135,524	$ 424,905	27%
12	69	$ 1,415,354	$ 975,179	$ 440,175	31%
13	92	$ 1,283,768	$ 1,181,067	$ 102,701	8%
14	90	$ 1,164,416	$ 1,047,974	$ 116,442	10%
15	87	$ 912,350	$ 793,745	$ 118,606	13%

*Pavement Section Information: Pavement Type AC, Area: 52250 m2, Replacement Unit Cost: $ 91.73, interest rate= 5%

The ratio indicates the total loss of asset replacement cost considering its current condition to that of a new asset. In other words, the ratio shows the loss of value due to the asset deterioration. If a preservation or rehabilitation is applied, the WDRC value increases as the condition improves and therefore reducing the loss ratio. The ratio allows for incorporating future values while addressing the challenges associated with economic fluctuations of unit costs. In addition, in the context of asset management decision making, the concept addresses the variation of value between sections due to the different area. Figure 1 graphically illustrates the value loss ratio over the analysis period. The proposed AV_L is a function of the WDRC valuation methods; which incorporate asset condition to write down the replacement cost. To establish the threshold for AV_L, equation 2 is rearranged and the $AV_{L,\ threshold}$ is calculated as follows:

$$[3]\ AV_{L,\ threshold} = 1 - \frac{Asset\ Condition\ minimum\ LOS}{Asset\ Condition\ (New)}$$

Where,

$AV_{L,threshold}$: The maximum Asset Value Loss acceptable for the specified Level of Service (LOS)

Asset Condition Minimum LOS: Established LOS or threshold in place for asset condition

Asset Condition (New): The asset condition if newly constructed

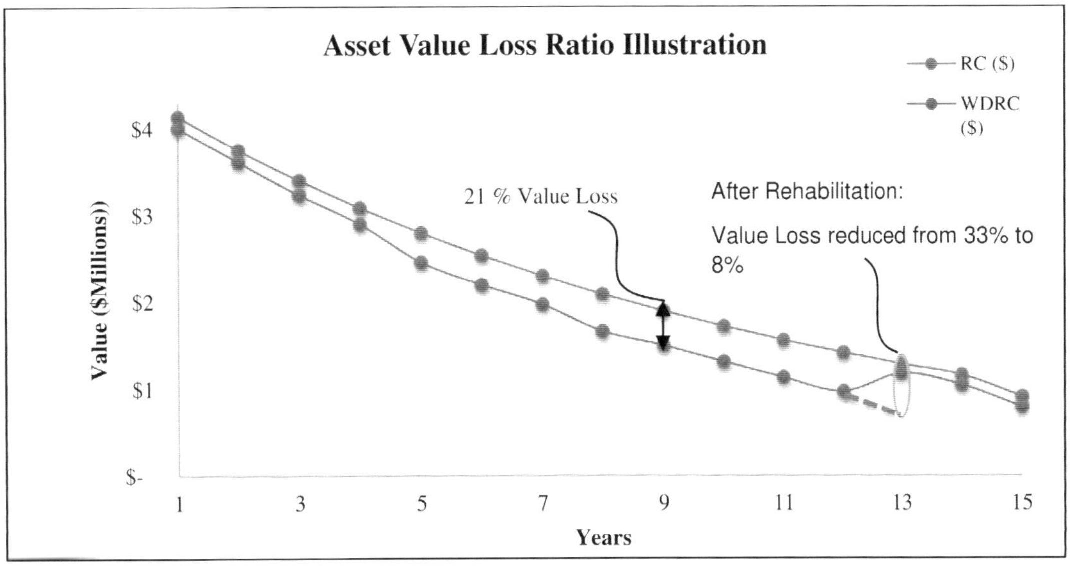

Figure 1: Graphical representation of value loss ratio over analysis period

As discussed earlier, one of the key challenges in reporting asset value using the current methods is predicting future asset values due to the instability of economic forces and the difficulty to predict future unit prices. Therefore, it is recommended that the AV_L is used to report the value as a percentage of the replacement cost (fixed to the analysis year unit cost) over the analysis period. The ratio can be presented to convey an increase of value loss due to lack of proper funding and asset management stewardship or vice versa. As presented earlier, the RC and WDRC methods are straightforward methods and easily understood and communicated. Using the ratio eliminates the impact changes to unit prices and inflation or discount rate as the percentage loss will remain constant regardless of any changes to unit prices. Figure 2 presents a systematic framework to establish the current and future AV_L for infrastructure assets.

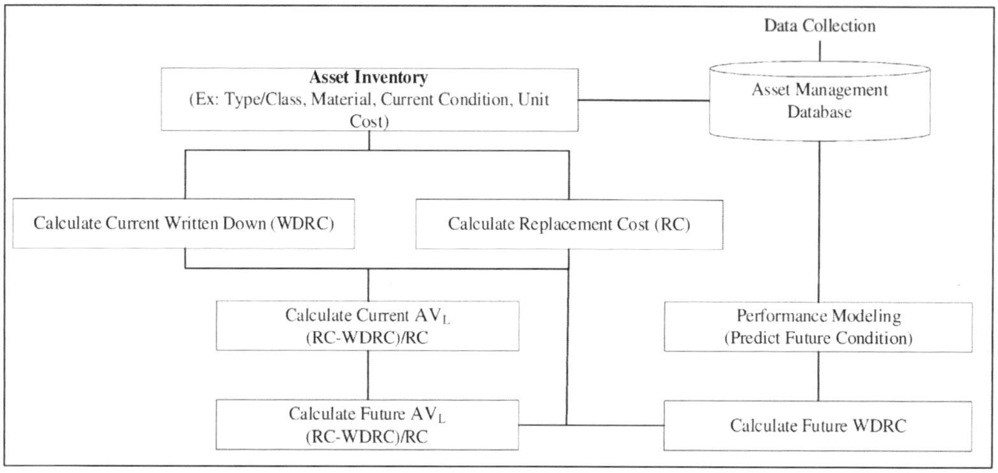

Figure 2: Asset value loss framework

5.2 Value-Based Life-Cycle Cost Analysis

As noted, the LCCA is used in asset management to evaluate different maintenance and rehabilitation strategies. In addition to maintenance and rehabilitation costs, agencies started to include other factors such as salvage value and user costs and environmental costs in the LCCA (Bryce et al. 2014; Mallela et al. 2011; Ozbay et al. 2004; Smith and Fung 2006; TAC 2013; Torres-Machí et al. 2015). However, to date, the impact of asset value on LCCA is not considered. Incorporating asset value in LCCA as a means of evaluating the return on investment is imperative. For example, the Transportation Research Board (TRB) issued a research needs statement with the objective to develop standard calculation methodologies to characterize the asset value of pavements for use in funding allocation, life cycle cost analysis and engineering evaluation.

The method proposed in the previous section can also be used in LCCA to evaluate different designs or the impact of different maintenance and rehabilitation strategies over the life cycle of the project on its value. The AV_L can be used to calculate the total loss of asset value over the analysis period and added as an incurred cost. Figure 3 shows an illustration of the proposed methods.

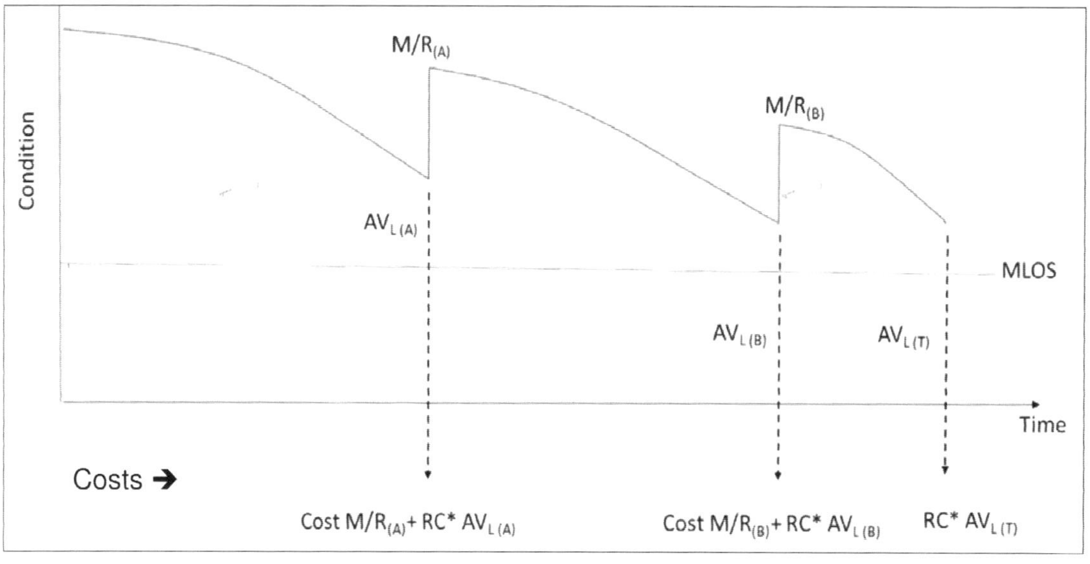

Figure 3: Incorporating asset value loss in LCCA illustration

As shown in Figure 3, the AV_L up to the time maintenance or rehabilitation is applied and at the end of the analysis period is used to calculate the loss in value up to that point. Mathematically, the calculation shown in Equation 4.2 can be modified to include asset value loss as follows:

[3] NPC = $IC + \sum_{i=1}^{k}(M\&R_j \; x \; (\frac{1}{1+i})^{n_j}) + \sum_{t=1}^{T}(AV_{Lj} \; x \; RC)$

Where:

AV_{Lj} is the Asset Value Loss ratio before the jth $R\&M_j$ treatment is applied, and at the end of the analysis period (Year T)

RC is the present Replacement Cost ($)

t = Year *t* of *T* years analysis period

Incorporating asset value loss allows for quantifying the opportunity loss in investment to preserve the asset and the impact of delay in maintaining assets.

5.3 Example

To further illustrate the proposed methodology, consider the pavement rehabilitation strategies example shown in Table 4.2. The pavement performance over the analysis period is illustrated in Figure 4.

Table 4.2: Pavement rehabilitation strategies example

	Option A	Option B
Rehabilitation 1	Hot Mix Overlay, Year 9	Year 12: Hot Mix Overlay
Rehabilitation 2	Mill + Hot Mix Overlay, Year 18	Year 20: Mil + Hot Mix Overlay

*Area = 21,491m^2, Replacement Unit Cost = $91.73, Rehabilitation Unit Costs = $19.16/m^2, Mill + Hot Mix Overlay = $19.5/m^2, Analysis Year = 2010, Analysis Period = 20 years, Discount Rate = 5%
Following Equation 3, the NPW of the alternatives are calculated as follows:

<u>Option A:</u>

[($ 19.16 / m^2) * (21,491 m^2) * (1/1.05^9)] + [($ 19.5 / m^2)*(21,491 m^2)*(1/1.05^18)] +

[(0.25+0.28+.012) * ($ 91.73 / m^2)*(21,491 m^2) = $ 1,508,257.53

<u>Option B:</u>

[($ 19.16 / m^2) * (21,491 m^2) * (1/1.05^12)] + [($ 19.5 / m^2)*(21,491 m^2)*(1/1.05^20)] +

[(0.3+0.26+0.1) * ($ 91.73 / m^2)*(21,491 m^2) = $ 1,688,525.61

Based on the calculation above, option A is preferred. It is worth noting that without taking asset value loss into consideration, option B is more preferred as the NPW is $387,421 while the NPW for option A is $ 439,775. Therefore, by taking into account the asset value loss due to the delay in maintaining the asset allows for a justifiable and quantifiable need for funding and opportunity loss in investment.

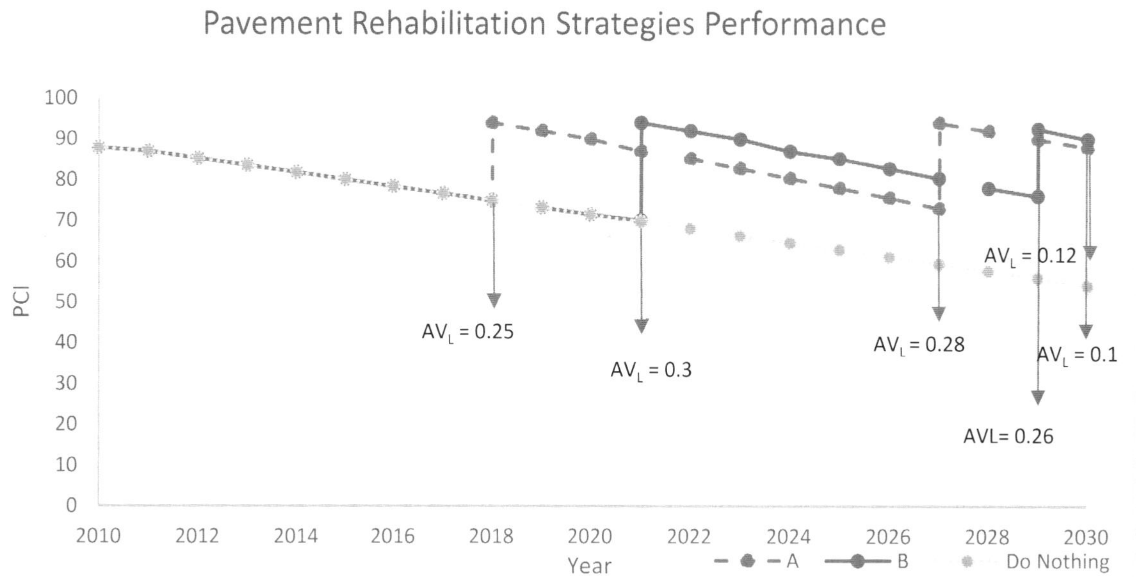

Figure 4: Value based life cycle cost analysis example for rehabilitation strategies

5.4 Summary and Conclusions

LCCA is used in asset management to evaluate different maintenance and rehabilitation strategies. In addition to maintenance and rehabilitation costs, agencies are using other factors such as salvage value and user costs and environmental costs in the LCCA. The paper presents a methodology to incorporate asset value as a component of LCCA as a means of evaluating the return on investment. The paper presented a methodology to establish asset value loss ratio that can be incorporated as a factor in LCCA. Incorporating asset value provides a tool for decision makers to evaluate different alternatives while taking into account the loss or add of value for a more justifiable and quantifiable need for funding and opportunity loss in investment.

References

AIREA. (1987). *The appraisal of real estate*. American Institute of Real Estate Appraisers.

Alyami, Z., and Tighe, S. (2016). "A Methodology for Integrating Asset Valuation in Transportation Asset Management." *Resilient Infrastructure, Canadian Society for Civil Engineering (CSCE) Proc.*, London, Ontario.

Amekudzi, A., Herabat, P., Wang, S., and Lancaster, C. (2002). "Multipurpose Asset Valuation for Civil Infrastructure: Aligning Valuation Approaches with Asset Management Objectives and Stakeholder Interests." *Transportation Research Record*, 1812(1), 211–218.

Bryce, J. M., Flintsch, G., and Hall, R. P. (2014). "A multi criteria decision analysis technique for including environmental impacts in sustainable infrastructure management business practices." *Transportation Research Part D: Transport and Environment*, 32, 435–445.

Byrne, R. (1994). "Asset valuations." *2nd Annual Conference, Institute of Municipal Engineering Australia, Queensland Division (IMEAQ)*, Townsville, Queensland, Australia.

Cowe Falls, L., Haas, R., and Tighe, S. (2004). "A Comparison of Asset Valuation Methods for Civil Infrastructure." *Coordinating Pavement and Maintenance Management with Asset Management Session Of the 2004 Annual Conference of the Transportation Association of Canada (TAC)*, Quebec City, Quebec.

Dewan, S. A., and Smith, R. E. (2005). "Valuing Pavement Network Assets and Use of Values as

Decision Supports." *Journal of Infrastructure Systems*, 11(4), 202–210.
FHWA. (1999). *Asset Management Primer.* Federal Highway Administration, U.S. Department of Transportation and Office of Asset Management, Washington, D.C.
FHWA. (2002). *Life-Cycle Cost Analysis Primer.* Washington D.C.
Haas, R., and Raymond, C. (1999). "Long Term Performance Specified Contract Framework for Road Netwroks." *Transpotation Asscoiation of Canada Annual Conference*, St. John.
Lugg, M. (2005). *Guidance Document for Highway Infrastructure Asset Valuation.* The Stationery Office, London UK.
Mallela, J., Sadasivam, S., Quintus, H. L. Von, Darter, M. I., Hallin, J. P., and Hein, D. K. (2011). *Guide for Pavement-Type Selection.* National Academies Press, Washington, D.C.
Markow, M. J. (2012). *NCHRP Synthesis 424: Engineering Economic Analysis Practices for Highway Investments.* Transportation Research Board.
Marston, A., Jean, C. H., and Robley, W. (1963). "Engineering Valuation and Depreciation." *Iowa State University Press.*
McNeil, S., Tischer, M. L., and DeBlasio, A. J. (2000). "Asset management: What is the fuss?" *Transportation Research Record: Journal of the Transportation Research Board*, 1729(1), 21–25.
OECD. (2000). "Asset Management for the Roads Sector." *Organisation for Economic Co-Operation and Development*, 1, 111.
Ozbay, K., Jawad, D., Parker, N., and Hussain, S. (2004). "Life-Cycle Cost Analysis: State of the Practice Versus State of the Art." *Transportation Research Record: Journal of the Transportation Research Board*, Transportation Research Board of the National Academies , 1864, 62–70.
Smith, T., and Fung, R. (2006). "Review of the Alternative Bid Tenders for Canadian Highway Construction Projects with Life Cycle Cost Component." *2006 Annual Conference and Exhibition of the Transportation Association of Canada: Transportation Without Boundaries*, Transportation Association Canada (TAC), Ottawa, ON.
TAC. (2001). *Measuring and Reporting Highway Asset Value, Condition and Performance.* Transportation Association of Canada, Ottawa.
TAC. (2013). *Pavement Asset Design and Management Guide.* Transportation Association Canada (TAC), Ottawa, Canada.
Torres-Machí, C., Chamorro, A., Pellicer, E., Yepes, V., and Videla, C. (2015). "Sustainable Pavement Management." *Transportation Research Record: Journal of the Transportation Research Board*, Transportation Research Board of the National Academies , 2523, 56–63.

Building Tomorrow's Society
Bâtir la Société de Demain

Fredericton, Canada
June 13 to 16, 2018 / *13 au 16 juin, 2018*

INCORPORATING FAST-TRACK PAVEMENT REPAIR METHODS INTO CURRENT INFRASTRUCTURE MANAGEMENT PRACTICES

Malek, Dahlia K.[1,2], Speller, Victoria[1], and Tighe, Susan L.[1]
[1] University of Waterloo, Canada
[2] dkmalek@uwaterloo.ca

Abstract: Pavement infrastructure deteriorates due to environmental conditions, traffic loading, and material aging. Regular maintenance and repair is required to maintain an acceptable level of service. High-traffic volumes and heavy traffic can increase the rate of pavement deterioration, necessitating more frequent rehabilitation or maintenance on these roadways. However, construction is often limited to a short window to minimize user impacts. Another common problem in asphalt roadways is pavement cracking; it is important to repair cracking to prevent further deterioration of the roadway. Accelerated construction or "fast-track" methods provide a solution to repair and extend the service life of pavement infrastructure, quickly, easily, and with minimal construction-related delays. Throughout this paper, two examples of fast track pavement repair methods are discussed and compared to traditional repair methods. Precast concrete inlay panels are used to rapidly rehabilitate structural rutting on high traffic volume highways, and infrared heating technology is used to quickly repair cracking of asphalt pavement. The objective of this paper is to examine the innovation, benefits, and drawbacks of these fast-track pavement rehabilitation methods and to present their role in pavement management. This paper investigates the project-level decision-making process of selecting between these new fast-track repair techniques and the existing traditional repair methods by examining various decision criteria.

1 INTRODUCTION

Roadways are essential infrastructure that allow the movement of people and goods. Road quality affects the social and economic development of the community and impacts individuals in their day to day commutes. Across Canada, there are approximately 415,000 km of paved highways and roadways ("Length of Canada's Public Road Network, 2003" 2015), so it is a large undertaking to maintain pavement quality and service.

Conventional asphalt-concrete pavements have a design life of approximately 20 to 25 years, while Portland Cement Concrete (PCC) pavements have a longer design life of 40 to 50 years. For the service life of pavement to reach its full design life, maintenance and possible rehabilitation is required. Without the proper maintenance/rehabilitation work, pavement failure or an inadequate level of service may occur before it reaches its full design life.

Today's infrastructure is subjected to heavy traffic loading, high traffic volumes, environmental conditions, and other factors that cause pavements to deteriorate. Construction and maintenance is needed to keep the pavement in adequate condition, but the task of performing the repair work can be made difficult by time limitations and traffic constraints. The challenge of performing construction within a short time period while maintaining a certain level of service drives the development of accelerated construction or "fast-track" methods. These are methods used to perform repairs rapidly to extend the service life of pavement.

For a given pavement repair or rehabilitation project, there are often multiple feasible treatment strategies. The major benefit of fast-track methods is the speed at which they can be performed to minimize the impact to roadways users. Many different fast-track methods for performing rapid roadway repairs and rehabilitation have been developed and are continuing to be researched. Some examples of fast pavement repair methods include the use of rapid-setting concrete pavements, precast concrete pavements, or partial depth asphalt milling and replacement.

Two examples of fast-track pavement repair technologies are discussed in detail in this paper. The first is a method of rapidly repairing structural rutting issues in asphalt pavements on high-traffic volume highways by partially milling the asphalt and inlaying precast concrete panels. The traditional repair method is to mill-and-replace the asphalt with a new hot-mix asphalt layer; however, this treatment strategy does not provide an adequate service life. The precast concrete inlay panels (PCIP) can be constructed overnight to minimize lane closures, and this strategy is expected to have a longer service life than the asphalt repair. The second fast-track pavement repair technology presented is a technique to rapidly repair cracking of asphalt pavements using an innovative infrared heating technology. This technology provides longer-term crack repair than traditional repair methods such as crack sealing/filling and mill-and-replace. Using infrared heating technology provides a very fast repair solution, with less opportunities for delay than more current methods of repair. Further research in infrared heating technology has produced some specialized fast-track repair processes for longitudinal crack repairs.

Both the PCIP and infrared heating technology are new methods of rapid pavement repair that have not been used extensively. The objective of this paper is to present the benefits and drawbacks of these two fast-track repair methods and the criteria which would be used to select between fast-track methods and the traditional repair methods.

1.1 Background

Several factors contribute to the deterioration of pavement over its service life, including construction quality, material aging, traffic loading, and environmental conditions.

One of the factors that causes a faster rate of deterioration of pavement is high-traffic volumes and heavy traffic loads. Some of the 400-series highways in Southern Ontario are the busiest in North America with annual average daily traffic (AADT) as high as 416,000 near the city of Toronto ("Ontario Provincial Highways Traffic Volumes On Demand" 2016). High-traffic volumes cause the pavement to deteriorate more rapidly than low-volume roads, requiring more frequent rehabilitation and maintenance. Furthermore, lane closures on high-volume highways have a greater user impact because there are thousands of roadway users. The effects of construction-related closures include increased traffic congestion and delay and increased safety risks for roadway users and workers (Abdelmohsen and El-Rayes 2016; Tayabji, Ye, and Buch 2013a). To reduce these effects, some departments of transportation require that highway lane closures for construction occur during an overnight window, typically around 10 p.m. to 6.a.m., with full reinstatement following this period. In these cases, fast-track pavement repairs can be advantageous to accomplish the repair within the tight construction window.

On all roadways, no matter how well designed and constructed, cracking will occur eventually. Cracking on pavement surfaces is most often caused by fatigue, and constant expansion and contractions of the pavement during freeze and thaw cycles (Rhino Asphalt Solutions, 2015). No severe problems regarding roadway performance or safety tend to occur after minor cracking propagates, but when not treated, cracked areas can quickly expand and cause further, more severe damage to roadways. Since cracking is such a common problem on all roadways, it is hard to keep up with the large volume of crack repair projects. In addition to the new cracks propagating throughout the roadway systems, many roads which have already been repaired continue to crack due to ineffective repair methods, or poorly implemented repairs. A fast track method of repairing pavement cracking can be beneficial to keep up with repair demand, but also needs to be more effective in creating a long-term repair on the roadway to increase its service life and eliminate the need for constant re-repair on the same sections of roadway.

2 FAST-TRACK REPAIR EXAMPLES

2.1 PCIP Rapid Repair for Highways

2.1.1 PCIP Technology

Precast concrete pavement (PCP) is a relatively new technology that is advantageous for projects requiring rapid construction or repair and high strength pavement. While only a small percentage of pavements are precast concrete, in the past 15 years PCP has become a more feasible and cost-effective alternative for certain projects (Tayabji, Ye, and Buch 2013b). Precast concrete inlay panels (PCIP) are a new, innovative application of precast concrete pavement. PCIP are being investigated as a rapid repair method to address deep-seated rutting issues observed on some high-traffic volume asphalt highways near the city of Toronto in Ontario, Canada. Precast concrete inlay panels during a trial construction are shown in Figure 1.

Figure 1: Precast concrete inlay panel installation in third lane of highway (Pickel and Tighe 2017)

Previously, the sections of highway exhibiting this problem were rehabilitated by partial-depth milling and replacement of the asphalt with a new layer of hot mix asphalt. However, this strategy has an observed service life of only 3 to 5 years and did not address the underlying problem (Pickel and Tighe 2017). Structural rutting is occurring, which is caused by permanent deformations in one or more layers below the asphalt surface layer (White et al. 2002). Structural rutting can typically only be repaired by performing a full-depth reconstruction of the layers causing problems. However, the high-traffic volumes deter the use of a full-depth repair strategy that would require a longer construction shutdown and have a greater user impact than the quicker alternative of a partial-depth repair. PCIP is a fast-track construction method developed to address this problem. The precast panels can be installed rapidly to minimize construction time, and concrete is a stiffer material that will reduce the stresses in the lower pavement layers to address the deep-seated issues without requiring a full-depth reconstruction

Repair of asphalt highways using PCIP involves partial-depth milling of the asphalt, preparing a support condition for the panels, installing the panels, and placing dowel, edge, and bedding grout. Providing a stable, uniform support condition for the panels is essential for a successful precast concrete pavement construction. Non-uniform support results in increased tensile stresses and cracking at the bottom of the panels in unsupported locations (Merritt et al. 2000). Three different support conditions were developed, designed, and constructed in the PCIP trial section; the alternatives are referred to as asphalt-supported, grade-supported, and grout-supported. The asphalt-supported condition consists of precision-milling the asphalt, placing the precast panels directly on top, and grouting. For the grade-supported alternative, the asphalt is milled, and a cement-treated base material (CTBM) is placed, levelled, and compacted before installing the panels on this base, followed by grouting. For the grout-supported alternative, the asphalt is milled, followed by placement of the panels. The panels are then raised to the correct elevation using leveling feet cast integrally with the panel, followed by grouting. Future research includes evaluating these three alternatives for constructability and performance to determine the optimal type of support for PCIP (Pickel and Tighe 2017).

2.1.2 Treatment Alternatives and Decision criteria

There are many criteria to consider when deciding between a repair strategy using PCIP or other traditional repair methods, such as asphalt mill-and-replace or cast-in-place concrete (CIPC). The general advantages and drawbacks of precast concrete pavement also apply to precast concrete inlay panels. Some of the primary criteria to consider when selecting between PCIP and a traditional repair method are traffic conditions, strength requirements, fabrication, construction considerations, installation and production rate, and costs.

One of the main benefits of using PCP over traditional repair methods is that precast concrete can be installed and the roadway can be re-opened to traffic rapidly to minimize construction-related lane closures for construction. Impacts of lane closures include increased traffic congestion and construction-related delays, increased safety risks for workers and roadway users, greater vehicle emissions, and higher vehicle-operating costs. (Merritt et al. 2000; Abdelmohsen and El-Rayes 2016). In overnight conditions, asphalt requires approximately 1 hour ± 20 minutes, depending on wind conditions, to cool before it can be reopened to traffic (Wang et al. 2018). CIPC typically requires 2 to 4 days after construction to gain sufficient strength to support traffic (Delatte 2008). High early strength concrete pavement can be used to reduce the time required for strength gain, however these are often accompanied by durability issues. In comparison, PCP can carry traffic immediately or shortly after installation. For the PCIP design, this depends on the support condition that is used. For the grade-supported and asphalt-supported condition, the panels can support traffic loads immediately after being placed, and the dowel, edge, and bedding grouts can be installed during the next construction shift. However, for the grout-supported case, the panels raised into position by the leveling screws cannot support traffic loads without the bedding grout; the bedding grout must be installed and requires approximately one hour to cure before the roadway can be re-opened to traffic. Although extra time is required to allow the grout to cure for the grout-supported case, this is offset by the time required to mill the asphalt to a higher precision for the asphalt-supported condition and the time to place and level the base material for the grade-supported case. In all cases, the construction can be performed overnight and the required construction activities can be completed in time to reopen the lane to traffic during the day (Pickel and Tighe 2017).

If a longer construction period is feasible and not too disruptive, then some of the conventional repair alternatives may be preferable depending on other criteria. For example, the problem that is addressed by PCIP is rutting of the asphalt pavement caused by deep-seated issues in the pavement structure; if longer construction shutdowns were allowable then this problem could also be addressed by improving the quality and performance of the lower layers through a full-depth reconstruction.

An inherent benefit of PCP is the high stiffness of Portland Cement Concrete (PCC) compared to asphalt concrete. Concrete pavement deflects less and distributes the wheel loads over a greater area to sub-surface layers than asphalt pavement (APCA 2013). This reduces the stresses transferred to the lower layers of the pavement structure. PCC pavements are well-suited to address structural rutting issues because they can carry heavy traffic and imparts less loads to the layers with deep-seated issues than flexible pavements.

Fabrication requirements may be an important decision criterion when selecting between treatment alternatives. PCIP is fabricated in a controlled environment and then transported to the site rather than being constructed on site. As a result, precast concrete is typically of better quality and durability than cast-in-place. In addition, shrinkage cracking which is a concern for CIPC and can lead to cracking, can be easily controlled in the fabrication of precast concrete (Merritt et al. 2000). High-early-strength CIPC is an alternative than can reduce the time required for strength gain to as little as 6 to 8 hours. However, the reduced setting time is often at the expense of the concrete's durability, as performance problems have been frequently observed with high-early strength concrete. Rapid-setting concrete can be successful, but they are achieved with more difficulty and there is greater risk of durability issues than in concrete with a longer setting time (Delatte 2008). The prefabrication of PCIP also makes this repair strategy less susceptible to weather conditions. The precast panels can be installed in temperature or precipitation conditions that may be prohibitive for placing CIPC or asphalt (Merritt et al. 2000). Prefabrication also eliminates other risks, such as construction delays due to the rejection of a poor-quality batch of material.

The construction or installation process weigh into the selection of a treatment strategy. For PCIP, the tasks of preparing the lane for panel placement and installing the panels may be more intensive than milling-and-replacing the asphalt or pouring cast-in-place concrete. The preparation for PCIP depends on the type of support condition being used and may include precision-milling of asphalt (asphalt-supported option), milling and placing a base material (grade-supported option), or milling, levelling and grouting (grout-supported option), before placing the panels. In addition, these procedures must be completed precisely so that the panels meet the desired crossfall and match the elevations of the adjacent panels and the existing pavement (Pickel and Tighe 2017).

The estimated production rate of precast concrete pavements, including PCIP, is 150 metres/day, assuming that construction only takes place overnight in a 10-hour window. The production rate of a conventional pavement is approximately 600 metres/day, assuming that construction takes place throughout the day (Merritt et al. 2000). This demonstrates that PCP installation for a given length of road takes longer than rehabilitation with conventional pavements. However, the overall traffic disruption is minimized because all construction is performed during the off-peak hours. It is also expected that the production rate of precast concrete pavements could increase as contractors become more familiar with this technique.

Since PCIP is a novel technique, the life-cycle costs are currently unknown. The initial cost and maintenance/rehabilitation cost of PCIP are not yet known quantitatively. However, in relative terms the initial cost of PCIP is higher than that of conventional repair strategies. PCIP is estimated to have a lifespan similar to other precast concrete pavements that have an expected service life of 20 years for repairs (Tayabji, Ye, and Buch 2013b). In comparison, the asphalt mill-and-replace strategy has been observed to have a service life of only 3 to 5 years. It is predicted that repair using PCIP will require significantly lower maintenance and rehabilitation costs in comparison to the asphalt-mill-and replace for a comparable time period. Furthermore, a primary motivation for using this fast-track repair method is to reduce the user costs. User costs are indirect costs such as increased commute time and vehicle-operating costs, caused by construction-related delays (Merritt et al. 2000). User costs are minimized by performing PCIP construction overnight when traffic volumes are at their lowest, in comparison to alternative repair methods which require 24-hour traffic diversions.

2.2 Infrared Heating for Crack Repair

2.2.1 Infrared Heating Technology

Infrared heating is a relatively new pavement repair technology used primarily for fixing cracking and potholes in roadways. Infrared heating machinery has been used in the pavement repair field for almost 30 years, but has only recently gained some popularity (Kieswetter, 2013).

Infrared light consists of longer wavelengths then all visible light. When exposed to infrared radiation, objects absorb the rays in the form of heat (Ray-Tech Infrared Corp., 2017). Infrared heating machines use this infrared radiation to heat asphalt until it is malleable enough at a specified depth to rework. To produce infrared radiation, machines feed propane through a device which mixes the correct ratio of air to propane into a heater plenum which creates ignition. Ignition occurs inside a ceramic cloth which heats up to temperatures ranging from 800 to 1000 degrees Celsius (Kieswetter, 2013). This hot surface of the ceramic cloth is what emits the infrared radiation. This surface is placed near the pavement surface so that the infrared radiation is emitted to and absorbed by the asphalt. Figure 2 below displays an infrared heating machine.

Figure 2: Left: Infrared Heating Machine in Use; Right: Bottom of Machine in Use (Kieswetter 2013)

The process used when repairing a section of pavement with an infrared heating machine is as follows: First, the affected area is heated using the infrared heating machine to approximately 350ºC, making sure the pavement is evenly heated past the depth of the cracked areas. Workers then rake the now malleable asphalt to remove damages and add new asphalt to bring the area back up to grade (mostly during pothole repairs). During this stage, a product consisting mostly of light oil contents called a rejuvenator is often added to the pavement to restore some of the old pavement's intrinsic properties. Old pavements tend to lose these chemicals over time and become more stiff, hard, and prone to cracking; adding the rejuvenators when the pavement is soft makes the pavement more flexible. Next, the area is compacted, making sure a strong thermal bond is created between the pavement surrounding the affected area and the rejuvenated asphalt (Parker, 2007).

Infrared heating machines are manufactured in many different sizes and shapes, depending on the jobs they were designed for. These can range from small, hand pushed machines for minor repairs to large, driver operated machines which can span a whole lane of traffic. One particular machine designed by Heat Design Equipment Inc. in Waterloo, Ontario, was specifically created to provide an effective, fast-tracked repair solution for longitudinal cracking. This machine has 14.6 metres (48 feet) in length of infrared heating mechanisms, which is towed at a rate which can exceed 1.5 metres/minute (5 feet/minute) (Heat Design Equipment Inc., 2017). By the time the affected pavement has been heated by the entire length of the train, it is at the perfect temperature to be scarified, reworked, and compacted by workers. This machine is very effective in treating longitudinal cracking repairs which tend to occur where joints were created during pavement construction while laying the asphalt layer. For the case of longitudinal crack repairs, this continuously moving "train" style of infrared heating machinery provides an improved performance with the benefits of a fast-tracked repair system. Figure 3 below displays this machine in action.

Figure 3: Infrared Heating Train (Heat Design Equipment Inc. 2017)

2.2.2 Treatment alternatives and Decision criteria

Some current, more commonly used methods for crack and pothole repairs include crack sealing, crack/pothole filling, and milling and patching. Crack sealing consists of cleaning the crack using compressed air, then filling it with a rubberized high modulus crack filling material. This method is generally used on small, lower severity cracks. For larger cracks or potholes, the affected areas are instead filled with asphalt emulsion to grade, and this is called crack or pothole filling. When crack and potholes are more severe, the affected area of asphalt is removed by milling, replaced by virgin materials and compacted (Uzarowski et al. 2011).

When compared to these more common crack and pothole repair methods, several performance benefits of the infrared heating method can be observed. Cracks repaired via infrared heating process tend to last longer without further deteriorating then these other methods mainly because of the thermal bond created between the repaired/rejuvenated area and the surrounding pavement. This thermal bond is created because both the repaired area and the pavement surrounding that area are heated when the joint between them is compacted. Compared to the cold joint created when using the more commonly used methods, the thermal bond will not allow any water or debris into the joint (Ray-Tech Infrared Corp., 2017).

When heating asphalt without using infrared heating, current methods use heat by flame. This method tends to overheat the top layer of asphalt, causing it to burn and damage the material, while also often not heating the base of the asphalt layer enough. Burnt materials will perform poorly if recycled in the future because they lose much of their innate oils and damage the aggregates. Materials which do not get heated enough fail to become properly malleable and do not mix well or properly absorb the rejuvenators (Ding et al. 2016). When using infrared heating, one does not have to worry as much about burning materials or not heating deep enough, as infrared heat penetrates the asphalt better and heats the asphalt evenly to full depth.

One main factor when selecting a repair option for any project is cost. Crack sealing is by far the cheapest repair option, costing about $8 per linear metre (Jeong, 2015). Prices of infrared heating repair and mill and patch repairs can range from $10-12 and $10-15 per linear metre respectively (Uzarowski et al. 2011). When completing a job using the infrared heating method, all of the in-place material is recycled, but when mill and resurfacing is implemented, the original asphalt layer is completely replaced with virgin materials. This results in a decrease of the relative price of using infrared heating compared to mill and resurfacing as the size of a repair job increases. Also, since infrared heating provides a longer lasting repair, using this method will pay itself off over time.

Another factor when deciding what kind of repair method to choose for a specific project is the availability of that repair method. Commonly used crack repair methods are readily available in all regions of the world, while new methods like infrared heating are not. When a method is not readily available, the machinery would have to be transported a longer distance to the construction site, dramatically increasing the cost of that option.

An example in infrared heating being chosen as the optimal repair method is in the Region of Waterloo Bleams road reconstruction project. For this project, several low to medium severity cracks had propagated on the binder course of a recently reconstructed portion of road. The Region of Waterloo required a cost effective, long term repair solution with minimal time increase since construction was already being delayed due to these cracks. Infrared heating repair was chosen because it met these criteria, while not requiring replacement of any in-place materials, which were only two years old at the time (Uzarowski et al. 2011). Also, the Region of Waterloo had local infrared heating machinery available for the job, eliminating long transportation costs associated with using this new method.

When infrared heating machines are used in the constantly moving "train" formation described earlier, it provides a very fast-tracked repair method desirable for high priority longitudinal crack repair projects. Several high demand roadways in Toronto, Ontario, and Washington, D.C. have used this method for repairs. This method is also very effective when used to repair airport aprons and runways (Heat Design Equipment Inc., 2017).

3 CONCLUSIONS

Two fast-track repair technologies were presented to demonstrate the advantages and disadvantages of rapid pavement repair methods and to examine the decision criteria that play a role in choosing a pavement treatment strategy.

The first technology is precast concrete inlay panels (PCIP), use for the rehabilitation of structural rutting problems on high-traffic volume asphalt highways. The structural rutting problem has been observed to occur on some of the 400-series highways in Ontario, and in the past, milling and replacing the asphalt has been performed to try to rehabilitate the pavement. However, the observed service life of this treatment strategy is only 3 to 5 years. Concrete pavement is a stiffer material that is well suited to carry heavy traffic loads and would be more resistant to structural rutting than asphalt pavement. Therefore, conventional cast-in-place concrete pavement is another potential treatment alternative. However, longer lane closures are required to allow concrete hardening before it can support traffic loads. Therefore, precast concrete pavement is proposed to minimize the construction-related lane closures, thereby reducing user costs. Table 1 summarizes the advantages and disadvantages that were previously described for the three pavement treatment strategies.

Table 1: Comparison of treatment alternatives for structural rutting on highways

Criteria	Precast Concrete Inlay Panels (PCIP)	Cast-in-place Concrete (CIPC)	Asphalt Milling and replacement
Time before re-opening to traffic	Immediately after installation	2-4 days (typical) or 6-8 hours (rapid-setting)	40 – 80 minutes
Supporting heavy traffic loads	Higher strength	Higher strength	Lower strength
Fabrication	Prefabricated; higher concrete quality and durability	On site construction; lower concrete quality and durability	On site construction; less controlled environment
Sensitivity to weather conditions during construction	Less susceptible to weather conditions	More susceptible to weather conditions	More susceptible to weather conditions
Installation Process	More intensive process	Simpler process	Simpler process
Production Rate	150 metres/day	600 metres/day	600 metres/day
Initial Cost	Higher	Lower	Lower
Maintenance & Rehabilitation Costs	Expected to be lower than asphalt replacement	Comparable to PCIP	Higher

Based on these decision criteria and the priorities of the specific project, the most appropriate treatment strategy can be selected. If longer lane closures do not have significant adverse impact on traffic, then a conventional cast-in-place concrete or full-depth reconstruction of the pavement may be suitable options. PCIP's main advantage is for rapid construction or rehabilitation applications, when reducing construction time and user impact is a priority.

The second technology is a new method for repairing asphalt cracking utilizing infrared heating to evenly heat asphalt to the required depth to resurface the affected area. This method provides a very fast and effective repair due to the thermal bond created when the repaired asphalt is bonded to the in-place asphalt while they are both heated. A very efficient fast-tracked repair method can be created by lining together a "train" of infrared heaters to constantly move over the affected areas. Table 2 summarizes the comparison of infrared heating to existing crack repair methods including crack sealing, crack filling, and asphalt milling and replacement.

Table 2: Comparison of treatment alternatives for crack repair on asphalt pavements

Criteria	Crack Sealing	Crack Filling	Asphalt Milling and Replacement	Infrared Heating
Typical Crack/pothole Size to Repair	Very small	Small	Large	Small-Large
Repair Time	Very low	Very low	Higher	Low
Availability	High	High	High	Low
Initial Cost	Lowest	Lower	Moderate	Moderate
Maintenance & Rehabilitation Frequency	High	High	Low	Expected to be Lower

Based on the above table, one can select the most appropriate crack sealing method for their project. For smaller projects, one may not want to invest in infrared heating due to its increased cost, however, if this technology is available in the area required, selecting this pavement repair strategy will provide a fast, effective method that will not require as much future maintenance causing increased delay for road users.

These are two examples of fast-track techniques, however, there are many other existing fast-track pavement repair methods. Due to high-traffic volumes experienced on many highways and the ongoing need to maintain and rehabilitate roadways, fast-track techniques are becoming necessary for certain projects because they offer faster construction times and a reduced user impact. The use of fast-track repair methods should be evaluated on a project-by-project basis using the decision criteria presented and other relevant criteria to select the optimal treatment strategy.

Acknowledgements

The authors of this paper gratefully acknowledge the Ministry of Transportation of Ontario, Cement Association of Canada, and the Natural Sciences and Engineering Research Council of Canada (NSERC) for providing funding for the PCIP project. We would also like to thank Heat Design Equipment Inc. for their continued support and assistance with infrared heating crack repair research, and Daniel Pickel for his support and expertise on the precast project. For their guidance and instruction in the Infrastructure Management course at the University of Waterloo, we thank Professors Chris Raymond and Vimy Henderson.

References

Abdelmohsen, Ahmed Z, and Khaled El-Rayes. 2016. "Optimal Trade-Offs between Construction Cost and Traffic Delay for Highway Work Zones." Journal of Construction Engineering and Management 142 (7). doi:10.1061/(ASCE)CO.1943-7862.

APCA. 2013. "Differences Between Concrete and Asphalt Pavement." http://overlays.acpa.org/Concrete_Pavement/Technical/Fundamentals/Differences_Between_Concrete_and_Asphalt.asp.

David Jeong, H, Omar G Smadi, Douglas D Gransberg, and H David. 2015. "InTrans Project Reports Institute for Transportation Iowa Pavement Asset Management Decision- Making Framework Iowa Pavement Asset Management Decision-Making Framework" 143. http://lib.dr.iastate.edu/intrans_reports.

Delatte, Norbert. 2008. Concrete Pavement Design, Construction and Performance. New York, NY: Taylor & Francis.

Ding, Yongjie, Baoshan Huang, and Xiang Shu. 2016. "Characterizing Blending Efficiency of Plant Produced Asphalt Paving Mixtures Containing High RAP." Construction and Building Materials 126 (126):172–78. https://doi.org/10.1016/j.conbuildmat.2016.09.025.

Inc., Heat Design Equipment. n.d. "HDE SEAMLESS JOINT REPAIR." Accessed February 2, 2018. https://d3ciwvs59ifrt8.cloudfront.net/87660493-6976-419d-bbf6-2d1314d01b82/24c7a7b6-2a3b-40e2-9f41-3ca7927d5167.pdf.

Kieswetter, B. 2013. "Use of Infrared Heaters in Asphalt Pavement Construction & Repair." Heat Design Equipment Inc. http://www.captg.ca/docs/pdf/13presentations/SWIFT/Tues_PM_-_Infrared_Heaters_-_Kieswetter.pdf.

"Length of Canada's Public Road Network, 2003." 2015. Statistics Canada. http://www.statcan.gc.ca/pub/16-002-x/2009001/tbl/transpo/tbl001-eng.htm.

Merritt, David K, B. Frank McCullough, Ned H Burns, and Anton K Schindler. 2000. "The Feasibility of Using Precast Concrete Panels to Expedite Highway Pavement Construction." Austin, TX. http://ctr.utexas.edu/wp-content/uploads/pubs/7_1517_1.pdf.

"Ontario Provincial Highways Traffic Volumes On Demand." 2016. Ministry of Transportation of Ontario. http://www.raqsa.mto.gov.on.ca/techpubs/TrafficVolumes.nsf/tvweb?OpenForm&Seq=4.

Parker, S. 2007. "The Infrared Repair Process - Its Uses and Advantages | Paveman Pro." Paveman Pro. Walkie Talkie Marketing Communications, LLC. http://www.pavemanpro.com/article/the_infrared_process_-_uses_and_advantages/.

Pickel, Daniel, and Susan Tighe. 2017. "Precast Concrete Inlay Panel Installation on Highway 400: Construction Report." Waterloo, ON.

Ray-Tech Infrared Corp. 2017. "The TRUTH About Infrared | Ray-Tech Infrared Corporation - Infrared Asphalt Maintenance Equipment." 2017. http://www.raytechinfrared.com/content/truth-about-infrared.

Solutions, Rhino Asphalt. 2015. "Causes of Cracking in Roads." 2015. https://www.rhino-uk.com/jointmaster/30-cracking-in-roads.

Tayabji, Shiraz, Dan Ye, and Neeraj Buch. 2013a. "Precast Concrete Pavement Technology." SHRP 2 Rep. S2-R05-RR-1. Washington, D.C.: Transportation Research Board. doi:10.17226/22710.

Uzarowski, L., Henderson, V., Henderson, M., Kiesswetter, B. 2011. "Innovative Infrared Crack Repair Method." Transportation Association of Canada. http://conf.tac-atc.ca/english/annualconference/tac2011/docs/m1/uzarowski.pdf.

———. 2013b. "Precast Concrete Pavements: Technology Overview and Technical Considerations." PCI Journal 58 (1): 112–28. https://www.pci.org/PCI_Docs/Design_Resources/Transportation_Resources/JL-13-Winter-13.pdf.

Wang, Yuhong, Songye Zhu, M Asce, and Alvin S T Wong. 2018. "Cooling Time Estimation of Newly Placed Hot-Mix Asphalt Pavement in Different Weather Conditions." Accessed February 28. doi:10.1061/(ASCE)CO.1943-7862.0000832.

White, Thomas D., John E. Haddock, Adam J. T. Hand, and Hongbing Fang. 2002. "Contributions of Pavement Structural Layers to Rutting of Hot Mix Asphalt Pavements." Washington, D.C. http://onlinepubs.trb.org/onlinepubs/nchrp/nchrp_rpt_468-a.pdf.

Building Tomorrow's Society
Bâtir la Société de Demain

Fredericton, Canada
June 13 to 16, 2018 / *13 au 16 juin, 2018*

INCORPORATING PHYSIOLOGICAL STRESS MEASURES INTO ASSESSMENT OF DRIVER PERFORMANCE: IMPLICATIONS FOR METHODOLOGY AND DESIGN

Sutherland, Jessica E.[1,3], Hassein, Udai[2], Easa, Said[2] and Day, David M.[1]
[1] Department of Psychology, Ryerson University, Toronto, Canada
[2] Department of Civil Engineering, Ryerson University, Toronto, Canada
[3] jessica.sutherland@psych.ryerson.ca

Abstract: Despite the evidence suggesting that physiological stress can compromise cognitive functioning and decision-making, there is a paucity of literature on the effects of physiological stress on driver behaviour and performance. This paper reviews the literature on physiological stress and decision-making, prior research on physiological stress and driving, and methodological considerations for effectively incorporating physiological data with driving performance. In particular, we will highlight implications and recommendations for future studies on driver stress and performance. These recommendations include selecting driver behaviours and experimental manipulations to improve ecological validity. As well, we discuss the benefits of incorporating physiological stress measures into experimental driving research to capture human factors related to driver decision-making. Finally, we include as an example the preliminary results of a study incorporating physiological stress into a model exploring peer passenger effects on young driver performance. We measured physiological stress via wearable biopotential recording sensors while male and female participants, ages 16 to 25 years, completed tasks either alone or with a peer, before driving. We use these preliminary findings to describe how to select stressors appropriate for the population and stress measurement tools appropriate to the methodological design. **Keywords** – Physiological stress; driver stress; driver performance; human factors

1 Introduction

Driver performance may be affected by a number of external variables, including the behaviour of other drivers on the road, road conditions, and visibility. However, a significant body of research supports the inclusion of internally-oriented variables when assessing driver performance. Specifically, the effects of physiological stress on decision-making suggest that stress should be a key variable of consideration when exploring driving outcomes sensitive to human error. Mather and Lighthall (2012) proposed the *stress triggers anticipated reward salience* (STARS) model of how physiological stress affects decision-making. The STARS model argues that stress sensitizes individuals to anticipated rewards and minimizes attention to, or learning of, possible negative outcomes. In the context of driving behaviour, stress may precipitate dangerous driving due to the various perceived rewards the driver might access. This might include, for example, speeding up to get to a destination faster, despite the higher likelihood of a collision due to speeding.

This paper reviews the relationship between physiological stress and decision-making and prior research on physiological stress and driving outcomes. We make specific methodological recommendations for future research on driver performance. Specifically, we note that driving outcomes, types of stress, and the unique context of the population being sampled are key variables for further consideration when designing studies examining stress effects on driver performance.

2 Literature Review

2.1 Physiological stress and decision-making

The "stress response" to an event is underpinned by physiological, endocrinal, cognitive, and affective changes that help the body respond effectively. Stress-inducing events can be acute (an immediate, sudden stressor such as an accident) or chronic (a long-term or repeated stressor, such as abuse). In the face of a stressor, the brain activates the "fight-or-flight" response to help the body respond by 'fleeing' (removal from the stressor) or 'fighting' (adapting to the stressor). Sterling and Eyer (1988) referred to the process through which the body adapts to acute, daily stressors and maintains homeostasis as *allostasis*. When individuals do not adapt to a repeated stressor or manage their responses to repeated stressors, their allostatic load increases, creating conditions for adverse health outcomes and behavioural consequences of repeated, chronic stress (e.g. poor cardiovascular health, poor diet; McEwan, 2008).

A major side effect of both chronic and acute stress is impaired decision-making. Chronic stress has been found to result in habit-based, rather than goal-oriented, decision-making (Dias-Ferreria et al., 2009). That is, when under chronic stress, choosing the less cognitively taxing option is more appealing, even when it makes goal-directed behaviour more challenging. Radenbach et al. (2015) found that a history of chronic stress compounded the effect of acute stress on decision-making, such that individuals with longer histories of stressors made less adaptive decisions after acute stressors. Wemm and Wulfert (2017) also found evidence for a reward bias after an acute stressor, whereby young adults were more likely to make risky choices after experiencing a social stressor. This effect of stress on decision-making is also particularly pronounced for women. Chronically stressed adults took more risks on an in-laboratory risk task after an acute stressor, and this was especially true for women (Ceccato, Kudielka, & Schwieren, 2016). Similarly, Lighthall et al. (2012) found that acute stressors are sensitive to gender. After a cold stressor (where individuals place their hands in ice-cold water), stress hormones measured in salivary cortisol increased and affected adaptive decision-making for female participants on a subsequent gambling task (Lighthall et al., 2012).

2.2 Prior studies of physiological stress and driving

A recent meta-analysis conceptualizing chronic exposure to driving as a stressor revealed mixed results (Antoun, Edwards, Sweeting, & Ding, 2017). Antoun et al. (2017) reviewed 7 studies of both on-road and simulated driving for their effects on cortisol and cardiovascular stress responses. Some studies found no evidence for driving as a stressor while others did; however, no particular pattern (i.e. between type of driver, on-road or simulated) emerged for the conditions under which driving increased physiological stress. A major limitation of this meta-analysis is the small sample size of participants (N = 162) and the lack of inclusion of variables relevant to physiological stress in driving contexts, such as emotional arousal.

For example, driver aggression, anger, and anxiety are linked to increased physiological stress and dangerous driving after sudden external stressors and internal perceived stressors. Morris and Pilcher (2016) had drivers complete a simulated drive while exposed to cold inside the car cabin to simulate winter weather conditions. Despite participants reporting that they felt cold but not uncomfortable, the cold stressor was still associated with slower braking time, harder braking, and giving less headway between cars, indicating a more aggressive driving style. Barnard and Chapman (2016) further propose that fear and anxiety about driving can increase physiological stress, negatively affecting hazard detection. Specifically, fear and anxiety may create a U-shaped relationship between physiological stress and hazard detection, such that at very low and very high levels of fear and anxiety, hazard detection is improved via deceleration of physiological responses to allow for effective visual scanning of the road. However, moderate levels of fear and anxiety can compromise hazard detection by increasing the physiological stress response to hazards on the road (Barnard & Chapman, 2016).

Perceived stress due to work- and personal-life stressors also predict driver errors and traffic violations. Ge, Qu, Jiang, Du, Sun, and Zhang (2014) found that individuals who perceived more stress (any type) in their lives also tended to drive more dangerously on the road, and this effect was compounded by higher reported levels of anger. Rowden, Matthews, Watson, and Biggs (2011) report that both work- and personal life-related stressors were associated with increased collision risk, but that driver traits (i.e. anger) interacted with stressors to increase or decrease the relationship between stress and driving outcomes. More daily hassles (i.e. minor acute stressors but chronic exposure) in particular were also predictive of both driver errors and traffic violations (Rowden et al., 2011). Ge et al. (2014) propose that working memory is limited, and that stress and its associated emotional consequences may limit one's capabilities to attend to changing road conditions.

3 Methodological Implications

Similar to Ge et al. (2014), Antoun et al. (2017) suggest that the effects of stress on driving outcomes may be due to the cognitive load of managing distractions while driving. Further, the type of stressor experienced and its relationship to driving outcomes may also be driver- and context-specific (Antoun et al., 2017). Similar to the transactional model of stress, which states that individual sensitivity to stress predicts responses to stressors, individual differences in how much a driver likes driving, feels comfortable on the road, or has traits associated with aggressive driving will intersect with how they perform as a driver and their response to driving situations (Rowden et al., 2011).

For this reason, research that incorporates stress physiology into driver assessment must consider the contributing roles of emotional arousal and the context of the driver sample. While *emotional arousal* can refer to any emotion that may affect behaviour, anger, anxiety, and fear can have independent and additive effects on driving outcomes and physiological stress due to their effects on individuals' ability to self-regulate. Other cars preventing a driver's intended behaviour or making them wait can elicit anger, possibly due to drivers feeling less in control of their travel, thereby resulting in changes to acceleration and speed (Roidl, Frehse, & Hoger, 2014). Further, higher levels of self-reported anxiety and fear around driving are associated with higher speed and acceleration, as well as faster braking (Barnard & Chapman, 2016; Roidl et al., 2014).

The context of the driver sample can also influence the types of stressors that will elicit stress responses. Professional drivers, such as taxi drivers or bus drivers, face particular occupational stressors that in turn affect driver performance. In long-haul truck drivers, the quality of relationship between employees' dispatchers and an occupational culture of support for safe driving ("psychological safety") was associated with fewer hard braking incidents and more safe driving behaviour (Zohar, Huang, Lee, & Robertson, 2014). A poor safety climate around professional drivers is also associated with higher occupational stress and more fatigue-related near misses while driving. That is, poor safety culture increases drivers' occupational stress, translating into fatigue and more dangerous driving (Strahan, Watson, & Lennon, 2008).

The need to consider individual differences in stress responses and the context of the driver sample means that methodological decisions should be made with ecological validity in mind. This means that the experimental design should reflect the real conditions drivers experience as closely as possible to effectively measure their physiological stress in response to driving stressors. As Antoun et al. (2017) found, the mixed results for driving as a stressor suggests that, in and of itself, driving may not always induce physiological stress, but that experimentally manipulated stressors sensitive to the sample do. Professional drivers, for example, face unique stressors that impact dangerous driving. Bus drivers have a combination of high-demand, low-control occupational strains, which in turn affects risky driving behaviour (Useche, Gomez Ortiz, & Cendales, 2017). This effect is compounded by fatigue (Useche et al., 2017). In non-professional drivers, a high degree of self-reported distress after a collision is associated with more anxious driving behaviours (hard braking, slower speed) for individuals with a longer history of stressful life events (Clapp et al., 2011). A longer history of stressful life events also distinguishes between individuals with histories of collisions who experience driving anxiety and those who do not. That is, sensitivity to stressors via chronic life stress increases the effect a previous collision has on future driving behaviour (Clapp et al., 2011). In practice, this means that the ability of a sudden hazard in the road to elicit physiological stress would depend on whether bus drivers', for example, were tested after a long shift, which would prime their fatigue (Useche et al., 2017). Alternatively, physiological stress may not appear to occur after a sudden hazard in a sample with a history of collisions, unless they are also asked about distress, anxiety, and chronic life stressors (Clapp et al., 2011).

4 Incorporating stress into a model of adolescent risky driving: an example study and 'lessons learned'

As an example of some of the above considerations, we describe below how sample-specific stressors were incorporated into an experimental study exploring mechanisms of peer passenger influence on young drivers. Peer passengers increase the likelihood of driver distraction, errors, and fatal accidents for adolescent drivers (Ouimet, Pradhan, Brooks-Russell, Ehsani, Berbiche, & Simons-Morton, 2015; Curry, Mirman, Kallan, Winston, & Durbin, 2012; Simons-Morton et al., 2011; Simons-Morton, Lerner, & Singer, 2005). Peer presence can induce physiological stress via elevated heart rate, pulse, and sweat production in non-driving contexts (Stroud et al., 2009; van den Bos, de Rooji, Miers, Bokhorst, & Michiel Westenberg, 2014), but physiological stress has not been explored for its effects on adolescent risky driving in the presence of a passenger. We conducted a driving simulator experiment as an initial effort to test these relationships. The simulation scenarios consisted of a 5-kilometre drive where we measured drivers' speed, lateral displacement, and acceleration through straight driving segments and intersections, including a left turn into oncoming traffic. We measured physiological stress via wearable biopotential recording sensors (BIOPAC, Goleta, CA) throughout the entire study, which transmit wirelessly to recording software. Participants were 75 young men and women between the ages of 17-25 completing tasks alone or with a peer before the driving scenarios. We measured heart rate via a three-lead electrocardiogram (ECG) and respiration rate via a respiration band worn around the chest.

Evidence suggests that peer presence is physiologically stressful among adolescents because they are especially sensitive to social evaluation and strongly desire social acceptance (Somerville, Jones, Ruberry, Dyke, Glover, & Casey, 2013). To use this as an ecologically valid stressor, we manipulated whether the participant was socially accepted or socially rejected by the peer passenger, as played by a research confederate who followed an acceptance or rejection script. We referred to this manipulation as a 'social stressor,' reflecting a type of stressor to which adolescents are particularly sensitive. We anticipated that social rejection would be especially stressful for young people desiring acceptance, and that this type of social stressor would serve as a distraction from the road and increase risky driving. Our results supported previous findings that peer presence increased physiological stress and affected driving behaviours relative to control participants who drove with no passenger. However, social acceptance, not rejection, seems to have a stronger effect on participants' stress and driving, particularly for young women (for the full study, see Sutherland, Hassein, Easa, & Day, 2018). This may be due to the fact that young women find social cohesion particularly rewarding, which enhances their physiological response to the presence of an accepting peer, as demonstrated in previous research (Voroboyev, Kwon, Moe, Parkkola, & Hämäläinen, 2015).

This initial test of physiological stress as having a role in young drivers' risky driving revealed several 'lessons learned.' During the pilot phase of data collection, we measured sweat production (skin conductance) and pulse in addition to heart rate and respiration. However, we had to alter our physiological data collection in response to the limits of these particular physiological measures within a driving simulator. Sweat production (as measured by skin conductance) and pulse are measured via sensors placed on the non-dominant fingers. It was possible to collect sweat and pulse data during study components prior to driving, but sensors had to be removed because the steering wheel interfered significantly with data collection. Given that we could not collect stress physiology data from these measures during variables of interest (the drives), we omitted them after piloting five participants.

However, physiological measures worn on the torso did not interfere with participants' ability to drive. The ECG leads reliably produced clean, analyzable data for almost all participants and was not affected by movement of the participant. The respiration band was particularly sensitive to anatomical differences in participants, such as chest depth and width, which required us to place the band above the breastbone for some participants and over the diaphragm for others. Since our experimental methodology had participants moving around at certain times (e.g. turning to face in different directions, getting in and out of the car), the respiration band would occasionally drop the signal, typically when a participant slouched in the sear. Unlike the ECG sensors, respiration bands may be most effective when study methodologies require minimal movement and participants can stay stationary in proper posture. Finally, participants putting on a seatbelt similarly interfered with the respiration band and ECG leads. Despite our interest in tracking whether or not participants put on their seatbelts, we amended our study script to instruct the participant not to put on the simulator's seatbelt because it interfered with the physiological data collection. Based on our experiences, wireless ECG apparatuses may be most effective for measuring physiological stress in a driving simulator when participants are also completing other tasks as part of the study, with respiration bands being similarly effective when the study is primarily stationary.

5 Conclusions

In summary, the key methodological consideration when designing experimental studies that measure driver stress is ensuring ecological validity of the selected stressor (e.g. passenger presence, concerns about safety climate) and measuring co-occurring variables, such as drivers' fear, anxiety, and history of chronic stress. It is likely that different populations of drivers experience varying degrees of stress while driving depending on the stressors they have been exposed to. Additionally, the measurement devices used to capture physiological stress must be carefully selected to minimize their impact on drivers while they complete study (e.g. avoidance of measurement sensors on the hands or fingers) and be adaptable to the driving environment (e.g. mobile, wirelessly recorded).

Our review of the literature indicates that physiological stress and individual sensitivities to stressors have significant effects on driver performance. Individual differences, such as anger or anxiety, and individual group differences, such as professional and non-professional drivers, have their own unique contributors to not only the stress response, but also to the effect on driver performance. Future research on driver performance and human factors related to driving should consider incorporating stress physiology as a unique contributor when measuring driver errors and violations. Using a machine learning approach, recent research suggests that driver behaviour during certain road conditions and road shape may predict upcoming physiological stress reactions while driving (Munoz-Organero & Corcoba-Magana, 2017). While preliminary, this indicates that physiological stress may be one human factors component that could be integrated into collision or hazard avoidance systems, and inform road and research designs to minimize the effect of physiological stress on driver performance.

The broader implication of measuring driver stress is that physiological stress is a modifiable risk factor for collisions and driver injury. Recent evidence suggests that young drivers respond positively to receiving feedback on their driving (i.e. being told when they are speeding) and adjust their driving behaviour accordingly (Paaver, Eensoo, Kaasik, Vaht, Maestu, & Harro, 2013). Further, event-triggered driving feedback (i.e. being video recorded when certain risky driving behaviours occur) also resulted in subsequent reduced rates of risky driving amongst newly licensed drivers (Carney, McGehee, Lee, Reyes, & Raby, 2010). These findings suggest that populations at higher risk for physiological stress while driving, such as young or professional drivers, may reduce the likelihood of stress interfering with their decision-making if biofeedback (i.e. information about their physiological stress) was incorporated into driver training. It is also possible that collision or hazard avoidance systems could be designed to measure drivers' physiological state via sensors in the steering wheel and/or driver's seat and integrate this data with traffic and road conditions to more cohesively warn drivers of possible threats (Munoz-Organero & Corcoba-Magana, 2017). Future research on human factors implicated in driver injury and collision risk should consider developing and testing driver training interventions and biofeedback systems to determine the optimal methods of identifying and modifying physiological arousal while driving.

Acknowledgements

The financial support of the Social Sciences and Humanities Research Council, the Harry Rosen Institute for Stress Research, and the Ryerson University Career Boost Research Assistant program is acknowledged. Two Research Tools and Instruments grants from the Natural Sciences and Engineering Research Council of Canada have financially supported the driving simulator.

References

Antoun, M., Edwards, K. M., Sweeting, J., & Ding, D. (2017). The acute physiological stress response to driving: A systematic review. *PLoS One, 12*(10), e0185517. 10.1371/journal.pone.0185517
Barnard, M. P., & Chapman, P. (2016). Are anxiety and fear separable emotions in driving? A laboratory study of behavioural and physiological responses to different driving environments. *Accident; Analysis and Prevention, 86*, 99-107. 10.1016/j.aap.2015.10.021
BIOPAC Systems, Inc. [Computer hardware/software]. Goleta, CA.
Carney, C., McGehee, D. V., Lee, J. D., Reyes, M. L., & Raby, M. (2010). Using an event-triggered video intervention system to expand the supervised learning of newly licensed adolescent drivers. *American Journal of Public Health, 100*(6), 1101-1106. 10.2105/AJPH.2009.165829
Ceccato, S., Kudielka, B. M., & Schwieren, C. (2016). Increased risk taking in relation to chronic stress in adults. *Frontiers in Psychology, 6.* doi: 10.3389/fpsyg.2015.02036
Clapp, J. D., Olsen, S. A., Danoff-Burg, S., Hagewood, J. H., Hickling, E. J., Hwang, V. S., & Beck, J. G. (2011). Factors contributing to anxious driving behavior: The role of stress history and accident severity.*Journal of Anxiety Disorders, 25*(4), 592-598. 10.1016/j.janxdis.2011.01.008
Curry, A. E., Mirman, J. H., Kallan, M. J., Winston, F. K., & Durbin, D. R. (2012). Peer passengers: How do they affect teen crashes? *The Journal of Adolescent Health, 50*(6), 588-594. doi:10.1016/j.jadohealth.2011.10.016
Dias-Ferreira, E., Sousa, J. C., Melo, I., Morgado, P., Mesquita, A. R., Cerqueira, J. J., . . . Sousa, N. (2009). Chronic stress causes frontostriatal reorganization and affects decision-making. *Science, 325*(5940), 621-625. 10.1126/science.1171203
Ge, Y., Qu, W., Jiang, C., Du, F., Sun, X., & Zhang, K. (2014). The effect of stress and personality on dangerous driving behavior among chinese drivers. *Accident; Analysis and Prevention, 73*, 34-40. 10.1016/j.aap.2014.07.024
Lighthall, N. R., Sakaki, M., Vasunilashorn, S., Nga, L., Somayajula, S., Chen, E. Y., . . . Mather, M. (2012). Gender differences in reward-related decision processing under stress. *Social Cognitive and Affective Neuroscience, 7*(4), 476-484. 10.1093/scan/nsr026
Mather, M., & Lighthall, N. R. (2012). Risk and reward are processed differently in decisions made under stress. *Current Directions in Psychological Science, 21*(1), 36-41. 10.1177/0963721411429452
McEwen, B. S. (2008). Central effects of stress hormones in health and disease: Understanding the protective and damaging effects of stress and stress mediators. *European Journal of Pharmacology, 583*(2), 174-185. 10.1016/j.ejphar.2007.11.071
Morris, D. M., & Pilcher, J. J. (2016). The cold driver: Cold stress while driving results in dangerous behavior. *Biological Psychology, 120*, 149 155. 10.1016/j.biopsycho.2016.09.011
Munoz-Organero, M., & Corcoba-Magana, V. (2017;2016;). Predicting upcoming values of stress while driving. *IEEE Transactions on Intelligent Transportation Systems, 18*(7), 1802-1811. 10.1109/TITS.2016.2618424
Ouimet, M. C., Pradhan, A. K., Brooks-Russell, A., Ehsani, J. P., Berbiche, D., & Simons-Morton, B. G. (2015). Young drivers and their passengers: A systematic review of epidemiological studies on crash risk. *Journal of Adolescent Health, 57*(1), S24-S35. doi:10.1016/j.jadohealth.2015.03.010
Paaver, M., Eensoo, D., Kaasik, K., Vaht, M., Mäestu, J., & Harro, J. (2013). Preventing risky driving: A novel and efficient brief intervention focusing on acknowledgement of personal risk factors. *Accident Analysis and Prevention, 50*, 430-437. 10.1016/j.aap.2012.05.019
Pradhan, A. K., Li, K., Bingham, C. R., Simons-Morton, B. G., Ouimet, M. C., & Shope, J. T. (2014). Peer passenger influences on male adolescent drivers' visual scanning behavior during simulated driving. *Journal of Adolescent Health, 54*(5), S42-S49. doi:10.1016/j.jadohealth.2014.01.004
Radenbach, C., Reiter, A. M. F., Engert, V., Sjoerds, Z., Villringer, A., Heinze, H., . . . Schlagenhauf, F. (2015). The interaction of acute and chronic stress impairs model-based behavioral control.*Psychoneuroendocrinology, 53*, 268-280. 10.1016/j.psyneuen.2014.12.017
Roidl, E., Frehse, B., & Höger, R. (2014). Emotional states of drivers and the impact on speed,

acceleration and traffic violations - a simulator study. *Accident; Analysis and Prevention, 70*, 282-292. 10.1016/j.aap.2014.04.010
Rowden, P., Matthews, G., Watson, B., & Biggs, H. (2011). The relative impact of work-related stress, life stress and driving environment stress on driving outcomes. *Accident Analysis and Prevention, 43*(4), 1332-1340. 10.1016/j.aap.2011.02.004
Simons-Morton, B.G, Lerner, N., & Singer, J. (2005). The observed effects of teenage passengers on the risky driving behavior of teenage drivers. *Accident Analysis and Prevention, 37*(6), 973-82. doi: 10.1016/j.aap.2005.04.014
Simons-Morton, B. G., Ouimet, M. C., Zhang, Z., Klauer, S. E., Lee, S. E., Wang, J., . . . Dingus, T. A. (2011). The effect of passengers and risk-taking friends on risky driving and Crashes/Near crashes among novice teenagers. *Journal of Adolescent Health, 49*(6), 587-593.
doi:10.1016/j.jadohealth.2011.02.009
Simons-Morton, B. G., Bingham, C. R., Falk, E. B., Li, K., Pradhan, A. K., Ouimet, M. C., . . . Shope, J. T. (2014). Experimental effects of injunctive norms on simulated risky driving among teenage males. *Health Psychology, 33*(7), 616-627. doi:10.1037/a003483
Somerville, L. H., Jones, R. M., Ruberry, E. J., Dyke, J. P., Glover, G., & Casey, B. J. (2013). The medial prefrontal cortex and the emergence of self-conscious emotion in adolescence. *Psychological Science, 24*(8), 1554-1562. doi:10.1177/0956797613475633
Sterling, P. & Eyer, J. (1988). Allostasis: A new paradigm to explain arousal pathology. In S. Fisher, & J. Reason (Eds.), *Handbook of life stress, cognition, and health* (p. 629-649). Oxford, England: Wiley.
Strahan, C., Watson, B., & Lennonb, A. (2008). Can organisational safety climate and occupational stress predict work-related driver fatigue? *Transportation Research Part F: Psychology and Behaviour, 11*(6), 418-426. 10.1016/j.trf.2008.04.002
Stroud, L. R., Foster, E., Papandonatos, G. D., Handwerger, K., Granger, D. A., Kivlighan, K. T., & Niaura, R. (2009). Stress response and the adolescent transition: performance versus peer rejection stressors. *Development and Psychopathology, 21*(1), 47-68. doi:10.1017/S0954579409000042
Sutherland, J.E., Hassein, U., Easa, S., & Day, D.M. (2018). Proposed model of adolescent risky driving: Integrating social stressors and physiological stress as predictors of driver performance. Paper submitted to the Canadian Society for Civil Engineering, Transportation Subspecialty Conference. Fredericton, New Brunswick.
Useche, S. A., Ortiz, V. G., & Cendales, B. E. (2017). Stress-related psychosocial factors at work, fatigue, and risky driving behavior in bus rapid transport (BRT) drivers. *Accident Analysis and Prevention, 104*, 106-114. 10.1016/j.aap.2017.04.023
van den Bos, E., Rooij, M., Miers, A. C., Bokhorst, C. L., & Westenberg, P. M. (2014). Adolescents' increasing stress response to social evaluation: Pubertal effects on cortisol and Alpha-Amylase during public speaking. *Child Development, 85*(1), 220-236. doi:10.1111/cdev.12118
Vorobyev, V., Kwon, M. S., Moe, D., Parkkola, R., & Hämäläinen, H. (2015). Risk-taking behavior in a computerized driving task: Brain activation correlates of decision-making, outcome, and peer influence in male adolescents. *PLoS One, 10*(6), 1-20. doi:10.1371/journal.pone.0129516
Wemm, S. E., & Wulfert, E. (2017). Effects of acute stress on decision making. *Applied Psychophysiology and Biofeedback, 42*(1), 1-12. 10.1007/s10484-016-9347-8
Zohar, D., Huang, Y., Lee, J., & Robertson, M. (2014). A mediation model linking dispatcher leadership and work ownership with safety climate as predictors of truck driver safety performance. *Accident; Analysis and Prevention, 62*, 17-25. 10.1016/j.aap.2013.09.005

Building Tomorrow's Society
Bâtir la Société de Demain

Fredericton, Canada
June 13 to 16, 2018 / *13 au 16 juin, 2018*

INVESTIGATING THE RISKS IMPOSED BY DIFFERENT DRIVER GROUPS ON OTHER ROAD USERS

Philip, Thomas[1], Dabbour, Essam[2,3], and Easa, Said[1]
[1] Department of Civil Engineering, Ryerson University, Toronto, Canada
[2] Department of Civil Engineering, Abu Dhabi University, United Arab Emirates
[3] essam.dabbour@adu.ac.ae

Abstract: Multi-vehicle traffic collisions usually result in increased injury severities to the more vulnerable drivers involved in those accidents. An obvious example is the case of an accident between a heavy vehicle and a motorcycle, which usually results in an increased injury severity to the motorcyclist. However, there is lack of quantified measures that estimate the level of risk imposed by different driver groups on other road users. As an example, there are no quantified measures available to estimate the level of risk imposed by impaired drivers on other road users. This paper provides a quantitative investigation regarding the level of risk imposed by different driver groups on other road users. The research is based on analyzing accident records from North Carolina for all two-vehicle collisions that occurred from January 1, 2012 until December 31, 2013. Different driver groups are investigated based on driver's gender, age, impairment and the age of the vehicle involved in the collision. To eliminate the random effect of the number of passengers in the vehicle, only driver's injuries are considered in the analyses. For each pair of groups, the number of collisions where the driver was at least visibly injured is compared between the two groups and based on that an estimate is made to the odds of every driver group to increase the severity of injuries for the other driver groups. The findings of this research may help decision makers identify driver groups that are more dangerous to other road users so that more resources might be allocated to improve awareness programs and licensing procedures dedicated to those groups. The findings of this research also provide quantified measures for insurance companies and law enforcement agencies regarding the potential liability imposed by different driver groups on other groups.

1 INTRODUCTION

Traffic collisions result in approximately 1.3 million deaths every year (World Health Organization 2015). For collisions involving two different types of vehicles, such as collisions between heavy vehicles and motorcycles, certain driver groups (in this case, the motorcyclist) are at more risk of being severely injured because of those collisions. However, there is lack of quantified measures that help us estimate the risk imposed by different driver groups on other road users, such as the risk imposed by impaired drivers on other road users. Daniels et al. (2010) examined risk externalities at roundabouts using logistic regression to analyze the severity level of drivers' injuries related to 1491 collisions that occurred at 148 roundabouts in Flanders-Belgium. They found that vulnerable road user groups (pedestrians, bicyclists, moped riders and motorcyclists) are more severely affected when colliding with other vehicle types. They also found that older road users (above the age of 45 years) are more severely injured when colliding with younger road users. There were other research studies that utilized logistic regression to investigate the significance of different factors that increase the severity of injuries in traffic collisions. Yau (2004) conducted a study to identify the various factors that can contribute to higher severity of single vehicle collisions in Hong Kong during the years 1999 and 2000. Factors pertaining to three types of vehicles, namely private vehicles, goods vehicles and motorcycles were identified using logistic regression models. Various factors such as vehicle age, gender of the driver, roadway lighting conditions, and seatbelt usage were found to be major contributors to the severity of collisions. Yan, et al. (2005) studied multi-vehicle rear-end collisions at signalized intersections in Florida for the year 2001 by using multiple logistic regression where they identified several factors that contributed to the severity of those collisions. Those factors include the number of lanes, road surface condition, driver's gender, speed limits, and alcohol/drug use.

Chang and Yeh (2006) analysed single vehicle crash data in Taiwan for the year 2000 and they were able to identify fatality risk factors for motorcyclists and non-motorcycle drivers. Based on the factors identified, they recommended several measures to reduce the fatality rates in collisions, including improving the quality of the roadway surface, proper speed management techniques, and enforcement of seatbelt use. Harb et al. (2008) analysed freeway work-zone crashes in Florida for the years from 2002 to 2004 and found that the geometry of the road, lighting conditions, driving under the influence, age, and gender are all factors associated with work zone crashes. Bham et al. (2012) analysed single vehicle and multi-vehicle collisions on urban U.S. highways. A multivariate analysis was used to identify various factors that result in collisions. Five types of collisions (angular, head-on, rear-end, sideswipe – same direction, sideswipe – opposite direction) were identified. Apart from this, various factors such as wet road conditions, poor lighting conditions, driving under the influence, driver's behaviour (decision making), and geometry of the road were found to increase collision severity. Another interesting study is the work done by Wenzel (2013), where he explored the possibilities of improved vehicle designs and their results in reducing accident fatalities. Various factors such as presence of side air bags, better alignment of light truck bumpers etc. were found to reduce the fatality rates during collisions. Yu et al. (2014) studied the effect of microscopic traffic, weather and road geometry on specific crash types on the freeway section I-70 in Colorado, USA. With the help of the automatic vehicle identification and weather detection systems installed in the corridor, the study provided valuable insights on how intelligent transport systems can give more focus on traffic safety improvement and effective traffic management. Dabbour (2017) analysed the effects of various factors on the severity of drivers' injuries in single-vehicle collisions involving light vehicles by using collision data from North Carolina during the period from 2007 to 2013. Factors related to the driver, the roadway, the vehicle, and the environmental conditions were identified, and temporal stability of these variables were also analysed.

This study provides a quantitative investigation on the level of the risk imposed by different driver groups on other road users. The research is based on analysing the accident data for all two-vehicles collisions that occurred in North Carolina from January 1, 2012 until December 31, 2013. To eliminate the random effect of the number of passengers in the vehicle, only driver's injuries are considered in the analysis. The results from this study could help decision makers identify those driver groups that increase the severity of injuries of other road users and this in turn can help in increasing the awareness among people and take necessary precautionary measures.

2 METHODOLOGY AND DATA COLLECTION

All two-vehicle collisions are analyzed to investigate the level of risk imposed by one driver group on other road users. Due to the significant differences between vehicle and driver characteristics of heavy vehicles and those of light vehicles, only collision data pertaining to light vehicles were considered for the analysis. The vehicle classes considered as light vehicles are passenger cars, pickup trucks, light trucks, panel trucks, sports utility vehicles, and vans.

The risk imposed by the driver groups is found out using logistic regression, which is a generalized linear model where the occurrence of an event is predicted by fitting the data to a logit function. The event is coded as a dichotomous outcome variable and its value is dependent on the various explanatory variables (Dabbour 2017). The logit function takes the general form:

[1] $f(z) = e^z / (1 + e^z)$

Where z denotes the logit function, and $f(z)$ represents the dichotomous variable which is assumed to follow Bernoulli distribution. This represents the probability that the driver is killed or seriously injured provided that the collision has occurred and has resulted in an injury to the driver. It can take a value of "1" if the driver injury is fatal or serious and "0" if it is a minor injury. The corresponding logit function can be represented as:

[2] $z = \beta_0 + \beta_1 x_1 + \beta_2 x_2 + \ldots. + \beta_k x_k$

Where $x_1, x_2, \ldots x_k$ represent the explanatory variables, $\beta_1, \beta_2, \ldots. \beta_k$ represent the regression coefficients and β_0 is the intercept. The collision data corresponding to the years 2012 and 2013 were collected from the Highway Safety Information System (HSIS). The HSIS is managed by the University of North Carolina Highway Safety Research Center under contract with the Federal Highway Administration. (University of North Carolina Highway Research Center, 2010). Explanatory variables related to both the driver and the vehicle were considered for the analysis. The different variables considered for the analysis are given below:

1. Driver-related variables: include driver's age, driver's gender and impairment level of driver; and

2. Vehicle-related variables: include vehicle's age.

The outcome variable is the injury level that is coded "1" if the driver is at least visibly injured, and coded "0" otherwise. The gender of the driver is coded "1" for female and "0" for male. The impairment level is coded "1" for impaired (which represents the impairment due to medications, drugs, alcohol) and "0" for not impaired. Once the explanatory variables were selected, those records which have incomplete/irrelevant data were eliminated. In this analysis, the odds ratio (OR) is used to interpret the significance of the different explanatory variables considered. Also, the 95% confidence interval (CI) is utilised to define the upper and lower limit values of the odds ratio with 0.05 significance level.

3 RESULTS AND DISCUSSION

Table 1 shows the frequencies and percentages of single-vehicle collisions, two-vehicle collisions, and collisions involving three (or more) vehicles. As shown in the table, two-vehicle collisions constitute more than 56% of the total collisions occurred; and hence it is worthwhile trying to understand the level of risk imposed by different driver groups on other road users in this category.

Table 1: Number [and percentage] of collisions based on the number of vehicle involved *

Number of vehicles involved	2012	2013
Single-vehicle collisions	57,789 [37.9%]	60,070 [37.8%]
Two-vehicle collisions	85,449 [56.1%]	89,386 [56.3%]
Collisions involving three (or more) vehicles	9,139 [6.0%]	9,374 [5.9%]
All collisions	152,377 [100%]	158,830 [100%]

*Note - Above percentages are obtained after eliminating outlier data pertaining to number of vehicles only.

Table 2 shows the coefficients associated with the developed logistic regression models along with the standard errors and the lower and higher 95% confidence intervals. The corresponding forest plot summary for the logistic regression performed is shown in Figure 1. All the explanatory variables considered for analysis (except the gender of the driver who is severely injured) were found to be significant during the analysis period at 0.05 significance level. Out of the different explanatory variables considered, impairment level was found to have the most significant coefficient with the highest odds ratio when compared to the other explanatory variables. Driving under the influence of alcohol or illicit drugs is therefore a significant factor that increases the risk of being seriously injured and also to cause severe injuries to other road users. As shown in Table 2, an impaired driver is five times more likely of getting injured when compared to other drivers. This result is similar to the findings of Behnood et al. (2014). The possible explanation for higher risk of injury for an impaired driver could be that the driver may not be reacting fast enough (for example normal driver might go to a bracing position faster at the time of accident compared to an impaired driver). Another finding which we see from Table 2 (related to impairment level) is that an impaired driver is 3.83 times more likely to increase the injury severity for other road users. This could possibly be explained by the fact that impaired drivers are less likely to adopt corrective maneuvers (in terms of braking or steering away) and correspondingly the impact speeds can be higher causing higher injury severities to the injured driver.

Table 2: Parameters of the developed logistic regression models

Explanatory Variable	Coefficient	Standard Error	Odds Ratio	95% CI	
				Low	High
Driver's age[(a)]	0.0067	0.0005	1.0068	1.0057	1.0078
Driver's age[(b)]	0.0025	0.0005	1.0025	1.0014	1.0035
Vehicle's age[(a)]	0.0175	0.0010	1.0177	1.0157	1.0197
Vehicle's age[(b)]	0.0107	0.0011	1.0108	1.0086	1.0130
Driver's gender[(a)]					
Female	NS[(c)]	NS[(c)]	NS[(c)]	NS[(c)]	NS[(c)]
Male (base category)	-	-	-	-	-
Driver's gender[(b)]					
Female	-0.2410	0.0190	0.7858	0.7571	0.8156
Male (base category)	-	-	-	-	-
Driver's impairment[(a)]					
Impaired	1.6302	0.0475	5.1051	4.6512	5.6033
Not impaired (base category)	-	-	-	-	-
Driver's impairment[(b)]					
Impaired	1.3428	0.0523	3.8298	3.4566	4.2432
Not impaired (base category)	-	-	-	-	-

[(a)] Characteristics of the driver who is more severely injured.

[(b)] Characteristics of the driver and the vehicle that caused more injury to the other driver.

[(c)] Not significant during the observation period (p value greater than 0.05)

Figure 1: Forest plot of the parameters associated with the developed logistic regression models

When it comes to the gender of the driver involved in the accident, we see from Table 2 that female drivers are less likely to cause more serious injuries to other road users. This can be explained by the fact that women tend to be more cautious drivers compared to their male counterparts. Generally, aggressive driving behaviour, road rage etc. is exhibited more by males than female drivers. When we look at previous studies related with the gender of the driver, we see discussions related with "which gender is more severely injured?". Behnood and Mannering (2015) analysed the effect of male drivers on injury severity for the years 2004 to 2012, where they found that for the year 2007 and 2009, with regard to severe injuries, male drivers have a greater likelihood of severe injuries. In all the other years it was found that males have a less likelihood of severe injuries. Another study by Kim et al. (2013) showed that male drivers are associated with a greater probability of fatal injuries and it was attributed to the fact that male drivers are usually over-represented in the fatal injury category. Further analysis will be required to see if the variable "driver gender who is severely injured" is significant during other years and also check the temporal stability of the trends shown during the analysis period 2012-2013 for a greater period of time.

The ages of the colliding vehicles were also found to have significant influence on the severity of drivers' injuries. It can be seen that as the vehicle age increases by one year, the odds of being seriously injured increase by 1.77%. So, we can say that, using a 10-year old vehicle can increase the odds of serious injury by 17.7%. Another result related to vehicle's age is that as the age of the colliding vehicle increases by 1 year, it increases the odds of serious injuries by 1.08%. So, in short, the older the vehicle, the greater the injury severity for the injured driver. This result is similar to previous studies (Wenzel 2013) and it might be explained by the fact that modern vehicles have more safety features than older vehicles. Also, due to the higher safety standard requirements, modern vehicles are subject to more rigorous crash tests prior to being commercially available.

The age of the driver was also found to increase the odds of driver's serious injuries. We see that for an increase in the driver age by one year, the odds of being seriously injured increase by 0.68%. This is logical because an 80-year-old driver is more likely to be injured because the physical condition deteriorates as the years go by. Another result related with the driver age is that as the age of the driver of the colliding vehicle increases by 1 year, it increases the odds of serious injuries by 0.25%. This result is similar to previous studies (Kim et al. 2013) and it could be explained by the fact that the responses of an older driver (in terms of braking or steering away) could be slower when compared to a young driver and correspondingly it can result in greater injury severity for the injured driver.

By understanding the risk imposed by the different explanatory variables considered for the above analysis, it is important to look at what we can to do minimise the risks for different road users in the future.

As we saw earlier, driver impairment is the major factor that has to be targeted so as to make the roads safer for all road users. The first step is by proper enforcement of laws which eliminate drunk driving and also the enforcement of strict measures such as license suspensions/cancellations depending on the severity of the offense. Another method of law enforcement is by setting up sobriety checkpoints for breath checks to see if the driver is impaired or not. A third option is by installing ignition interlocks in vehicles. These are devices installed in vehicles that measure the driver's breath before each journey and if the driver is impaired, he/she won't be able to start the vehicle. Generally, these devices are installed in vehicles of drivers who have been convicted multiple times for driving under the influence (DUI). However, making it mandatory for all vehicles can act as a deterrent for people who drive while being impaired.

When it comes to the gender of the driver, we saw that female drivers are less likely to cause serious injury to the other driver compared to their male counterparts. The best option here is to educate drivers the significance of being calm behind the wheel and to take necessary precautionary steps to avoid aggressive driving. Usually, drivers get agitated when they are late. So, planning ahead and giving yourself enough time to reach your destination can definitely help. Traffic congestions is another reason for aggressive driving behaviour. So, the driver should identify alternate routes with less traffic and be ready to use them so as to avoid aggressive driving. Even though such precautionary measures are adopted, the drivers should come to terms with the fact that traffic congestions are part of driving and its best handled with a peaceful attitude.

When it comes to the age of the vehicle, we saw that as the age of the vehicle increases, the severity of injuries increases as well. The average life span of a passenger car ranges from 8 to 10 years. However, there are older/classic cars on the road which have been brought back to life by restorations. While performing restorations/improvements, importance should be given for the safety aspects as well. Many of these classic cars do not even have seatbelt provisions which is a disadvantage. Combine that with the lack of modern features like airbags and dedicated crumple zones during impact, older vehicles are far behind in terms of modern safety standards. Specific rules should be set apart for older vehicles such that at least the basic safety requirements are met during their restoration.

When it comes to the age of the driver, we saw that as the age increases, the chances of severe injury increases. We also saw that aged drivers can cause more injuries to the other driver. Frequent senior driver assessments can ensure that the senior drivers are physically and cognitively fit to be on the road. Vehicle related improvements can also be made by car manufacturers such that driver restraint systems like seatbelts, airbags etc. are designed specifically for senior citizens so as to reduce injuries. Road infrastructure changes such as better road lighting facilities can help senior citizens see vehicles better during night time driving reducing the chances of accidents. Improved educational and refresher courses related with driving should be made available to old drivers before their license renewals. All the above-mentioned measures can definitely help make our roads much safer.

4 Conclusions

Logistic regression was used to identify and quantify the effects of various explanatory variables that increase the risk of injury in drivers associated with two-vehicle collisions for light duty vehicles. The study was conducted by analysing the accident records of all collisions that occurred in North Carolina from January 1, 2012 to December 31, 2013. Explanatory variables related to the drivers as well as those related to the colliding vehicles were considered for the analysis and all variables taken into consideration (except the gender of the driver who is severely injured) were found to be statistically significant at 0.05 significance level. The explanatory variables related to the driver were the age of the injured driver, the gender and impairment levels of both colliding drivers. Vehicle's ages were the explanatory variables related to the vehicles. It was found that an impaired driver is most likely to be seriously injured and also to cause serious injuries to other road users. Female drivers were found to cause less injury to the injured driver compared to males. The age of the injured driver and the age of the driver in the colliding vehicle were found to increase the odds of serious injuries. The ages of the two colliding vehicles were also found to increase the odds of serious injuries.

Traffic collisions that result in serious injuries constitute a large burden on the healthcare system. Identifying various factors that increase the risk of drivers' serious injuries can help decision makers take necessary precautionary measures to reduce those risks or to improve legislative policies. The findings of this research can also help insurance firms and law enforcement agencies in estimating the potential liability imposed by different driver groups on other groups.

References

Behnood, Ali, and Fred L. Mannering. 2015. "The Temporal Stability of Factors Affecting Driver-Injury Severities in Single-Vehicle Crashes: Some Empirical Evidence." *Analytic Methods in Accident Research* 8: 7–32.

Behnood, Ali, Arash M. Roshandeh, and Fred L. Mannering. 2014. "Latent Class Analysis of the Effects of Age, Gender, and Alcohol Consumption on Driver-Injury Severities." *Analytic Methods in Accident Research* 3–4: 56–91.

Bham, Ghulam H., Bhanu S. Javvadi, and Uday R. R. Manepalli. 2012. "Multinomial Logistic Regression Model for Single-Vehicle and Multivehicle Collisions on Urban U.S. Highways in Arkansas." *Journal of Transportation Engineering* 138(6): 786–97.

Chang, Hsin-Li, and Tsu-Hurng Yeh. 2006. "Risk Factors to Driver Fatalities in Single-Vehicle Crashes: Comparisons between Non-Motorcycle Drivers and Motorcyclists." *Journal of Transportation Engineering* 132(3): 227–36.

Dabbour, Essam. 2017. "Investigating Temporal Trends in the Explanatory Variables Related to the Severity of Drivers' Injuries in Single-Vehicle Collisions." *Journal of Traffic and Transportation Engineering (English Edition)* 4(1): 71–79.

Daniels, Stijn, Tom Brijs, Erik Nuyts, and Geert Wets. 2010. "Externality of Risk and Crash Severity at Roundabouts." *Accident Analysis and Prevention* 42(6): 1966–73.

Harb, Rami et al. 2008. "Freeway Work-Zone Crash Analysis and Risk Identification Using Multiple and Conditional Logistic Regression." *Journal of Transportation Engineering* 134(5): 203–14.

Kim, Joon Ki, Gudmundur F. Ulfarsson, Sungyop Kim, and Venkataraman N. Shankar. 2013. "Driver-Injury Severity in Single-Vehicle Crashes in California: A Mixed Logit Analysis of Heterogeneity due to Age and Gender." *Accident Analysis and Prevention* 50. 1073–81.

University of North Carolina Highway Research Center. 2010. "Highway Safety Information System." http://www.hsisinfo.org/index.cfm (December 20, 2017).

Wenzel, Tom. 2013. "The Effect of Recent Trends in Vehicle Design on U.S. Societal Fatality Risk per Vehicle Mile Traveled, and Their Projected Future Relationship with Vehicle Mass." *Accident Analysis and Prevention* 56(2013): 71–81.

Yan, Xuedong, Essam Radwan, and Mohamed Abdel-Aty. 2005. "Characteristics of Rear-End Accidents at Signalized Intersections Using Multiple Logistic Regression Model." *Accident Analysis and Prevention* 37(6): 983–95.

Yau, Kelvin K W. 2004. "Risk Factors Affecting the Severity of Single Vehicle Traffic Accidents in Hong Kong." *Accident Analysis and Prevention* 36(3): 333–40.

Yu, Rongjie, Mohamed A. Abdel-Aty, Mohamed M. Ahmed, and Xuesong Wang. 2014. "Utilizing Microscopic Traffic and Weather Data to Analyze Real-Time Crash Patterns in the Context of Active Traffic Management." *IEEE Transactions on Intelligent Transportation Systems* 15(1): 205–13.

Building Tomorrow's Society
Bâtir la Société de Demain

Fredericton, Canada
June 13 to 16, 2018 / *13 au 16 juin, 2018*

METHODOLOGY FOR PERFORMANCE EVALUATION OF PRECAST CONCRETE INLAY PANELS AS A PAVEMENT REPAIR TECHNIQUE

Malek, Dahlia K.[1,2], Pickel, Daniel J.[1] and Tighe, Susan L.[1]
[1] University of Waterloo, Canada
[2] dkmalek@uwaterloo.ca

Abstract: Deep-seated rutting of asphalt pavements has been observed on some high-traffic volume highways in Ontario, Canada. The current rehabilitation technique of milling and replacing the asphalt has been observed to last for only 3 to 5 years. The use of Precast Concrete Inlay Panels (PCIP) is a new rehabilitation strategy that provides a longer service life and can be installed rapidly during overnight lane closures to minimize construction-related impacts on roadway users. A trial construction of PCIP was completed in 2016 on Highway 400 near the city of Toronto. Evaluating the long-term performance of PCIP is necessary to ensure that this rehabilitation strategy is feasible and provides the expected level of service. The objective of this paper is to establish a methodology to assess the performance of the PCIP based on monitoring, testing, and inspections of the trial site and by modelling PCIP using ABAQUS finite element analysis (FEA). The trial site will be inspected for panel and joint condition and distresses. Instrumentation installed on site is used to collect data remotely. Deflection testing, roadway surface friction testing, profile measurements, and noise testing will also be completed. The performance of PCIP will be evaluated for various traffic, temperature, site and construction conditions using FEA. Deflection results can be used to calculate theoretical joint faulting and load transfer efficiency of the joints. Stresses and strains in the concrete and dowel grout will be used to predict whether concrete/dowel grout crushing or tensile cracking will occur.

1 INTRODUCTION

Precast concrete pavement (PCP) technologies have been in use throughout the past 50 years, however they have only recently become a more viable alternative for certain pavement projects. PCP is a high-strength pavement that is long-lasting and can be constructed rapidly within overnight construction windows. Despite these advantages, a lack of long-term performance data is one of the factors that has deterred more widespread use of this technology. Due to the novelty of PCP, there are not many aged PCP systems in place and the performance of existing installations has not been well documented.

Performance monitoring of future PCP projects is essential to develop a database and to further the understanding of PCP behaviour. This information will be useful to calibrate models and to create design guides and will encourage greater industry usage of PCP.

One application of PCP currently being researched is the use of precast concrete inlay panels (PCIP). The installation of PCIP is a method to rehabilitate structural rutting issues on high-traffic volume asphalt pavements. This rehabilitation strategy is expected to provide a longer service life and superior performance to the existing repair strategy which is to mill and replace the asphalt. Performance evaluation of PCIP will be completed as a part of this research. This will contribute to the documentation of PCP performance and will be beneficial to evaluate project-specific aspects of PCIP, and to identify deficiencies or recommend improvements in PCIP design, materials, or construction. One method of evaluating PCIP performance will be through monitoring of a PCIP trial site which was constructed in 2016. In addition, finite element modelling of the PCIP will be performed to assess the behaviour and long-term performance of PCIP when subjected to various loading and site conditions, or design variations.

The objective of this paper is to identify the requirements and methods for long-term performance evaluation of PCIP. This paper discusses methods for monitoring and evaluating the performance of the PCIP based on general recommendations for PCP, plans for in-field monitoring of the PCIP trial site, and finite element modelling.

1.1 Background

Precast concrete inlay panels are currently being researched to address structural rutting issues on high-traffic volume asphalt highways. Structural rutting is caused by issues in subgrade or base layers of the pavement structure; deformation occurs in these lower layers which is then reflected up to the surface layer. The traditional rehabilitation method to repair the rutted asphalt has been to mill a layer of the asphalt and replace it with a new layer of hot mix asphalt (HMA). However, the service life of this strategy has been observed to be only 3 to 5 years because it does not address the root cause of the problem and the high-traffic volumes cause the pavement to deteriorate rapidly. Performing frequent maintenance or rehabilitation on high-traffic volume highways is expensive, and the required lane closures for construction are disruptive to the roadway users. Often, transportation agencies limit highway lane closures to a window between 8 p.m. and 6 a.m. to minimize the impacts on traffic. Therefore, a rapid rehabilitation technique with a longer service life is required to address the structural rutting problem. Precast concrete inlay panels (PCIP) are a rehabilitation strategy which could potentially meet these requirements. The PCIP can be constructed overnight, and use Portland Cement Concrete (PCC) which is stiffer than asphalt pavement, to reduce the stresses in the lower pavement layers where the structural rutting problems originate. PCIP are precast reinforced concrete slabs, 4.57m long by 3.66m wide (lane width). There are dowels, facing in the direction of traffic, on one transverse edge of the slab, which are covered by dowel slots in the following slab (Figure 1).

Figure 1: PCIP being installed on site (Pickel and Tighe 2017)

A trial section of PCIP was constructed on Highway 400, near Toronto in Ontario, Canada. The trial section is 100 metres long, and three different support conditions were constructed within this section. The three support conditions are referred to as asphalt-supported, grade-supported, and grout-supported. For the asphalt-supported case, the existing asphalt was milled precisely and then the panels were placed directly on top. For the grade-supported case, the existing asphalt was milled, and then a cement-treated base material was placed, levelled, and compacted (Figure 2), followed by placement of the panels on this base.

Figure 2: Grade-supported support condition (Pickel and Tighe 2017)

For the grout-supported case, the existing asphalt was milled, the panels were placed and then raised to the correct elevation by levelling feet cast into the panels. A rapid-setting bedding grout was then pumped underneath to support the panels. In each of these cases, additional grouting of the dowels, edges, and/or bedding was also performed after panel installation (Pickel and Tighe 2017).

A finite element model of the PCIP will also be created as part of this research. The main components of the finite element model for PCIP are the concrete panel, the support layer, existing asphalt, base, and subgrade layers. More than one panel will be modelled to assess the behaviour and performance of the transverse joint between panels, which will be modelled with the dowel bars and dowel slot grout.

1.2 Purpose of Performance Evaluation

Over the past 50 years, precast concrete pavement (PCP) has been used infrequently, but research and trials with PCP began to take off in the early 2000s. Within the last 15 years there have been sufficient technological advancements to make PCP a feasible and cost-effective option for selective projects. Recognizing the potential of this technology, research is being pursued in many aspects of PCP, including identifying current practices, developing specifications and guidelines for selection, fabrication, and installation of PCP systems. Monitoring and performance evaluation is another aspect that is required to continue to advance the state of PCP technology. There is limited information on the long-term performance of PCP because there are few in-service installations older than 15 years and the performance of existing installations has not been well documented. The lack of long-term performance data is one of the factors that have deterred more widespread use of this technology. Data collection and monitoring of PCP projects is especially important, not only to evaluate the performance of the individual project, but also to contribute to building a performance database for PCP. In the future, this database could be used to calibrate design procedures for different types of PCP and to identify successes and problems with existing PCP installations to improve future projects (Tayabji, Ye, and Buch 2013).

PCIP is a type of precast concrete pavement, so monitoring and performance evaluation of this project is an important component of this research. Furthermore, PCIP is accompanied by its own unique performance considerations that have not been investigated in previous PCP installations. Previous PCP projects for HMA rehabilitation have been undertaken, however the purpose and installation processes differed from that of PCIP. One previous project involved completely removing the HMA, grading and compacting the base, and adding new material, before placing the panels. A second project involved removing the HMA, cement-treating the subgrade, and placing a new HMA layer before placing the panels (Tayabji and Tyson 2017). PCIP is unique in that its purpose is to address deep-seated rutting issues while only milling the asphalt to partial depth, and without modifying the base, subbase, or subgrade layers in any manner. Since providing a uniform, stable support for the panels is essential for the successful performance of PCP, three types of support conditions were constructed for the PCIP trial section. Therefore, unique characteristics of PCIP that must be evaluated include the performance of the three support conditions and performance of the PCIP given the existing pavement structure upon which it is installed. Assessing how well PCIP performs in these conditions is necessary to establish the feasibility of this rehabilitation technique for addressing the structural rutting issues.

2 MONITORING IN-FIELD PERFORMANCE OF PCIP

This section discusses recommendations for evaluating the in-field performance of PCIP based on studies of precast concrete pavement systems and planned monitoring of the PCIP trial section.

2.1 Pavement Performance Categories

Monitoring and testing of pavement is undertaken to evaluate its performance in four main categories: physical condition, structural integrity/capacity, safety, and service (Hudson, Haas, and Uddin 1997). Physical condition is evaluated by the severity and extent of distresses such as cracking, patching, rutting, and spalling. The structural integrity of the pavement is evaluated by performing tests to determine the load-carrying capacity of the pavement and properties of the in-situ materials. The main characteristic used to provide an indication of safety is friction, however there are other characteristics that affect safety, including the pavement surface condition (presence of potholes, spalls, faults, etc.), light reflectivity of the surface, condition of the lane markings, and the presence of debris or foreign objects on the roadway. Service is usually measured by riding comfort which is based on surface roughness (Haas et al. 1997).

These categories of pavement performance are interrelated since the performance measures in one category often provide insight into the pavement's performance in other categories as well. For example, surface distresses are a measure of the pavement's condition, but also cause roughness which will be reflected in the service category and a severe distress could be a safety concern. Poor structural integrity or capacity could be the cause of surface distresses, as demonstrated by the PCIP rutting problem in which deep-seated rutting is the cause of the surface rutting. The relationships between these performance measures and categories continues. Data collection, in-service monitoring, and modelling of pavement are methods used to assess the performance of pavement in each of these categories.

2.2 Monitoring PCP Performance

Recent studies on PCP recommend that performance data should be collected every 2 years and should include identification and measurement of distresses, joint width, deflection, road surface profile, and skid resistance (Tayabji, Ye, and Buch 2013).

The panels should be inspected for distresses that include cracking and poor surface condition, and the joints should be inspected for spalling and to assess the condition of the joint sealant and dowel slots. Furthermore, the differential elevations at the joints and the joint widths should be monitored and measured. The slabs' horizontal movements due to thermal expansion and contraction, which cause changes in joint width, should be measured in both the summer and winter season. Deflection testing using a falling weight deflectometer (FWD) is recommended to evaluate the adequacy of loads transfer at the joints. The test should be performed on either side of each joint and in the middle of the panel. Friction testing along the inner wheelpath should be performed to evaluate the skid resistance of the surface, and profile testing should be completed with the results reported as an International Roughness Index (IRI) value. (Tayabji, Ye, and Buch 2013)

Depending on the specific application of PCP, other tests or inspections may be required. For the PCIP trial installation, each of these recommended tests will be performed as well as some additional tests and monitoring; these are described in the following section.

2.3 Monitoring of PCIP Trial Site

The precast concrete inlay panel (PCIP) trial section on Highway 400, is the only in-service application of this rehabilitation strategy. For this reason, monitoring the section is crucial for gaining insight into its behaviour and performance under service conditions.

The monitoring activities related to the trial section are generally divided into two parts: Closure and Non-Closure. Closure activities are those that require the right lane of highway 400 to be closed to traffic to undertake testing. Due to the high traffic volumes on the site, these closures are infrequent and only occur during the overnight period when traffic is at its lowest daily level. Closure activities include: falling weight deflectometer (FWD) testing, T2Go and British pendulum friction testing, and visual inspections.

FWD testing involves imparting impact loads on the pavement and measuring the deflection of the pavement. These loads are applied on one side of a transverse joint and deflections are measured on either side of this joint. A ratio between the two deflections is determined which is referred to as the load transfer efficiency (LTE_{δ}), which serves as a measure of joint performance. High LTE_{δ} indicates good load transfer and a properly functioning joint, while low LTE_{δ} indicates the load is not being transferred to adjacent slabs. Poor load transfer results in deflections and stress concentrations at joints which can lead to premature failure. Scheduling FWD testing over time tracks the performance of joints.

Friction testing provides insight into the way the PCIP is interacting with traffic on its surface. Providing a safe surface for vehicle traffic is a primary function of any pavement, and therefore tracking the behaviour of tire/pavement interaction is a performance indicator for the pavement.

Visual inspections provide an opportunity for close observation of the panels in-situ. This is when issues in the panels can be identified and quantified, including cracking and panel surface condition, joint spalling, condition of the joint sealants, and measuring of the joint widths and faulting.

Non-closure activities include those activities which can take place without closing any traffic lanes. Because these activities are less intrusive, they can occur more frequently and at any point during the day. Non-closure activities include monitoring instrumentation installed beneath the panels, brake-tester friction testing, and noise testing.

Instrumentation was installed beneath the PCIP trial during construction. The instrumentation includes pressure cells, moisture probes, and temperature sensors which were installed at the interface between the PCIP and the underlying HMA support. The pressure cells track the pressures due to warping and curling on the underlying material indicating the severity of these effects on PCIP support. The moisture sensors tracked the initial infiltration of moisture beneath the slabs and the levels thereafter. The support beneath precast pavements is a primary predictor of the pavement's performance, and therefore tracking the conditions at this level of the pavement structure is an important component of the PCIP trial analysis.

The noise produced by the interaction between a pavement and a vehicle's tires is a performance consideration, and particularly for pavements located in urban corridors with large populations. Noisy pavements can be detrimental to the health of nearby residents, and therefore must be monitored for performance.

3 FINITE ELEMENT MODELLING

A finite element analysis (FEA) of the PCIP will be completed to evaluate the behaviour and performance of the panels and joints when subjected over time to various loading and site conditions, and for each support condition. A three-dimensional finite element model will be created using ABQAUS software and calibrated. The methodology for evaluating the performance of the precast concrete inlay panels can be divided into three main phases:

1. Determine the range of conditions to be applied to the model,

2. Apply the conditions and obtain results for each case, and

3. Analyze the results and compare to failure or performance criteria.

In addition, part of the research will be to investigate possible modifications in the design, materials, or construction of PCIP to recommend improvements.

3.1 Variations in Traffic, Site, and Support Conditions

The performance of the PCIP will be evaluated for a variety of traffic and temperature loading, site conditions, and different support conditions for the panels. Each different combination of traffic, temperature, site and support conditions is referred to as a new "case" throughout this paper.

Three different support conditions were developed and constructed in the trial installation of PCIP. Providing a uniform and stable support is one of the main factors for a successful precast concrete pavement project (Tayabji, Ye, and Buch 2013). Therefore, an important part of the research involves assessing the performance of PCIP for each type of support condition to be able to make recommendations for future projects. Different variations of the finite element model will be created to represent each of the three support conditions, and the results will be analysed for each case.

The FEA will also consider the effect of varying site conditions. For example, the model will be varied to investigate the impacts on PCIP performance of different subgrade materials and properties. This can be used to determine whether PCIP does adequately resist the effects of deep-seated issues, such as poor subgrade, as intended.

Another variation will arise from changing traffic and environmental loading combinations. The types of loading to consider include the slab self-weight, thermal gradients for daytime and nighttime conditions, and traffic loading. The applied thermal gradient will vary depending on the season and time of day, and the magnitude and position of traffic loads on the panels will also vary. Various load combinations will be considered, and the results will be analyzed for the load combinations producing the worst-case effects.

Finally, PCIP performance must be evaluated for all combinations of these changing conditions over a specified time period. The cases that produce the worst-case effects will be analyzed. To determine the long-term performance of PCIP using the FEM, the results will be analyzed to determine how the pavement response changes over time.

3.2 Expected Distresses and Analysis of Results

3.2.1 Panels

The performance of the concrete panels can be established from the type and extent of distresses on its top and bottom surface, and in localized areas (such as around the joints). Panel distresses are a measure of pavement performance in each of the four main categories previously described; panel distresses relate to the condition of the pavement structure and its structural integrity, surface distresses directly impact the panel roughness, and major distresses can be safety concerns.

Cracking and spalling of concrete are the primary distresses expected to occur in PCP. Types of cracking that can occur in PCIP are longitudinal and transverse cracks or durability cracks (also referred to as D-cracking). D-cracking is caused by using aggregates which are susceptible to fracturing in freeze/thaw conditions, causing cracks in the panel. This can cause progressive failure in PCC pavement. However, D-cracking will not be evaluated by the FEA and can be avoided by using durable aggregates. Longitudinal cracks are not expected to occur in the PCIP panels, but transverse cracks can form at the bottom of the panel and propagate upwards or they may form at the top of the panel and propagate downwards.

Transverse cracking in the PCIP would be caused by a combination of traffic loads, thermal loads, and panel self-weight. During the daytime, the top of the panel is warmer than the bottom and will expand, causing the panel to curve upwards such that the middle of the panel is unsupported. During the nighttime, the top of the panel cools to a lower temperature than the bottom and will contract, causing the panel to curl upwards such that the ends of the panel are unsupported. This curling effect is counteracted by the self-weight of the slab which worsens the situation. When the middle of the slab curls up, it is unsupported, and the self-weight causes tensile stresses at the bottom of the slab. When the ends of the slab curl up, the self-weight produces tensile stresses at the top of the slab. Furthermore, if the position of the wheel loads on the panel also counteracts the slab curling, this increases the tensile stresses formed at the top or bottom of the panel. Under these conditions, the critical response is transverse cracking in the middle of the panel, originating at the top or bottom surface (Merritt et al. 2000; National Cooperative Highway Research Program 2004).

The stress and strain results from the ABAQUS model will be analyzed to evaluate the behaviour and predict the condition of the panel. Without comparing to a failure criterion, the magnitudes of the tensile stresses in the panel can be compared for each case, where each case is a different combination of traffic, temperature, site, and support conditions. In general, it is desirable to reduce the tensile stresses in the panel to minimize the amount of cracking that will form over time. In addition, the ABAQUS damaged plasticity model will be used to model the non-linear behaviour of the concrete material for the panel (Simulia 2014). Using this model, one of the results obtained from the FEA will be the maximum principal plastic strains, which can be used to determine the locations where cracks will initiate and to visualize the crack patterns (Genikomsou and Polak 2015). By evaluating the stresses and strains in this manner, the extent of cracking can be predicted and compared for each of the cases.

3.2.2 Joints

The performance of the joints can be determined from the type and extent of distresses in the panel area near the joints, joint faulting, distresses in the joint load transfer mechanism, and measurements of the joint load transfer efficiency (LTE).

If the joints become filled with incompressible materials, it can prevent the panels from expanding and this causes stress concentrations around the joints. This may cause spalling or "blow-ups", which is crushing of concrete caused by the upwards movement of the pavement in a localized area (Delatte 2008).

Joint faulting is a difference in elevation between panels and is an indication of poor load transfer. Faulting can be caused by differential settlement of the panels, curling of the panels, or a phenomenon referred to as pumping. Pumping is when heavy traffic loads on a panel cause it to deflect and force loose material, or if there is water accumulation under the slab, pressurized water carrying loose material from underneath the slab to the adjacent slab (Delatte 2008). This only occurs if there is poor joint load transfer because then the traffic load will cause the loaded panel to deflect significantly more than the adjacent panel, allowing space for the movement of materials.

The dowels are grouted into dowel slots to embed them into the panel on one side of the joint. On the other side of the joint, the dowels are free to move along their axis to allow expansion and contraction of the panels. Crushing of the dowel grout around the dowel bar creates voids and can result in reduced load transfer at the joints. This can be tested by determining the LTE of the joint, and if low load transfer is occurring this may contribute to joint faulting.

Deflection results from the FEM can be used to determine the faulting across the joints. This can be compared to the recommended maximum allowable joint faulting for jointed PCP over its service life, 3.81 mm (0.15 in), to determine if this performance criterion is met (Tayabji, Ye, and Buch 2013). The LTE can also be calculated from the deflection results obtained from the FEA and compared to the LTE calculated from the FWD testing of the PCIP trial site. This can be used to calibrate the finite element model. Furthermore, for the long-term performance of concrete pavement, it is recommended that the LTE be maintained at 70% or greater over its service life (Chan and Lane 2005; Tayabji, Ye, and Buch 2013). The theoretical LTE calculated from the FEA deflections can be compared to this criterion to determine whether the PCIP meets this requirements for all the simulated cases.

Distresses in the panels near the joints and in the dowel grout material can be evaluated by analyzing the stress results of the FEM. When a load is imparted on the slab and it deflects, a shear force is exerted on the dowel bar and it bends, which transfers the force to the concrete or dowel grout through bearing.

Around the dowel bar, there are compressive stresses in the concrete/dowel grout below the bar, and tensile stresses above the dowel bar. By modelling the concrete with the damaged plasticity model, the maximum principal strains can be evaluated to determine if concrete crushing or cracking occurs. It can be determined if the dowel grout will fail in compression or tension by comparing the FEM stresses with the maximum allowable stresses of the surrounding material.

The potential for concrete spalling or blow-ups could be evaluated by reducing the joint gap in the finite element model and observing the effect on the stresses in the concrete near the joint.

4 CONCLUSIONS

The use of precast concrete inlay panels (PCIP) is currently being researched as a method of addressing structural rutting problems exhibited by asphalt pavements on high-traffic volume highways. This rehabilitation technique can be performed rapidly during overnight lane closures to minimize the impacts of construction on the roadway users. A trial construction of PCIP has been completed on Highway 400 near Toronto in Ontario, Canada.

Long-term performance evaluation of the PCIP is required to establish this technology as a feasible strategy for rapid, long-lasting pavement rehabilitation and to promote industry usage. Other applications and types of precast concrete pavement (PCP) have been in use throughout the past 50 years, and within the past 15 years PCP has become a feasible and economical alternative on a project-by-project basis. However, there is a limited amount of information on the long-term performance of PCP, and there are few aged in-service installations that can be monitored and tested to obtain this data. Therefore, monitoring and evaluating long-term performance is a recommendation for any new PCP project.

The long-term performance evaluation of PCIP will be completed by testing, inspecting, and remote monitoring of the trial construction site and by performing finite element analysis (FEA) using ABAQUS software to predict PCIP performance for various conditions.

To monitor the performance of PCP installations, it is generally recommended to monitor panel cracking and surface condition, joint widths and faulting, and joint condition. It is also recommended to perform deflection, profile, and friction testing. These inspections and tests will be performed for the PCIP trial site to collect data and monitor its performance over time. In addition, noise testing will be performed, and pressure cells and moisture probes were installed in the trial section to collect data remotely. Table 1 summarizes the evaluation techniques that will be used to monitor the in-field performance of PCIP in each of the pavement performance categories.

Table 1: Technique for PCIP Performance Evaluation

Evaluation Technique	Condition	Capacity	Service	Safety
Visual Inspections	Panel and joint distresses	-	Roughness and riding comfort*	-
British Pendulum (friction)	Pavement surface texture*	-	-	Skid resistance
T2GO (profile)	Surface distresses*	-	Surface roughness	-
FWD (deflection)	Distresses in joint load transfer mechanism*	Joint LTE	-	-
Noise testing	-	-	Noise-level comfort*	-

Pressure Cells Instrumentation	-	Pressure on underlying layer due to slab curling/warping	-	-
Moisture Probes Instrumentation	Moisture infiltration below slab	-	-	-

* The evaluation technique does not directly measure pavement performance in this category, but it may provide an indirect indication of performance.

Furthermore, FEA will be performed to evaluate the performance of PCIP for various applied traffic loads, thermal gradients in the panel, panel support conditions, and subgrade properties. The performance will be evaluated over time for the cases producing the worst-case effects.

Table 2 summarizes the distresses that are expected to be observed in PCIP over its service life, their causes, and the FEA results that will be used to evaluate the extent of these distresses. The distresses will provide an indication of PCIP performance.

Table 2: Evaluation of PCIP using Finite Element Analysis

Distresses	Cause	FEA Results
Panel transverse cracking/spalling	Combination of self-weight, traffic and thermal loads	Magnitude of tensile stresses and plastic strains
Surface distresses in panel near joints	Incompressible material filling the joints, preventing slab expansion	Localized stresses in panel near the joint
Distresses in load transfer mechanism	Material strength exceeded under applied loads	Stresses and plastic strains in concrete and dowel grout
Poor joint Load Transfer Efficiency (LTE)	Distresses in load transfer mechanism	Theoretical LTE calculated from deflections; performance criterion: LTE $\geq$ 70%
Joint faulting	Pumping of materials under slab; poor joint LTE	Deflections; performance criterion: faulting $<$ 3.81 mm

The panels must primarily be evaluated for the occurrence of transverse cracking. For the varying conditions, the tensile stresses will be compared, and the extent and location of cracks can be predicted from the maximum principle plastic strains obtained from the ABAQUS analysis. The joints will be evaluated for distresses in the panel near the joints, distresses in the load transfer mechanism, and joint faulting. Stresses in the panels can be analyzed to assess the potential for localized crushing or spalling of the concrete near the joints. Crushing and tensile cracking can occur in the concrete or dowel grout surrounding the dowel bars. The stresses in the dowel grout obtained from the FEA can be compared to maximum allowable stresses to determine if failure occurs, and the principal plastic strains can be used to determine if concrete cracking occurs. In addition, deflection results can be used to calculate joint faulting and the theoretical load transfer efficiency (LTE) at the joints. These can be compared to the recommended performance criterion for PCP over its service life.

In-service monitoring of the trial site and finite element analysis will be used to establish the long-term performance of PCIP. Performance evaluation is a critical step to determine the feasibility and effectiveness of this rehabilitation technique.

Acknowledgements

The authors of this paper gratefully acknowledge the Ministry of Transportation of Ontario, Cement Association of Canada, and the Natural Sciences and Engineering Research Council of Canada (NSERC) for providing funding for this project. For their assistance and expertise in completing construction of the PCIP trial section, we thankfully acknowledge Rico Fung, Peter Smith, Tom Kazmierowski, Stephen Lee, Warren Lee, Susanne Chan, Mark Snyder, Sherry Sullivan, and Aaron Toth.

References

Chan, Susanne, and Becca Lane. 2005. *Falling Weight Deflectometer (FWD) Testing Guideline*. Downsview, ON: Ontario Ministry of Transportation.

Delatte, Norbert. 2008. *Concrete Pavement Design, Construction and Performance*. New York, NY: Taylor & Francis.

Genikomsou, Aikaterini S., and Maria Anna Polak. 2015. "Finite Element Analysis of Punching Shear of Concrete Slabs Using Damaged Plasticity Model in ABAQUS." *Engineering Structures* 98: 38–48. https://doi.org/10.1016/j.engstruct.2015.04.016.

Haas, Ralph, Ken Anderson, Assafm Gabriel, and Art Bergan. 1997. "Chapter 2: Data Requirements, Collection Methods and Data Base." In *Pavement Design and Management Guide*, 38–80. Ottawa: Transportation Association of Canada.

Hudson, W. Ronald, Ralph Haas, and Waheed Uddin. 1997. "Chapter 7: Uses of Monitoring Data and Examples of In-Service Evaluation." In *Infrastructure Management*, 143–65. New York: McGraw-Hill Press.

Merritt, David K, B. Frank McCullough, Ned H Burns, and Anton K Schindler. 2000. "The Feasibility of Using Precast Concrete Panels to Expedite Highway Pavement Construction." Austin, TX. http://ctr.utexas.edu/wp-content/uploads/pubs/7_1517_1.pdf.

National Cooperative Highway Research Program. 2004. "Part 3. Design Analysis." In *Guide for Mechanistic-Empirical Design of New and Rehabilitated Pavement Structures*. Champaign, IL. http://onlinepubs.trb.org/onlinepubs/archive/mepdg/Part3_Chapter4_RigidDesign.pdf.

Pickel, Daniel, and Susan Tighe. 2017. "Precast Concrete Inlay Panel Installation on Highway 400: Construction Report." Waterloo, ON.

Simulia. 2014. "Abaqus 6.14 Analysis User's Guide." Providence, RI: Dassault Systèmes. http://abaqus.software.polimi.it/v6.14/books/usb/default.htm.

Tayabji, Shiraz, and Sam Tyson. 2017. "Precast Concrete Pavement Implementation." *Concrete International* 39 (4): 41–46. https://global-factiva-com.proxy.lib.uwaterloo.ca/ga/default.aspx.

Tayabji, Shiraz, Dan Ye, and Neeraj Buch. 2013. "Precast Concrete Pavement Technology." SHRP 2 Rep. S2-R05-RR-1. Washington, D.C.: Transportation Research Board. https://doi.org/10.17226/22710.

Building Tomorrow's Society
Bâtir la Société de Demain

Fredericton, Canada
June 13 to 16, 2018 / *13 au 16 juin, 2018*

NOVEL ASSET MANAGEMENT FRAMEWORK FOR ROAD MAINTENANCE

Mohammadi, Alireza[1, 5], Igwe, Charles[1], Amador-Jimenez, Luis[1], and Nasiri, Fuzhan[1]
[1-4] Concordia University, Canada
[5] alireza.mohammadi@concordia.ca

Abstract: As assets begin to deteriorate either due to age or additional demands based on urbanization, it becomes imperative to renovate or replace them to ensure that they continue to deliver the level of service required by the end user. Asset management, therefore, becomes a very important consideration in the development of nations. However, there is another key factor worthy of consideration and this is based on the principles of lean construction. According to the lean construction paradigm, to increase value to the end user in any project, especially projects in urban areas where the effect of delays is more noticeable and profound due to the number of people requiring the use of these assets, waste must be eliminated or reduced to the barest minimum and doing this requires conformance to five major principles. Classical practices in the asset management domain; however, have not embraced the lean construction paradigm in road maintenance or rehabilitation plan. This paper suggests a framework to establish a formalized road management procedure with a focus on the principles of lean construction for avoiding non-value activities. To validate the proposed approach, a case study project is tested. The results of the case study revealed that by using the developed model, a 20% reduction in the project budget can be achieved while still ensuring that the required level of service is maintained. Based on this developed model, roads and other infrastructure decision-makers will be able to make better asset management plans that will ensure that project stakeholders including the end users get better value for their money and time.

1 INTRODUCTION

The competition for resources required to promote economic development and cater for the effects of urbanization has made it imperative to ensure that whatever budget or resources are allocated to projects be adequately utilized to provide value for time and money. One of the precursors to development is the availability of good and well-maintained road networks to meet the demands of road users and reduce inconveniences that may arise from bad and poorly maintained roads. The decision on which road to select for maintenance is not usually an easy one as there are several factors to consider including the selection of which asset (e.g. which road segments in a highway) requires the most intervention, the choice of the maintenance or rehabilitation actions, and the best time for actions. The aforementioned factors are mainly bothered by finance.

There is a rich literature existing in terms of Road Asset Management (RAM) efforts and several types of research have been conducted to provide recommendations and solutions aimed at improving the performance and safety of road systems subject to budgetary constraints. The traditional planning approach for providing interventions in RAM involves three distinct phases of planning in strategic level, translating this plan into tactical and operational projects, and finally implementation through assigning the plan to construction teams.

In the planning phase, decision-making models of cost-effectiveness analysis or mathematical optimization could be applied to find the optimum treatment/intervention actions across competing alternatives (trade-off) which is mostly done in a long-term horizon. Next phase involves the implementation of the selected intervention actions which is typically achieved through agencies outsourcing the work to be performed. However, before outsourcing, it is important to convert long-term (strategic) plan into tactical and operational projects which is less discussed and elaborated topics in the literature and possibly producing wasting values and disruption of services (Faghih-Imani and Amador 2013). Coordination could be applied by asset managers in this phase based on expert judgments; however, the feasibility of the plan lies in its ability to ensure the avoiding unnecessary costs (mobilization and demobilization) when carrying out the road maintenance actions plan as well as minimization of the disruption that will be suffered by the road users.

1.1 Lean Construction Principles

The focus of the lean paradigm is the elimination of wastes to promote the value. According to Fewings (2013), the lean thinking goes beyond the elimination of waste and extends its application to value delivery. Womack and Jones (2010) add further that lean thinking creates a means for specifying the value, differentiates value adding from non-value adding activities and helps in the sequential arrangement of value-adding activities. There are five major principles of lean thinking and these include value, value stream, flow, pull and continuous improvement. These principles are focused on the reduction in the share of non-value adding activities, reduction in variability and lead time, increased flexibility, transparency and simplicity of operations. Applying the lean construction paradigm to RAM ensures that stakeholders get value for their time and money by ensuring the proper selection of intervention measures for road assets, prioritizing the intervention measure to be applied while ensuring that intervention activities flow seamlessly from one selected road section to another.

1.2 Common Wastes in Road Asset Management (RAM)

The principles of lean thinking can be applied to RAM to facilitate value for main stakeholders including the transportation agencies, the road maintenance contractors, and the end users. From a lean perspective, asset manager could look for wasted value in the process of reinvestment for Maintenance, Rehabilitation, and Replacement (MRR). One source which contributes to wasting is identified in the classical process including those costs are involved to repetitive mobilization and demobilization. Thus, in the planning step, decision-maker could pick the alternatives in which minimize these kinds of costs in the construction phase. Implementing the lean idea during the planning phase in RAM helps to improve MRR plan for the construction phase by ensuring the elimination of non-value adding activities as will be demonstrated later in the case study section. Figure 1 is developed to present applying lean perspective in road asset management.

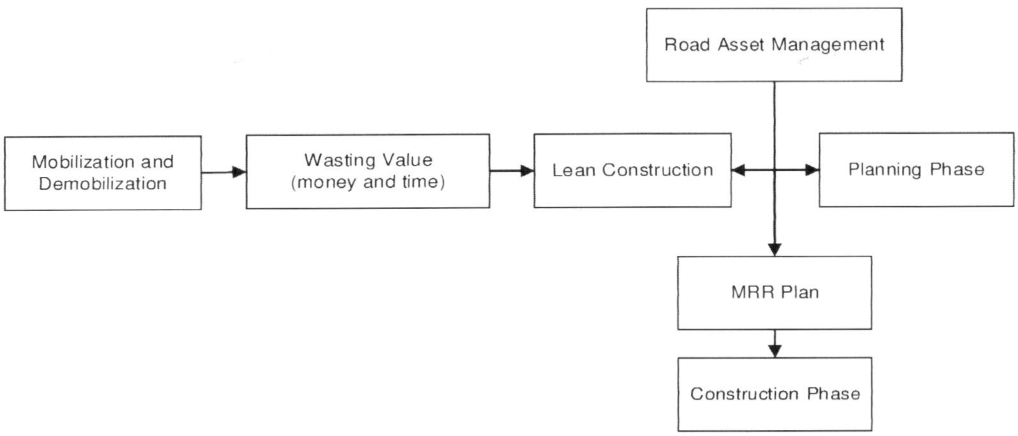

Figure 1: Addressing lean construction in RAM

1.3 Background

Mobilization and demobilization are common wastes usually experienced when performing road maintenance works for separate road segments geographically distributed along the same road network. Studies conducted by Amador and Magnuson (2011) and Osman (2016) generated coordinated programs of maintenance and replacement works for different types of infrastructure in the same location to avoid wastes, disruptions, and repetition. Attempts have been made by researchers to reduce this time and money waste. Faghih-Imani and Amador (2013) developed a framework with specific distance and temporal rules to coordinate road maintenance projects; however, clustering alternatives is still done in the separate step by decision-maker based on designed plan doubting to achieve an optimal solution. Galenko et.al. (2015) conducted a study to optimize created construction plan by modifying the maintenance schedule while respecting coordination. However, this study had some notable shortcomings, the developed model addressed coordination of the road segments for repairs by fast-tracking or postponing treatment actions. However, this method does not ensure acceptable and safe performance levels for all segments in each year. Also, the user cost associated with each project as a part of the decision-making process was also not defined.

Several types of research have focused on addressing coordination during the construction phase as a means of minimizing disruptions. Theiss et al. (2016) developed a guideline for the U.S. federal highway administration (FHA) to minimize work zone mobility impacts in the construction phase. Lee (2005) tested different construction closure times of weekday and weekend to find the lowest cost scenario for both user and agency. Morgado (2014) proposed a model for work zone planning in pavement rehabilitation to integrate costs, duration, and user effects and select the best alternative minimizing disruption. Oh et al. (2011) developed a scheduling framework to minimize road project disruptions while using existing transportation networks; however, this model may work in urban road networks with enough alternatives. Addressing coordination is typically neglected during the planning stage, which may give more flexibility. In the traditional planning models RAM, plan alternatives are analyzed through several indicators of costs, level of service, safety, gas emission and user costs. However, user costs mostly reflects the relationship between road conditions and costly attributes for users (such as car maintenance costs or fuel consumption) while delays caused when road sections are under maintenance works is often neglected in decision-making (Santos, 2005 and Moreira 2017). It might be resulted by difficulty in data collection and quantification.

The state of the art shows that there is a gap in the literature providing transportation agencies with the opportunity for avoiding time and money waste during the planning and construction phase. This paper suggests a framework to establish a formalized asset management procedure based on lean construction principles (eliminating waste to enhance value) in road management process. A decision-making model will be developed to provide guidance to road agencies in arriving at the optimum choice of asset MRR plan. Based on the result of the study, decision-makers will be able to make more efficient asset management decisions addressing coordination that will ensure project stakeholders get better value for their time and money.

2 METHODOLOGY

2.1 Mobilization and Demobilization

As a repetitive process, road agencies need to select several roads each year for maintenance and rehabilitation purposes. This kind of modeling has been advanced through time mostly from planning rather than construction point of view. Mature frameworks, objectively distribute available budget in order to maximize performance and safety. However, from a practical point of view, some factors may impact construction quality and cost. One time-consuming and costly step in any construction activity is the mobilization and demobilization of personnel/machinery and contractors always try to avoid repeating mobilization as it is a waste of time and very costly phase. The results from road management planning is commonly a bunch of segments geographically distributed along the road network, which is not necessarily near or next to each other. Meanwhile, transit agencies, particularly on low and medium-size roads, often avoid working with several contractors for a specific road. Therefore, the contractor must mobilize and demobilize several times to implement requested interventions and this increases the share of non-value activities with an attendant decrease in actual value-adding activities. In this study, the developed model tries to improve the overall performance of each road while respecting the coordination between the plan and the different road sections requiring intervention. By increasing the number of side by side segments in each year planning, this fixed cost could be decreased while time is saved too.

2.2 Mathematical Approach

The budget limitation is a common obstacle in pavement management. Transit agencies usually concentrate on utilizing the bulk of the budget on respecting safety and performance thresholds while utilizing the remaining for preventive maintenance works. A two-step optimization will be applied to model: in the first step, the minimum budget for satisfying all constraints will be estimated. This step could give a long-term view to decision makers. In the second step, a higher budget (subject to more available funds) could exceed the outcome of the first step optimization and will be applied to find an optimum solution and maximum performance. However, in this study, an innovative approach is proposed to classical one for the second step with the ability of the model to select side by side segments, saving funds, and reducing incoordination issues. The performance of a road segment ($P_{s,t}$) could be estimated based on any performance index such as international roughness index (*IRI*). The first step optimization model is shown in Equations 1 to 4 and Equations 5 to 6 present classical formulations for the second step optimization.

Step I:

$$[1]\ MIN(C_t) = \sum_{s=1}^{S} C_{s,t} X_{s,t} l_s$$

Subject to

$$[2]\ P_s < P_{s,t} \quad and \quad P_o < P_t$$

$$[3]\ P_t = \frac{\sum_{s=1}^{S}(l_s \times P_{s,t})}{\sum_{s}^{S} l_s}$$

Where the following time dynamic link applies.

$$[4]\ P_{s,t} = X_{s,t}\left(P_{s,t-1} + Pi_{s,t}\right) + \left(1 - X_{s,t}\right)\left(P_{s,t-1} - Pd_{s,t}\right)$$

Step II (Classical model):

$$[5]\ MAX\left(P_t\right) = \sum_{s=1}^{S} (l_s \times P_{s,t})$$

Subject to:

$$[6]\ 0 < \sum_{s=1}^{S} C_{s,t} X_{s,t} l_s \leq B_t \qquad B_t > C_t$$

Where:

$$[7]\; X_{s,t} = \begin{cases} 1 \; if \; action \; is \; taken \; on \; segment \; s, in \; the \; year \; t \\ 0 \; if \; no \; action \; taken \; on \; segment \; s, in \; the \; year \; t \end{cases}$$

P_t = *Overall performance in the network or road on year "t'*

$P_{s,t}$ = *Performance of segment "s" on year "t" on a 0 to 100 scale*

C_t = *Minimum budget ($) as given by Equation 1 on year "t"*

$C_{s,t}$ = *Unitary cost ($) of rehabilitation action of segment "s" on year "t"*

$Pi_{s,t}$ =*Improvement portion from the year (t-1) to year "t" for segment "s"*

$Pd_{s,t}$ =*Dropped portion of performance from the year (t-1) to year "t" for non-selected segment "s"*

P_s=*Minimum acceptable performance for one segment respecting safety threshold*

P_o=*Minimum acceptable performance overall a network or road*

l_s= *Segment "s" length*

In case of using *IRI*, due to nature of index, maximum acceptable levels as thresholds will be used and the objective would be minimizing overall *IRI*.

For this study, Equations 5 and 6 could be customized to reduce the cost for those segments which are selected sequentially in each time period plan. As it is mentioned in the cost estimating manual for projects, published by Washington State Department of Transportation (WSDOT) (2015) mobilization and demobilization cost is adjustable contractor's preconstruction expenses in transportation activities which could be changed from 5 to 12% of the construction cost depending on the contract value. However, working in adjacent segments drops the construction costs as well as accelerates activities which directly decrease the overall cost. Therefore, the contractor will gain even more than the suggested rates for the mobilization percentage. Typical decision rules are implemented in developed optimization model as follows:

If two sequential segments are assigned the same treatment action in the one-time period, 10% reduction of treatment costs in the smaller segment will be applied. Meanwhile, for different selected actions in two side by side segments, 5% drop could be used which is applied to less total price treatment. The above formulation also addresses coordination and user costs by reducing traffic flow disruptions and improving safety; however, future studies could add this factor (delay caused by road maintenance projects) in the decision-making process.

2 CASE STUDY

Highway 417 west (King's Highways) in Canada is selected as a case study. This freeway connects Montreal City in Quebec to Ottawa in Ontario. Around 180 km from Quebec border to the end of the highway in Ontario is analyzed in this study. Provincial Traffic Volumes (2016) indicates that traffic load in this highway could be categorized in three levels: low, medium and high based on 22000, 80000, 160000 annual average daily traffic volume (AADT), respectively. The highway is mostly including four lanes with low AADT; however, partially has six and eight lanes in medium and high traffic portions, respectively.

Deterioration trends for each traffic load group are developed based on Paterson and Attoh-Okine (1992) model (Equation 8) where IRI is predicted through equivalent single-axle loads (*ESAL*), thornthwaite moisture index (*m*), structural number (*SNC*) and initial as-built quality (IRI_0). The almost similar environmental condition is observed in this case and *SNC* is calculated based on AASHTO (1993). As-built quality (IRI_0) is set to 0.8 and the coefficient α is set to 265 (Zareie et al. 2016).

$$[8]\ IRI_t = e^{mt}[IRI_0 + \alpha(SNC)^{-5} \times ESAL_t]$$

Treatment action windows and pavement lifespans are developed based on pavement design and rehabilitation guidelines (Government of New Brunswick 2018; Ontario Ministry of Transportation 2018) in Table 1. Also, British Columbia ministry of transportation and infrastructure (2013) guideline for rehabilitation costs is used for developing this table; however, the costs are presented in $US. Maximum twice preventive maintenance is allowed to be applied to each segment to avoid the inefficiency of treatments. Based on these guidelines, overall *IRI* should be set on less than 1.5 m/km and 2.7 m/km is defined as the poor threshold for each segment corresponding to 15-year lifespan. Pavement condition observations for 2014 are collected from Ministry of Transportation (2018) and since the shortest segment in this report is 3.64 km, the highway is divided into 45 segments, each four kilometers long. Planning intervals could be set respecting road current condition, segments length, agency policy, deterioration rates, and seasonal construction duration in each year. In this study, since the current overall condition is good, the model is run for each 4-year time period to give flexibility for clustering. Also, results could be comparable with government rehabilitation plan for the same road.

Table 1: Treatment and Operational Windows for pavement

Treatment	Drop in IRI	Cost ($US/lane.km)	Lower IRI (m/km)	Upper IRI (m/km)
Do nothing	0.00	0	0.00	1.10
Preventive maintenance	0.54	4,000	1.11	1.56
Minor rehabilitation	0.90	40,000	1.57	2.08
Major rehabilitation	1.40	100,000	2.09	2.70
Reconstruction	Brand New	220,000	2.71	-

The model is run for this case study and the minimum budget was derived for each time period (Figure 2). The minimum required budget has been increased particularly after 20 years. Two groups of segments are observed in this highway, and the first group has a poor condition (more than *IRI* 2 m/km) and approaching the end of its life cycle. In the second group, several segments with *IRI* around 1 m/km are observed and this situation improved the overall *IRI*, which is 1.46 m/km and less than acceptable level. Therefore, the model in the first step concentrates more on critical segments (first group) and ignore the rests. Thus, preventive actions are ignored in this step and additional budget is needed for the second step of the optimization.

To test the model, the results are compared with the Ontario ministry of the transportation plan for 2014-2018 and 2015-2019 (Ontario Ministry of Transportation 2018). Seven segments were selected by the model for major rehabilitation while the Ontario ministry of transportation rehabilitation plan shows that 17 segments were picked for resurfacing as a major rehabilitation action in the 5-year plan in both directions of west and east. Analyzing the results indicates that same approach is applied by the government for east direction and very similar segments were selected for rehabilitation in the first plan (2014-2018). The model also assigned preventive maintenance to only 4 segments to respect the overall performance threshold (*IRI*<=1.5 m/km); however, as it can be seen in Figure 2, it is not a proper long-term approach and brings more future costs as a result of ignoring enough preventive actions and concentrating only on critical segments. In order to reduce this future cost, more than minimum required budget should be assigned to this highway.

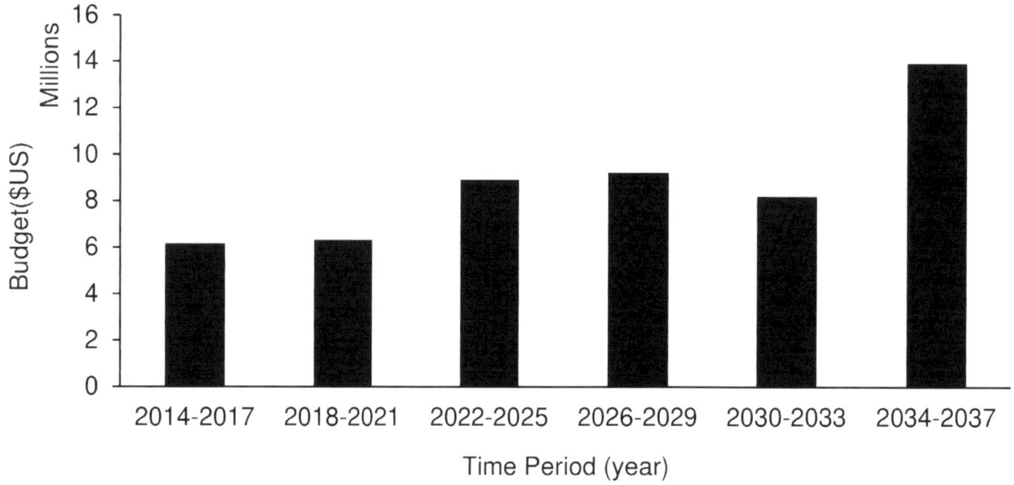

Figure 2: Minimum budget for respecting thresholds

Respecting the results of the first step, the budget is increased to 6,500,000 $US for each 4-year time period and is applied to both the classical and the developed model in this study to minimize the overall condition while respecting thresholds of condition and budget. Figure 3 presents overall *IRI* in both approaches and a slightly better condition is achieved by proposed approach in this study. Average IRI is 1.12 and 1.14 for the new and classical model, respectively.

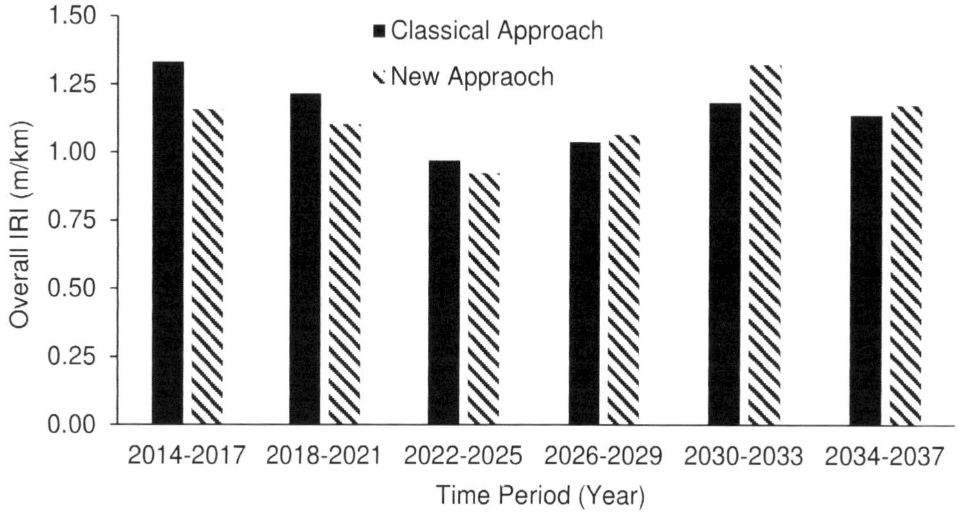

Figure 3: Overall IRI (m/km) in classical and new model

Figure 4 presents the total expenditures in each approach. The first interesting point in these figures is the very high efficiency of preventive maintenance actions. By adding less than $US 500,000 to the minimum budget at the beginning, the agency could reduce the total cost significantly while dropping required budget for the 2026-2033 periods with less than $US 500,000 due to an acceptable level of performance (IRI<=1.5 m/km) in all segments. The novel model even could gain more and only need the assigned budget ($US 6,500,000) for two cycles of 2014-2017 and 2018-2021. Comparing Figures 3 and 4 indicates that new formulation could save $US 5 million, while similar performance is achieved.

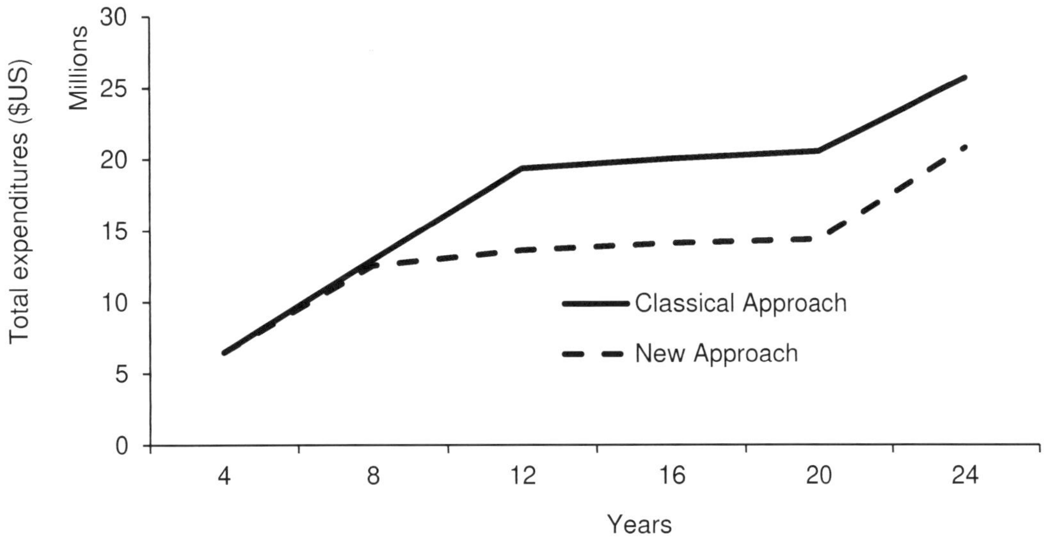

Figure 4: Total expenditure in the classical and new formulated model

3 CONCLUSION

The classical approach for providing intervention in RAM involves two distinct steps of planning and implementation which result in wasting money and time. This paper suggested a framework to establish a formalized asset management procedure with a focus on the principles of lean construction. The model focuses on eliminating wastes including mobilization and demobilization costs in the construction phase to enhance value in road management process. To test the model, the results were compared with the Ontario ministry of the transportation plan for 2014-2018 and 2015-2019 (Ontario Ministry of Transportation 2018) for Highway 417 in Ontario, Canada. Long-term analysis of this freeway showed that an approximate budget of $US 25,000,000 for 24 years is required to keep this freeway at an acceptable level of service. However, the developed formulation for this case study could drop the approximate budget by about 20% and still achieve a similar level of service.

References

AASHTO. (1993). *AASHTO Guide for Design of Pavement Structures, 1993* (Vol. 1). AASHTO.

Amador, Luis, and Sherry Magnuson. "Adjacency Modeling for Coordination of Investments in Infrastructure Asset Management: Case Study of Kindersley, Saskatchewan, Canada." *Transportation Research Record: Journal of the Transportation Research Board* 2246 (2011): 8-15.

British Columbia ministry of transportation and infrastructure. (2013). Construction and Rehabilitation Cost Guide. https://www2.gov.bc.ca.

Faghih-Imani, Ahmadreza, and Luis Amador-Jimenez. From Strategic Optimization to Tactical Plans: Coordinating Treatments of Road Infrastructure. No. 13-0632. 2013.

Fewings, P. (2013). *Construction project management: an integrated approach.* Routledge, Ney York.

Galenko, Alexander, Eric Perrone, and Tonya Scheingberg. "Project Coordination Model." *Procedia Computer Science* 52 (2015): 83-89.

Government of New Brunswick (2018), Capital Maintenance of Highways, chapter 5, http://www2.gnb.ca/content/gnb/en.html

Osman, Hesham. "Coordination of urban infrastructure reconstruction projects." *Structure and Infrastructure Engineering* 12, no. 1 (2016): 108-121.

Lee, Eul-Bum, C. William Ibbs, and David Thomas. "Minimizing total cost for urban freeway reconstruction with integrated construction/traffic analysis." *Journal of infrastructure systems* 11, no. 4 (2005): 250-257.

Morgado, João, and José Neves. "Work zone planning in pavement rehabilitation: Integrating cost, duration, and user effects." *Journal of Construction Engineering and Management* 140, no. 11 (2014): 04014050.

Moreira, André V., Tien F. Fwa, Joel RM Oliveira, and Lino Costa. "Coordination of User and Agency Costs Using Two-Level Approach for Pavement Management Optimization." *Transportation Research Record: Journal of the Transportation Research Board* 2639 (2017): 110-118.

Oh, Jun-Seok, Hyunmyung Kim, and Dongjoo Park. "Bi-objective network optimization for spatial and temporal coordination of multiple highway construction projects." *KSCE Journal of Civil Engineering* 15, no. 8 (2011): 1449-1455.

Ontario Ministry of Transportation (2018), Pavement condition for provincial highways, https://www.ontario.ca/data/pavement-condition-provincial-highways

Ontario Ministry of Transportation (2018), Pavement design and rehabilitation manual, second edition, www.bv.transports.gouv.qc.ca/mono/1165561.pdf

Ontario Ministry of Transportation (2018), Southern Highways Program (2014-2018), http://www.mto.gov.on.ca

Ontario Ministry of Transportation (2018), Southern Highways Program (2015-2019), http://www.mto.gov.on.ca

Santos, Bertha, Luis Picado-Santos, and Victor Cavaleiro. "Road User Costs in Pavement Management Systems: Methodologies and Costs." In *Proceedings of the Fourth International Conference on Maintenance and Rehabilitation of Pavements and Technological Control.* 2005.

Theiss, LuAnn, Gerald Ullman, and Amy Moinet. *Guide to Project Coordination for Minimizing Work Zone Mobility Impacts.* No. FHWA-HOP-16-013. 2016

Washington State Department of Transportation (WSDOT), 2015, Cost Estimating Manual for Projects, 3034.03, ttps://www.wsdot.wa.gov

Womack, J. P., and Jones, D. T. (2010). *Lean thinking: banish waste and create wealth in your corporation.* Free Press, New York.

Yang, Jidong. "Nested Markov Decision Framework for Coordinating Pavement Improvement with Capacity
Expansion." *Journal of Transportation Engineering* 138, no. 4 (2012): 387-394.

Zareie, Alireza, Md Shohel Reza Amin, and Luis E. Amador-Jiménez. "Thornthwaite moisture index modeling to estimate the implication of climate change on pavement deterioration." *Journal of Transportation Engineering* 142, no. 4 (2016): 04016007.

Building Tomorrow's Society
Bâtir la Société de Demain

Fredericton, Canada
June 13 to 16, 2018 / *13 au 16 juin, 2018*

OPTIMIZATION OF SINGLE-LANE ROUNDABOUT GEOMETRIC DESIGN: ENVIRONMENTAL SUSTAINABILITY

Ahmed, Hend[1, 3] and Easa, Said[2]
[1] Ryerson University, Canada
[2] Ryerson University, Canada
[3] hend.ahmed@ryerson.ca

Abstract: Intersections are a significant source of congestion which produces the majority of vehicle emissions; therefore, environmental sustainability is the primary objective of intersection design. Research has proven that roundabouts provide an effective solution to calm traffic and improve the environmental impact of intersections. This study presents a model that directly identifies the optimal geometric parameters of a single-lane roundabout in order to minimize vehicle emissions. The model is based on an optimization technique which uses environmental sustainability as a design objective. The Vehicle Specific Power (VSP) methodology was used to model the vehicle emissions of approaching vehicles based on an operation speed model. The traffic conditions, site limitations, and guideline recommendations were used as inputs and the optimal geometric parameters that minimize vehicle emissions were the decision values determined by the model. A sensitivity analysis was conducted and revealed that the model improved environmental sustainability, providing reductions in vehicle emissions.

1 INTRODUCTION

A roundabout is a type of intersection in which the traffic circulates counter clockwise around a central island, and the entering traffic must yield to the circulating traffic (Rodegerdts et al., 2010). Roundabouts play a key role in the improvement of the environmental impact of traffic compared to other forms of at-grade intersections. Roundabouts improve mobility by reducing the total delay compared to other controlled intersections. This improvement leads to environmental benefits and savings in fuel consumption (Federal Highway Administration, 2010). Mandavilli et al. (2008) examined the effect of converting AWSC intersections into modern roundabouts on vehicle emissions. The researchers evaluated 6 intersections indicating the environmental improvement of installing roundabouts. They reported reductions of 42%, 59%, 48%, and 65% for CO, CO2, NOX, and HC emissions, respectively. Another study conducted in Sweden indicated that converting a signalized intersection into a roundabout reduced vehicle emissions by 29% and 21% for CO and NOX (respectively), and fuel consumption by 28% (Várhelyi et al., 2002). Roundabouts are an effective traffic calming approach for intersection congestion, reducing delay and average queue length (Rao et al., 2014). Congested intersections are a major source of vehicle emissions. According to the Government of Canada, passenger cars are a major contributor to air pollution; these vehicle emissions account for almost 21% of the nitrogen dioxide (NO_x), 51% of the volatile organic compound (VOC), and 4% of the fine particle matter (PM 2.5)(Canada, 2017). Road designers need to include vehicle emissions as an important objective in order to achieve improved environmental impact and air quality. This research project involves the development of an optimization model based on minimizing vehicle emissions at single-lane roundabouts. Single- lane roundabouts have the highest safety performance due to the minimizing of conflict between traffic and because of their simplicity. The single-lane roundabouts that have been constructed in the U.S until the year of 2013 represent approximately 71% of the total number of roundabouts (Pochowski et al., 2016). The geometric features of roundabouts are linked to several design measurements such as speed, safety, and mobility and significantly affect the environmental impact. It is therefore necessary to develop a geometric design optimization model for single-lane roundabouts that is able to determine the optimal geometric parameter values that satisfies specific conditions and objectives of each site.

In the planning stage, designers determine the design objectives based on the unique conditions of each site. Each site has its own issues and specific objectives that require improvement. If the location is within a sensitive environmental area, more weight will be applied on minimizing vehicle emissions to improve the air quality in that location. Sometimes the implementation of a roundabout is required in order to improve the safety of a particular intersection due to increases in annual crashes. In this case, safety performance would be the primary design objective. The purpose of this study was to develop an optimum design model. The proposed model can determine the geometric design parameters of single-lane roundabouts that minimize vehicle emissions in order to improve the environmental impact of the intersection. This model improves the geometric design process for roundabouts and improves environmental sustainability. The model can also be expanded by adding more design objectives such as safety performance, mobility, cost-effectiveness, energy consumption, and sight distances to become a complete design tool.

2 METHODOLOGY

Geometric data and traffic data are required as inputs when modeling vehicle emissions. The data are used to model the operating speed of the roundabout, second by second. The vehicle specific power is determined for each speed profile using the modeled operating speed of the roundabout and the vehicle emissions are modeled. Using Solver, the model is solved minimizing the total vehicle emissions subjected to all the defined constraints to provide optimal geometric parameters including the inscribed circle diameter, the circulated width, and the entry width. The methodology is described in more detail in the following section.

2.1 Roundabout Data

2.1.1 Geometric Data

The maximum ranges of the geometric parameters are retrieved from an aerial photograph of the proposed location and verified through the site survey which includes utility information such as poles and drainage lines. The minimum ranges of the geometric parameters are retrieved from the guidelines that are used based on the design vehicle. In this study, "Roundabouts - An Informational Guide (2nd Edition)" was used as a design guideline (Rodegerdts et al., 2010).

2.1.2 Traffic Data

The actual traffic data are used if the model is used to evaluate an existing roundabout or convert an existing intersection into a roundabout in order to improve any of the design objectives. The predicted traffic data are used if the model is used to design a new roundabout. The total approach entering traffic is the sum of all movements at this approach while the conflicting traffic or the circulating traffic is the sum of all of the traffic that passes in front of this approach, given by Equation 1.

$$[1]\ Q_{conf,j} = Q_{UT,(j+1)} + Q_{UT,(j+2)} + Q_{LT,(j+2)} + Q_{UT,(j+3)} + Q_{LT,(j+3)} + Q_{TH,(j+3)}$$

Where, $Q_{conf,j}$ = the circulating traffic for approach j, Q_{UT} = the U-turn movement entering approach j (pce/hr.), Q_{LT} =the left turn movement entering approach j (pce/hr.), Q_{TH} = the through movement entering approach j (pce/hr.), and j = the approach number.

2.2 Modeling Vehicle Specific Power

Numerous researchers have used models based on vehicle specific power to measure vehicle emissions (Coelho 2006; Federal Highway Administration, 2010; Salamati at al., 2013; and Song, 2015). Vehicle specific power (VSP) is the engine power needed to move a vehicle unit mass. Researchers have simplified the VSP equation to be expressed in terms of the instantaneous speed and acceleration of the vehicle as given by Equation 2.

$$[2]\ \mathrm{VSP} = \mathrm{v} * (1.1\mathrm{a} + 0.132) + 0.000302 * \mathrm{v}^3$$

Where, VSP = the vehicle specific power, v = the vehicle instantaneous speed or second by second speed (m/s), a = the vehicle acceleration or deceleration (m/s^2),

In this study, the acceleration and deceleration values were 2.1 m/s^2 and 1.3 m/s^2 , respectively, as recommended by Rodegerdts (2010). The acceleration and deceleration are considered constant for simplicity and due to the unavailability of acceleration/ deceleration models for roundabouts.

The speed is the operating speed modeled by Bassani and Sacchi (2011) and given by Equation 3.

$$[3]\ \mathrm{V}_{85} = 0.4433 * \mathrm{D}_{\mathrm{INT}} + 0.8367 * \mathrm{C} + 3.2272 * \mathrm{E}$$

Where, *v* = 85th percentile speed (km/hr), *C* = the width of the circulatory roadway (m), *E* = the width of the entry lane (m), and $\mathrm{D}_{\mathrm{INT}}$ = the central island diameter (m).

2.3 Modeling Speed Profiles

The speed profile of roundabouts based on vehicle trajectories is divided into three cases with three different speed profiles, as modeled by Coelho et al. (2006).

1. Unstopped vehicles

The percentage of vehicles that cross the roundabout without stopping is given by Equation 4.

$$[4]\ \%(Unstopped\ vehicles) = 100 - 0.0000611(Q_{conf} + Q_{en})$$

2. One-Stop vehicles

The percentage of vehicles that stop only once at the yield line of the roundabout while waiting for an acceptable gap is given by Equation 5.

$$[5]\ \%(\mathrm{one-stop\ vehicles}) = 100 - \%(Unstopped\ vehicles) - \%(more\ than\ one - stop\ vehicles)$$

3. More than one stop vehicles

The percentage of vehicles that experience more than one stop through go and stop cycles before crossing the roundabout is given by Equation 6.

$$[6]\ \%(more\ than\ one - stop\ vehicles) = \exp[0.00123(Q_{conf} + Q_{en} - 300)^{1.2} - 1$$

Where, Q_{conf} = the conflict traffic or the corresponding circulating traffic of the calculated approach,

Q_{en} = the total traffic entering the approach entering traffic

The stop and go cycles experienced by the vehicles while stopping more than once were characterized into short stop and go cycles (SSG) and long stop and go cycles (LSG) based on the queue length. The numbers are given by Equations 7 and 8.

$$[7]\ SSG = 1.834 * \exp(0.0759QL) - 1$$

$$[8]\ \mathrm{LSG} = 1.997 * \exp(0.1124\mathrm{QL}) - [1.834 * \exp(0.0759QL) - 1]$$

Where, SSG= short stop and go cycles number, LSG= long stop and go cycles number, and QL = the queue length (veh.)

The typical short cycle maximum speed is 3.8 km/h, distance is 5.2m, and idle time before each cycle is 4.5 sec. The typical long cycle maximum speed is 6.6 km/h, distance is 15.1m, and idle time before each cycle is 5.2 sec. The idle time at the yield line was modeled theoretically based on the probability of accepting a gap between the circulating traffic. The average headway of the circulating traffic is given by Equation 9.

$$[9]\ \mathrm{AHW} = \frac{3600}{\mathrm{Q_{conf}}}$$

Where, AHW = the average headway time (sec), $\mathrm{Q_{conf}}$ =the circulated traffic in pce/hr

The probability of waiting for 0 gaps is calculated using Equation 10.

$$[10]\ P = \exp(-\frac{t_c}{AHW})$$

Where, P = the probability of waiting for 0 gaps, t_c = the critical gap specified in HCM6 to be 4.99 sec

The expected number of gaps is given by Equation 11.

$$[11]\ \mathrm{E_n} = (1 - \mathrm{P})/\mathrm{P}$$

$$[12]\ \text{Idle time} = \mathrm{E_n} * AHW$$

Where, $\mathrm{E_n}$ = the expected number of gaps

The vehicles decelerate from the approach downstream speed to the roundabout operation speed defined by Equation 2, crossing the roundabout at in the operation speed for a distance that differs for in each movement, and accelerate from the roundabout operation speed to the upstream approach speed. Each speed profile includes three basic movements: through movements, right turn movements, and left turn movements. The vehicles for each movement travel different distances through the roundabout. The three basic movements with the three speed profiles provides nine different speed profiles. In this study, three movements are modeled, through movement, right turn movement, and left turn movement in each speed profile of the three speed movement specified in the previous section. The three movements are characterized by the distance or the time traveled in the roundabout until exiting. Easa and Mehmood (2004.) drew have drawn the fastest path for of vehicles in of single-lane roundabouts with different inscribed circle diameters.They found that the through movement vehicles enter and exit the roundabout at around a 60° angle, the right turn movement vehicles enter and exit the roundabout at a in 30° and 60° angle (respectively), and the left turn movement vehicles enter and exit the roundabout at around a 45° angle, as shown in Figure 2-1. The vehicles fastest path is laying 1.5 m from the concrete curb (Rodegerdts et al., 2010).The distances $\mathrm{d_{th}}$, d_r and d_L can be derived from the geometry geometric of Figure 2-1 as follows:

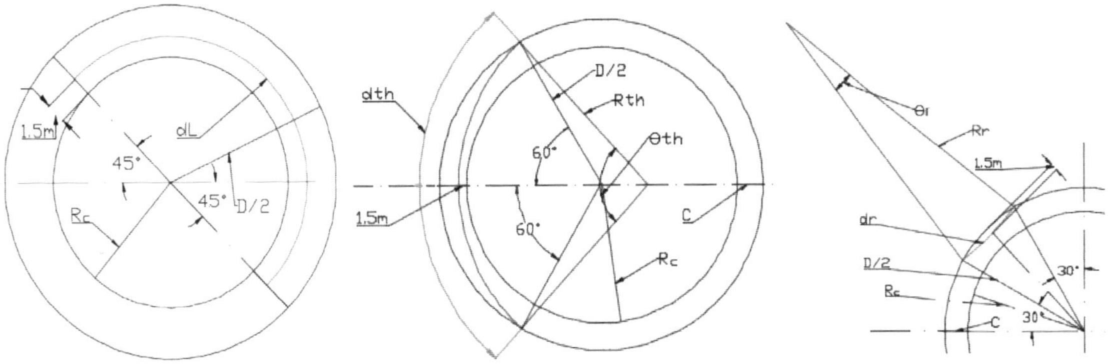

Left turn path Through path Right turn path

Figure 2-1: Movement paths modeling

1. Through path:

$$[13]\ \frac{D}{2}\cos 30 = R_{th}\sin(\frac{\theta_{th}}{2})$$

$$[14]\ \mathrm{R_{th}} = \mathrm{R_{th}}\cos\left(\frac{\theta_{th}}{2}\right) + 1.5 + R_c - \frac{D}{2}\sin 30$$

From Equations 13 and 14, $\mathrm{R_{th}}$ and θ_{th} can be expressed in terms of D and R_c as followings:

$$[15]\ \theta_{\mathrm{th}} = 4\tan^{-1}\left[\frac{R_c + 1.5}{0.433} - 0.57735\right]$$

$$[16]\ \mathrm{R_{th}} = \frac{0.433\mathrm{D}}{\sin\left(\frac{\theta_{th}}{2}\right)}$$

$$[17]\ \mathrm{d_{th}} = \mathrm{R_{th}} * \theta_{\mathrm{th}} * \pi/180$$

Where, D = the inscribed circle diameter (m), R_c = the central island radius (m), $\mathrm{R_{th}}$ = the through path radius (m), and θ_{th}= the through path angle (degrees).

2. Right turn path

$$[18]\ \mathrm{R_r}\sin\left(\frac{\theta_r}{2}\right) = D/2sin15$$

$$[19]\ \frac{D}{2}(1 - cos15) + R_r\left(1 - \cos\left(\frac{\theta_r}{2}\right)\right) = 1.5$$

From Equations 18 and 19 θ_r and $\mathrm{R_r}$ can be expressed in terms of D as given by Equations 20 and 21:

$$[20]\ \theta_r = 4\tan^{-1}\left(\frac{11.59111}{D} - 0.131651\right)$$

$$[21]\ R_r = \frac{0.1294D}{\sin\left(\frac{\theta_r}{2}\right)}$$

$$[22]\ d_r = R_r * \theta_r * \pi/180$$

Where, D = the inscribed circle diameter (m), R_r = the right turn path radius (m), and θ_r = the right turn path angle (degrees).

3. Left turn path

$$[23]\ R_L = R_c + 1.5$$

$$[24]\ d_L = \pi(R_c + 1.5)$$

Where, R_L = the left turn path radius (m), d_L= the distance traveled in the roundabout by the left turn vehicle (m)

The time spent in the roundabout for each movement can be determined from the distances traveled in the roundabout at with the roundabout operation speed, the time spent in the roundabout for each movement can be determined. Using a small optimizing model, the stop and go cycles in speed profile 3 are modeled based on the information given by (Coelho et al., (2006) and the Kinematic Equations considering acceleration and deceleration are the same value. The model determines the acceleration/deceleration of SSG and LSG to be 0.327 and 1.902 m/s^2 respectively.

2.4 Geometric Constraints

The geometric constraints are defined as follows:

$$[25]\ D_{min} \le D \le D_{max}$$

$$[26]\ E_{min,j} \le E_j \le E_{max,j}$$

$$[27]\ \mathrm{Max}(E_j) \le C \le 1.2 * \mathrm{Max}(E_j)$$

As recommended by "Roundabouts - An Informational Guide", the circulated width should be more than the maximum entry width of all legs and should not exceed 20% more than the maximum entry width.

$$[28]\ D = 2R_c + 2C$$

$$[29]\ E_{ex,min,j} \le E_{ex,j} \le E_{ex,max,j}$$

Where, D = the inscribed circle diameter (m), E = the entry width (m), C = the circulated width (m), E_{ex} = the exit width (m), R_c =the central island radius in (m), and j = the leg number (1, 2, 3 or 4)

2.5 Modeling Vehicle Emissions

The Vehicle Specific Power methodology was used to calculate vehicle emissions using Equation 1. The vehicle specific power is determined using Equation 1 (second by second) for each speed profile, and the vehicle emissions are identified based on the VSP value. Frey specified the corresponding vehicle emissions of NOx, HC, CO2 and CO (g/s) for each VSP range (Frey et al., 2002; Frey et al., 2003). The total approach emissions for NOx, HC, CO2, or CO are given by Equation 30.

$$[30]\ \mathrm{E_j} = \sum_{k=1}^{3} [\%P_k * (Q_{TH,j} * \sum_{i=1}^{n^{th}} E_{TH,j,k,i} + Q_{RT} * \sum_{i=1}^{n^{th}} E_{RT,j,k,i} + Q_{LT} * \sum_{i=1}^{n^{th}} E_{LT,j,k,i})]$$

Where, E_j = approach emissions, Q_{TH}, Q_{RT}, and Q_{LT} = the through, right turn and left turn approach traffic, respectively, $\%P_k$ = the vehicles percentage of vehicles from the total entry traffic that experience speed profile k, k = the profile number 1, 2 or 3, j = the approach number 1, 2, 3, or 4, I = the second number, E_{th}, E_{RT}, E_{LT} = the emissions of vehicles making through, right turn or left turn movements, respectively.

2.6 Objective Function

In this study, the design objective was maximizing environmental sustainability which was represented by vehicle emissions. The objective function is the total vehicle emissions given by Equation 31.

$$[31]\ \mathrm{OF} = \lambda_1 \sum_{j=1}^{4} NOX_j + \lambda_2 \sum_{j=1}^{4} HC_j + \lambda_3 \sum_{j=1}^{4} CO2_j + \lambda_4 \sum_{j=1}^{4} CO_j$$

Where, $\lambda 1 \ldots \lambda 4$ are the weight applied to each emission kind type according to the importance or the site needs, NOX represents is Nitrogen oxide emissions, HC represents is Hydrocarbon emissions, CO2 represents is Carbon dioxide emissions, and CO represents is Carbon monoxide emissions. In order to maximize environmental sustainability, the objective function, vehicle emissions, is are minimized. Solving the model provides module presents the optimal geometric parameters; such as, the inscribed circle diameter, the central island radius, the enter entry width, the exit width, and the circulated width.

3 APPLICATION EXAMPLE

The model was applied using a hypothetical intersection with assumed traffic values. The model can be used to evaluate an existing roundabout using the same stages as the design of a new one. In the planning stage, the sizing and space requirements should be assessed based on the actual traffic if it is an existing roundabout or the predicted traffic if it is a new design. The module provided the optimal approach decision variables (D, C, and E_{en}) that minimize the total vehicle emissions.

3.1 Data Preparation

The roundabout turning traffic should be calculated for a passenger car equivalent. The hypothetical traffic values are shown in Table 3-1.

Table 3-1: Application example input data

Heading	Q_e (PCE/h)	Q_c (PCE/h)	Q_{TH} (PCE/h)	Q_{RT} (PCE/h)	Q_{LT} (PCE/h)
West Approach	580	381	250	315	15
South	693	315	27	450	216

Approach					
East Approach	718	258	375	17	326
North Approach	105	917	35	20	50

3.2 Results and Discussion

Table 3-2 shows the optimal values for the inscribed circle diameter (D), circulated width (C), central island radius (in meters), and entering width that minimize vehicle emissions and crash rates for each roundabout approach, as shown in Table 2.

Table 3-2: Application example output data

WB-50 (WB-15)	Een (m)	NOX (g)	HC (g)	CO2 (kg)	CO (g)
West Approach	5.454	47.710	21.070	45.370	1305.644
South Approach	4.919	49.928	1177.917	839.112	1283.704
East Approach	4.919	34.250	471.440	632.835	803.352
North Approach	4.944	45.570	1175.081	836.724	1036.384

The inscribed circle diameter was 46.036 m, the circulating width was 6.760 m, and the central island radius was 16.259 m.

4 SENSTIVITY ANALYSIS

In order to evaluate the sensitivity of vehicle emissions to the geometric parameters of the roundabout, the model was applied to different ranges of the inscribed circle diameter, entry width, and circulated width. The sensitivity analysis was conducted for the west approach based on increasing and decreasing the optimal geometric parameters by 5% and 10%. The results are shown in Table 4-1 and illustrated by Figure 4-1.

Table 4-1: Sensitivity analysis results

West Approach.	NOX (g)	HC (g)	CO2 (kg)	CO (g)
-10% of the optimal	51.146	24.310	48.734	1512.574
-5% of the optimal	51.492	24.478	48.828	1526.238
Optimal parameters	47.710	21.070	45.370	1305.644
+5% of the optimal	52.830	26.373	48.110	1714.412
+10% of the optimal	52.855	26.436	47.815	1726.685

As shown in Table 4-1, increasing the optimal geometric parameters by 5% resulted in an increase of approximately, 10.7%, 25%, 6%, and 31% for the NOX, HC, CO2, and CO emissions, respectively. The results also revealed that decreasing the optimal geometric parameters resulted in an increase in the vehicle emissions. The sensitivity analysis revealed that the optimal geometric parameters identified by the model improved environmental sustainability, decreasing vehicle emissions.

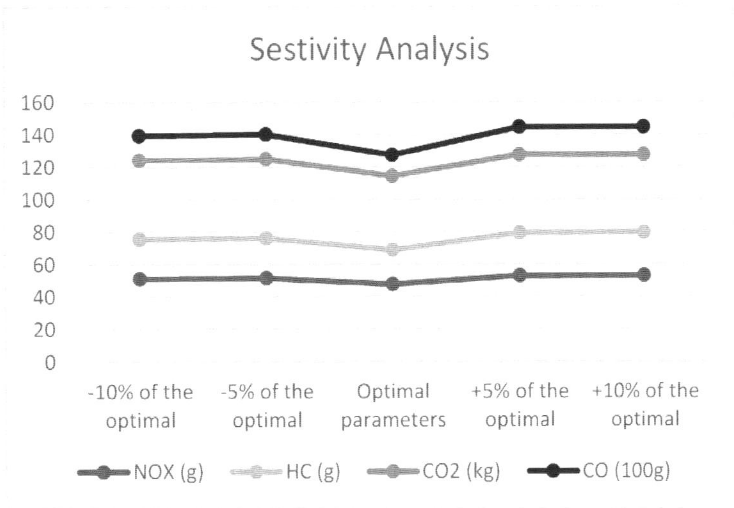

Figure 4-1: The sensitivity of vehicle emissions to the geometry

5 CONCLUDING REMARKS

Numerous researchers have linked the geometric parameters of roundabouts to important aspects of roundabout design including speed and safety performance. There is also evidence that good geometric design of roundabouts can improve environmental sustainability. This paper presents an optimization model for the design of a single-lane roundabout based on minimizing vehicle emissions in order to improve the environmental impact of the intersection. The model was applied to a hypothetical example to test its ability to determine the optimal geometric parameters. The results revealed that the model identified the geometric parameters that minimize vehicle emissions. A sensitivity analysis was conducted and revealed that using the optimal geometric parameters provided by the model improved the environmental sustainability of the roundabout. The use of the model resulted in reductions of 10.7%, 25%, 6%, and 31% for NOX, HC, CO2, and CO emissions (respectively) compared to increasing the optimal geometric parameters by 5%. The model can be improved by addressing the limitations of this study in future research. In this study, the acceleration and deceleration rates were assumed to be constant for emissions calculations, as recommended by the guidelines. Future research should model the acceleration and deceleration and relate them to the geometric parameters of the roundabout. In this study, the model used an operating speed model that was calibrated to the local conditions. Future research can feed the model with more realistic models calibrated using local data. More design objectives can be also be added to the model to suit the needs of any site. Lastly, the model can be modified for different roundabout types instead of limiting it to single-lane roundabouts.

Acknowledgements

This research project was financially supported by the Discovery Grant from the Natural Sciences and Engineering Research Council of Canada (NSERC). The authors are thankful to anonymous reviewers for their thorough and useful comments.

Reference

Ariniello, A. and Przybyl, B. (2010). "Roundabouts and sustainable design." Green Streets and Highways 2010: An Interactive Conference on the State of the Art and How to Achieve Sustainable Outcomes, 82–93.

Bassani, M., & Sacchi, E. (2011). The investigation into speed performance and consistency of urban roundabouts: an Italian case study. Proceedings of the TRB 3rd International Roundabout Conference, Carmel, Indiana.

Canada, E. (2017). "Air pollution from cars, trucks, vans and suvs 388 canada.ca, <https://www.canada.ca/en/environment-climate-change/services/air389 pollution/sources/transportation/cars-trucks-vans-suvs.html> (Nov. 5, 2017).

Coelho, M, Farias, T and Rouphail, N. 2006. "Effect of roundabout operations on pollutant emissions," Transportation Research Part D, vol. 11, (5), pp. 333-343.

Easa, Said M., and Mehmood, A. 2004. Optimizing the geometric design of single-lane roundabouts: Consistency analysis. Canadian Journal of Civil Engineering 31 (6): 1024-38.

Frey, H., Unal, A., Chen, J., Li, S., and Xuan, C. (2002). "Methodology for developing modal emission rates for EPA's multi-scale motor vehicle & equipment emission system." Ann Arbor, Michigan: US Environmental Protection Agency.

Frey, H., Unal, A., Chen, J., Li, S., 2003. Modeling Mobile-Source Emissions Based On In-Use and Second-by-Second Data: Development of Conceptual Approaches for EPA's New MOVES Model. Presented at Air and Waste Management Association 96th Annual Conference and Exhibition, San Diego.

Mandavilli, Srinivas, Margaret J. Rys, and Eugene R. Russell. 2008. Environmental impact of modern roundabouts. *International Journal of Industrial Ergonomics* 38 (2): 135-42.

North Carolina State University, 2002. Methodology for Developing Modal Emission Rates for EPA's Multi-Scale Motor Vehicle and Equipment Emission System. EPA Contract No. PR-CI-02-10493. Office of Transportation and Air Quality, US Environmental Protection Agency, Washington DC.

Pochowski, A., Paul, A., and Rodegerdts, L. A. (2016). Roundabout Practices. Number Project 349 20-05, Topic 46-02.

Rao, Q, Zhang, L, Yang, W, Fang, B. 2014. Analysis of Instantaneous Vehicle Emissions Models Based on Speed and Acceleration. 14th COTA International Conference of Transportation Professionals, American Society of Civil Engineers, pp 2736-2747

Rodegerdts, L, Justin, T, Christopher, K, Julia, M, Edward, J, Mark, M, Persaud, B, Lyon, C, Shauna I, Hillary, C, Barry, G, Bernard, O'Brien, Andrew. 2010. Roundabouts - An Informational Guide (2nd Edition): (NCHRP Report 672). Transportation Research Board. Online version available at http://app.knovel.com/hotlink/toc/id:kpRAIGENCA/roundabouts-an-informational/roundabouts-an-informational

Salamati K, Coelho M, Fernandes P, Rouphail N, Frey H, Bandeira J. 2013. Emission Estimation at Multilane Roundabouts: Effect of Movement and Approach Lane. Transportation Research Board 92nd Annual Meeting, Washington, DC.

Song, G, Zhou, X and Yu, L.2015. "Delay correction model for estimating bus emissions at signalized intersections based on vehicle specific power distributions," The Science of the Total Environment, vol. 514, pp. 108-118.

Várhelyi, András, Transport and Roads, Lund University, Trafik och väg, and Lunds universitet. 2002. The effects of small roundabouts on emissions and fuel consumption: A case study. *Transportation Research Part D* 7 (1): 65-71.

Building Tomorrow's Society
Bâtir la Société de Demain

Fredericton, Canada
June 13 to 16, 2018 / *13 au 16 juin, 2018*

PERFORMANCE EVALUATION OF A FOUR-LEGGED INTERSECTION USING MICROSCOPIC SIMULATION MODEL

Orvin, Muntahith[1,5], Ivy, Afia[2], Hadiuzzaman,Md[3] and Ahsan, Hasib[4]
[1] BSc, Military Institute of Science And Technology (MIST), Bangladesh
[2] BSc, Military Institute of Science And Technology (MIST), Bangladesh
[3] PhD, Professor, Bangladesh University of Engineering And Technology (BUET), Bangladesh
[4] PhD, Professor, Bangladesh University of Engineering And Technology (BUET), Bangladesh
[5] mmceorvin@gmail.com

Abstract: Traffic congestion is a widely spreading problem throughout the world. The gap between traffic demand and supply is increasing day by day that makes junctions, a traffic congestion spot. In this study, a four legged intersection has been evaluated that suffers from high congestion. The traffic within the study area is mostly heterogeneous with non-lane based flow. A microscopic simulation model has been used to assess the performance of the intersection. Using VISSIM, the on-field condition is replicated to observe the behaviour of the prevailing system. Several traffic data are collected from the study area. Then the simulation model is calibrated with the field data and validated with another day's field data, signifies that the developed model clearly mimics the reality. The driving behaviour parameters are tuned to replicate the lane indiscipline existing in the field. The model is analyzed for several conditions such as existing two phase, modified two phase and four phase signal control. It is found that travel time for modified two-phase is more favourable than the existing two-phase signal. Moreover, delay result is more conducive in modified two-phase system. It is to be noted that the four-phase signal produces maximum queue length. The signal design modification can be a great solution to improve the existing congestion condition prevailing in the studied intersection. The strategy used in the model can be incorporated for the improvement of any intersection adjusting the calibration standards. Increase of efficiency of intersections suffering congestion will lead to a better urban traffic system.

1 Introduction

On a typical four-leg intersection, one out of two intersecting roads has the higher traffic volume. This roadway is referred to as the major or arterial road. The second roadway, which services the lower traffic volume, is referred to as the minor or collector road. When the volume on either road nears capacity, queues begin to form, raising the potential for unsafe driving manoeuvres. For this reason, improving safety and operational efficiency at intersections on urban road intersections remain a constant goal (Naghawi 2014). Intersections are usually considered as the critical points within the network and the evaluation of their performance provides valuable understanding and useful indication about the performance of the system (Sunil 2013).

Signalised intersections are one of the key locations in any transportation network, hence improper planning of the intersection affects the performance and productivity of the whole road network. A thorough analysis has to be made to utilise the signalised intersections effectively. The analytical approach has the limitation of underlying assumption of homogeneity, which is having high variations of traffic characteristics when compared with a mixed traffic condition. Simulation models on the other hand, follow the dynamic nature of the traffic system and give a continuous view of the state of the traffic system over time. This characteristic of simulation model is an advantage over analytical models, since it gives more insight to what is happening in the system in detail. Hence, researchers concentrated on appropriate modelling technique through simulation, which is emerged as the most powerful and acceptable solution searching tool (Sheela 2014).

Dhaka is the capital of Bangladesh, one of the mega cities in the world. 'Banglamotor' of Dhaka is a busy four legged intersection during the peak hours. Non-lane based flow predominates and traffic is mixed with both motorized and NMV's in the minor road while only motorized vehicle ply on major road due to the restriction imposed by the road authority. This intersection need to be analyzed for improvement in terms of various traffic characteristics. In this study this intersection is modeled with microscopic simulation software. Out of numerous parameters, signal timing and phasing is modified to understand the traffic properties, how it changes and improves the present condition. The improvement will help the overall increase in performance of the urban intersection that will eventually help in building a congestion free society.

Figure 1: Four-legged busy intersection named 'Banglamotor' of megacity Dhaka

2 Literature Review

Traffic micro-simulation models are widely used to evaluate the benefits and limitations of traffic operation alternatives. Traffic simulation models are mainly classified as microscopic and macroscopic. Models that simulate individual vehicles at small time intervals are termed as microscopic, while models that aggregate traffic flow are termed as macroscopic. Microscopic simulation has been used for a long time to simulate project scale cases such as intersection design. Thus computer simulation is a valuable tool for the analysis and design of complex transportation systems.

Simulation models may be classified as being static or dynamic, deterministic or stochastic and discrete or continuous. (Keller and Saklas 1984) developed a procedure to estimate Passenger Car Equivalent (PCE) values using simulation model. (Banks et al. 2000) mentioned the choice of whether to use a discrete or continuous simulation model is a function of the characteristics of the system and the objectives of the study. (Hossain 2001) used micro simulation technique to model traffic operations at signalised intersections. (Arasan and Koshy 2005) developed a simulation model for urban roads. (Mallikarjuna and Rao 2006) developed a Cellular Automata based simulation model to estimate PCU. (Arasan and Krishnamurthy 2008) conducted a study on the effect of traffic volume and road width on PCU values of vehicles using Microscopic simulation. (Arasan and Vedagiri 2006) applied the simulation model to estimate the saturation flow rate of heterogeneous traffic. (Gowri and Sivanandan 2008) developed a simulation model and examined the effects of left turn channelization on vehicle waiting times. (Arasan and Arkatkar 2010) and (Arasan and Dhivya 2010) also have developed micro simulation models for midblock sections. (Radhakrishnan and Mathew 2011) developed a traffic simulation model integrating the concepts of cellular automata. Thus microscopic simulation models are used in recent times for numerous purposes. However, the effect of signal time and signal phase modification can alter a lot of changes to traffic properties of an intersection which has not been extensively studied.

Traffic signals are one of the most effective active controls of traffic. The advantage of traffic signal includes an orderly movement of traffic and increased capacity of the intersection. However, the disadvantages of the signalized intersection are large stopped delays and complexity in both design and implementation. For a signalized section, signal phase is the part of a cycle allocated to a stream or a combination of two or more streams having the right of way simultaneously. Traffic fixed time controller signal may be of different phasing options. Figure 2 illustrates the conventional signal phase systems. This study discuss various alternatives among signal phase and timing modification of fixed time controller and their comparative effects on traffic condition such as congestion.

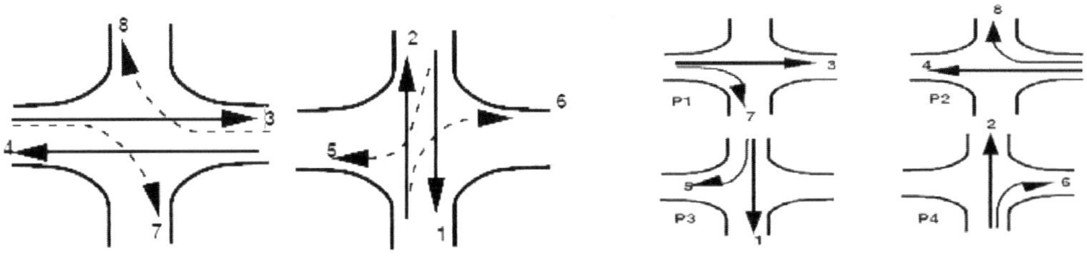

Figure 2: Traffic movements in two phase and four phase fixed-time controller signal system respectively (Mathew 2017)

3 Methodology

The efficiency of a transportation network depends on the performance characteristics of its various components. To improve the efficiency, proper planning and design of various facilities have to be implemented (Sheela 2014). This paper focuses on simulating traffic movements at signalised intersection by developing a micro simulation model using various signal phase system to improve the intersection performance.

The studied intersection is modeled with micro simulation software named 'VISSIM' version 5.40. As micro simulation encounters with individual track of vehicles, distance and time headway, lane changing behaviour, car following model, so extensive data are required. Basic data collected are the geometry of the intersection, number of lanes, lane width, median width, channel dimensions, traffic volume, directional count, signal phasing, cycle times, green, red and amber times, saturation flow, traffic composition, queue length, travel time, speed of vehicles etc. The traditionally simplest approach for conducting manual count is to record each observed vehicle with a tick mark in a tally sheet that has been used in this study. Indirect method of manual counting, the video camera is also utilized for the purpose of verifying the collected data (Manual of Transportation Engineering Studies 2010) as presented in Figure 3.

Figure 3: Manual method (direct and indirect both) is used to collect traffic volume and intersection counts

With these data input, the intersection model is generated. The network generation process includes importing and scaling overlay image after fixing units, determining driver behavior parameters, individual link behavior fixing, link and connector generation, determination of vehicle types and classes, vehicle composition selection, vehicle input in all the arrival points of the network, routing for various links, conflict area and priority implementation and signal head along with signal program implementation etc. Then, data collection points, travel time sections and queue counters establishments are done where necessary for evaluation purpose.

After the network design completion, the simulation is initiated with random seed value. If the random seed varies, the stochastic functions in VISSIM are assigned a different value sequence and the traffic flow changes. This allows stochastic variations of vehicle arrivals in the network. For meaningful results it is recommended to determine the arithmetic mean based on the results of multiple simulation runs with different random seed settings.

Calibration of the generated model is executed to check whether it replicate the existing situation of the intersection. Calibration is defined as the process of adjusting the parameters used in the model to ensure that it accurately reflects input data. The subsequent process of validation is to run an independent check on the calibrated model. Two sets of observed data are therefore required during the model development process. One is used to calibrate the model by adjusting the parameters to ensure that the output matches observed data and the second is used to verify that the aspects of the performance of the calibrated model are in agreement to the set of observed data (Sykes 2010). It is very significant that the output found for particular traffic parameters combinations are within the calibration target range. Calibration target is shown in Table 1.

Table 1: Calibration target

Criteria and measures	Acceptability targets
Hourly flows, model vs observed	
Individual link flows	
Within 15% for 700 vph < Flow < 2700vph	>85% cases
Within 100 vph for Flow < 700 vph	>85% cases
Within 400 vph for Flow > 2700 vph	>85% cases
Total link flows	
Within 5%	All accepting links
GEH statistic- Individual link flows	
GEH < 5	>85% cases
GEH statistic- Total link flows	
GEH < 4	All accepting links
Travel times, model vs observed	
Journey Times Network	
Within 15% (or 1 minute, if higher)	>85% cases
Visual audits	To analyst's satisfaction
Individual link speeds	
Visually acceptable speed-flow relationship	To analyst's satisfaction
Bottlenecks	
Visually acceptable queuing	To analyst's satisfaction

*Source: FHWA Freeway Model Calibration Criteria (Dion et al. 2012)

After the calibration and validation of the generated model is done, the signal phase and timing is redesigned and modified. The existing model with prevailing two phase signal system is checked for modified two phase and four phase signal system. For phase design, saturation flows at all the approaches and lost times are determined first. Subsequently, the green period of all phases and cycle time are calculated. Various traffic parameters are analyzed according to the changes. Alternatives are compared and discussed whether the modification or measure will be suitable and efficient in terms of performance increase of the Banglamotor intersection.

4 Data Collection

Banglamotor is a four legged intersection where Northbound, Southbound, Westbound and Eastbound roads are labelled as SAARC Foara road, Shahbag road, Link road and New Iskaton road respectively. East and Westbound roads have three lanes while South and Northbound roads having four lanes (Figure 4). It is to be mentioned that Eastbound road has a dedicated left turning lane. North and Southbound roads carrying larger traffic flow hence considered as major road arterial.

Figure 4: Geometry of Link road (Westbound), New Iskaton road (Eastbound), Shahbag road (Southbound) and SAARC Foara road (Northbound)

Typical data for the vehicle count for peak hour on a working day along with vehicle composition is shown in Table 2 for the Shahbag road. Table 3 represents the Passenger Car Equivalent (PCE) values which are used to calculate total volume in terms of PCE.

Table 2: Vehicle count for Shahbag road (Southbound)

Movement	Motorcycle	Bike	Car/Jeep/ Micro/Laguna	CNG	Bus	Utility Van	Rickshaw/ Van	Total	%
Left	18	5	41	15	0	0	0	79	3.72
Through	32	11	1074	374	428	2	0	1989	93.73
Right	2	1	25	14	10	2	0	54	2.54
Total =	55	19	1170	423	451	4	0	2122	
%	2.59	0.90	55.14	19.93	21.25	0.19	0		
PCE	41.25	9.5	1170	317.25	1353	4	0	2895	
Total PCE			2895 passenger car per hour						

Table 3: PCE value of vehicles (Partha 2009)

Vehicle Categories	MoC[6] (2001)
Passenger car	1.0
Truck	3.0
Light good vehicle (Utility van)	1.0
Bus	3.0
Auto-rickshaw (CNG)	0.75
Motorcycle	0.75
Pedal cycle (Bike/rickshaw/van)	0.50

In the field it is observed that two phase fixed controller signal system prevailed but it was implemented manually by traffic police. As a result, variation in cycle time often occurred. For the calibration purpose the signal data taken as average of three cycle time observed in the field within the studied hour.

Total cycle time is found 241 sec, among which 190 sec of green time allocated for major road. As per the calculation, the cycle time distribution for the considered phase systems is displayed in Table 4. The generated model is illustrated in Figure 5.

Table 4: Cycle time distribution for existing, modified two phase and four phase system

Phase	Cycle time (sec)	Effective green time (E-W) in sec	Effective green time (N-S) in sec	Total Lost Time, L in sec
Existing two phase	241	190	45	6
Modified two phase	200	162	34	4
Four phase	200	18(E) 18(W)	82(N) 72(S)	10

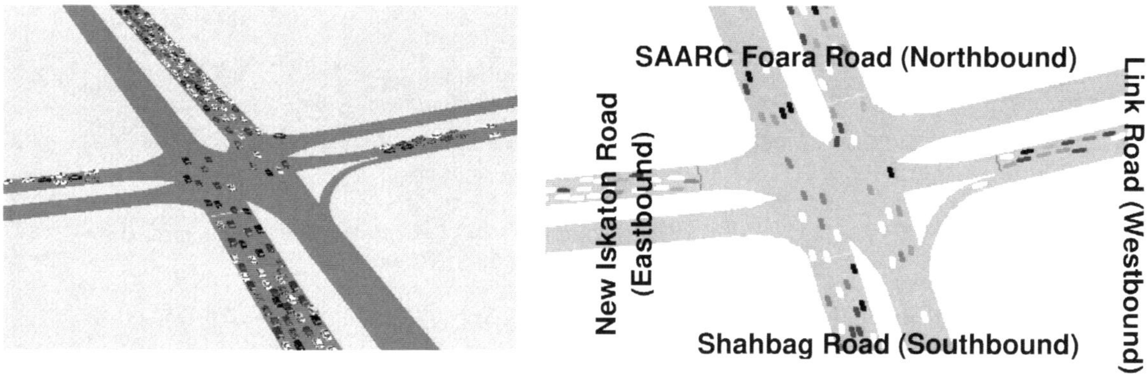

Figure 5: VISSIM 3-D and 2-D Model respectively used in the study

5 Model Calibration and Validation

To achieve high credibility for a simulation model, calibration and validation are of utmost importance (Park 2005). The studied model conform the calibration targets as shown in Table 1, where the limits are within the specified range. The heterogeneity of flow is eliminated by converting all traffic to PCE. The pedestrians are not simulated in the study. Non-lane based flow stands in sharp contrast to micro simulation and rigid in associating vehicles with specific lanes. So, driving behaviour parameters are modified since traffic flow does not follow lanes unlike North American condition. The main parameters for incorporating non-lane based flow are desired position at free flow, lateral distance, diamond shape queuing, consider next turning direction and overtake on same lane. The other parameter's sensitivity analysis is performed. Finally using trial and error method, the calibrated parameters set is found that mimics the field result in terms of flow and travel time. The parameter combination is displayed in Table 5. Comparison of real data observed in the field and simulated data from VISSIM is illustrated in Table 6(a) and (b). The calibrated model is validated for data collected from another working day.

6 Results and Discussion

After studying it is seen that at field condition there is considerable amount of average delay and queue length generation. The average travel time of the vehicles in between the various links is moderately high as well. Especially the right turn from Sahbag to New Iskaton road, face considerable amount of congestion due to the conflict situation with the through movement of traffic coming from SAARC Foara road. Link road and New Iskaton road traffic constitutes minor traffic flow so the signal timing modification for major and minor road may be one solution to improve the present situation.

To achieve better performance in this intersection, various road traffic parameters, characteristics and measures need to be experimented. In this study signal time modification has been implemented and results are compared. Finally the best alternative in terms of efficiency and of the intersection is suggested.

Table 5: Calibrated and modified parameters combination

Parameter	Default Value	Modified Value
Significant Driving Behaviour Parameters:		
Look ahead distance	Min: 0 m, Max: 250 m	30
Look back distance	Min: 0 m, Max: 150 m	30
Smooth close-up Behaviour	unchecked	checked
Average standstill distance	2 m	2.45 m
Additive part of safety distance	2 m	1.90 m
Multiple part of safety distance	3 m	2.98 m
General behaviour	Free lane selection	as it is
Waiting time before diffusion	60 s	as it is
Minimum headway (front/rear)	1 m	0.5 m
Desired position at free flow	Middle of lane	any
Keep lateral distance to vehicle on next lane	unchecked	checked
Diamond shaped queuing	unchecked	checked
Consider next turning direction	unchecked	checked
Collision time gain	2.6 s	3.0 s
Overtake on same lane	unchecked	checked
Minimum lateral distance	1.0	0.90

Table 6 (a): Observed data in field v/s data from simulation

Direction	Individual link flow		Diff. |M-C|	GEH statistic, $\sqrt{\frac{2*(M-C)^2}{(M+C)}}$	Remarks
	Field, C	Model, M			
From SAARC Foara	3336	3254	82	1.43	Difference within 400 veh/hr, for Flow > 2700 veh/hr and within 100 veh/hr, for Flow < 700 veh/hr for more than 85% of the cases.
To SAARC Foara	3152	3062	90	1.61	
From Shahbag	2895	2905	10	0.19	
To Shahbag	3668	3526	142	2.35	
From New Iskaton	504	526	22	0.97	GEH Statistic < 5 for Individual Link Flows for more than 85% of the cases
To New Iskaton	261	224	37	2.33	
From Link road	543	535	8	0.34	
To Link road	194	185	9	0.59	
Sum	14553	14217	336 (2.31%)	2.80	Sum of All Link Flows within 5% of sum of all link counts and GEH < 4 for sum of all link counts.

Table 6 (b): Observed data in field v/s data from simulation

Direction	Travel Times		Difference	Percentage	Remarks
	field	model			
From SAARC Foara	0.69	0.8	0.11	15.94	Road section considered: 10 meter
To SAARC Foara	0.79	0.9	0.11	13.92	
From Shahbag	0.89	0.8	0.09	10.11	
To Shahbagh	0.80	0.9	0.10	12.50	Journey Times, Network: within 15% (or 1 min, if higher)
From New Iskaton	0.76	0.8	0.04	5.26	

To New Iskaton	0.94	0.9	0.04	4.26	for more than 85% cases
From Link Road	1.67	1.9	0.23	13.77	
To Link Road	0.94	0.8	0.14	14.89	

With the existing two phase signal system, if average travel time is considered, it is observed from simulation that modified two phase signal system serves the intersection well for almost all the cases. Typical values are shown in Table 7. For instance, from Shahbag to New Iskaton, the travel time decreases 369 to 318 sec. On the other hand, the four phase signal system generates more time than the prevailing condition at present at a number of cases, making it an ineffective alternative. It is to be mentioned that VISSIM evaluation result of travel time is the average travel time (including waiting or dwell times) determined as the time required by vehicles between crossing the first and crossing the second cross sections (VISSIM manual 5.40, PTV, 2012)

Table 7: Typical comparison of travel time for the generated model

Origin-Destination	Distance in meter	Travel Time in sec		
		Existing two phase signal	Modified two phase signal	Four phase signal
SAARC Foara to Shahbag	886.4	92.10	84.70	227.90
Shahbag to New Iskaton	1371	369.00	318.60	295.00
New Iskaton to SAARC Foara	1363	235.00	220.50	231.40
Link road to Shahbag	622.1	146.30	134.50	322.40

* Travel time refers to the average travel time (including waiting or dwell times)

From simulation, average total delay per vehicle, average standstill time per vehicle and average number of stops per vehicle is calculated. The total delay is computed for every vehicle completing the travel time section by subtracting the theoretical (ideal) travel time from the real travel time. The theoretical travel time is the time that would be reached if there were no other vehicles and no signal controls or other stops in the network. Comparison of delay statistics for the alternative solutions studied are typically illustrated in Table 8. The modified two phase system appears to be a feasible improvement in terms of delay minimization.

Table 8: Typical Comparison of delay for the generated model

Direction	Existing two phase signal			Modified two phase signal			Four phase signal		
	Delay in sec	Stopd in sec	Stops	Delay	Stopd	Stops	Delay	Stopd	Stops
SAARC Foara to New Iskaton	9.20	5.50	0.36	6.50	3.50	0.22	153.60	134.90	1.96
From Shahbag to New Iskaton	252.10	192.20	9.44	201.20	152.70	8.28	177.50	161.90	1.79
Link road to New Iskaton	102.30	92.20	1.63	104.20	94.30	1.53	367.50	341.70	3.58
New Iskaton to SAARC Foara	118.90	102.70	1.97	104.40	89.00	2.03	115.00	105.20	1.27

*Delay is the average total delay per vehicle, Stopd is the average standstill time per vehicle and Stops is the average number of stops per vehicle

In VISSIM, queues are counted from the location of the queue counter on the link or connector upstream to the final vehicle that is in queue condition. For Link from SAARC Foara, queue length improves a lot as shown in Table 9 for modified two phase system. Although maximum queue length at link from Shahbag increases from to 337 to 391 m. The minor roads do not show significant improvement in terms of queue length generation. Four phase system yet again fails to satisfy improvement, as it creates longer queue for almost all approaches.

Table 9: Typical Comparison of queue length for the generated model

Direction	Existing two phase signal		Modified two phase signal		Four phase signal	
	Queue length	Max queue length	Queue length	Max queue length	Queue length	Max queue length
From SAARC Foara	37	264	17	115	345	505
From Shahbag	85	337	85	391	331	505
From New Iskaton	25	72	23	67	25	66
From Link road	43	115	38	116	182	472

* Queue length is the average queue length and queue length is output in units of length not in number of cars

Finally the network performance of the model is checked. The average delay time per vehicle, average speed, total delay and travel time are simulated as displayed in Table 10.

Table 10: Typical Comparison of network performance for the generated model

Term	Existing two phase signal	Modified two phase signal	Four phase signal
Average delay time per vehicle in sec	34.75	29.49	240.36
Average speed in km/hr	28.81	30.25	14.56
Total delay time in hr	69.69	59.14	479.66
Total travel time in hr	224.55	213.95	737.30

The comparison of various data extracted by simulation among the three alternatives (existing two phase, modified two phase and four phase signal system) discussed, it is obvious that four phase system will not be suitable at all and it will worsen the present condition. Most of the parameters suggest that modified two phase system will be more effective in elevating the performance of the studied intersection.

7 Conclusions

Signalized intersections are key elements in the urban transportation network. Increasing traffic demand can exceed the carrying capacity of the intersection during peak periods. Consequently, traffic condition deteriorates. The studied intersection has started facing congestion problem and measures should be taken to improve the prevailing condition. The complex traffic situation prevailed in this study is non-lane based and heterogeneous. The real field condition is simulated by modifying the default driving behavior characteristics of VISSIM. This modification is significant to represent the driving behavior that does not respect lane discipline. Also all the vehicles are converted to equivalent passenger car to induce homogeneity. As per study, it is evident that the modification of existing two phase signal system can be a great alternative to facilitate better performance to the intersection. Four phase signal system implementation will not be suitable as far as the study results suggest. Existing two phase signal's cycle time modification eliminates congestion problem to a great extent and flourish other parameters such as lesser travel time and queue length etc. The model can be tested for other traffic parameter changes other than signal time modification. However, it is obvious that signal time modification can be a feasible alternative to improve performance of the studied intersection. This technique to improve intersection proficiency can be implemented to any North American intersection. Although the driving behavior parameters need to be adjusted due to the fact that there generally exists lane based homogenous or heterogeneous flow. Congestion in intersections is a waste of money and resources. The studied strategy can be applied to fabricate efficacious intersection that would lead to an efficient urban traffic system.

Acknowledgements

Thanks to Almighty Allah for His graciousness. The author expresses gratitude to Prof Dr. Md. Hadiuzzaman, Prof Dr. Hasib Mohammad Ahsan and Anup Paul for their continuous assistances. Thanks to Head of the Civil Engineering department and Commandant of MIST for the encouragement. A very special gratitude is offered to the author's parents, wife and his younger sister for their support.

References

Arasan, V.T. and Arkatkar, S.S. 2010. Microsimulation study of effect of volume and road width on PCU of vehicles under heterogeneous traffic. *Journal of Transportation Engineering*, ASCE, **136**(12): 1110-1119.

Arasan, V.T. and Dhivya, G. 2010. Methodology for Determination of Concentration of Heterogeneous traffic. *Journal of Transportation Systems Engineering and Information Technology*, **10**(4): 50-61.

Arasan, V.T. and Koshy, R.Z. 2005. Methodology for Modeling Highly Heterogeneous Traffic Flow. *Journal of Transportation Engineering,* ASCE, **131**(7): 544-551.

Arasan, V.T. and Krishnamurthy, K. 2008. Study of the Effect of Traffic Volume and Road Width on PCU Values of Vehicles Using Microscopic Simulation, *Journal of the Indian Roads Congress*, **69**(2): 133-149.

Arasan, V.T. and Vedagiri, P. 2006. Estimation of saturation flow of heterogeneous traffic using computer simulation. *Proceedings of the 20th European Conference on Modelling and Simulation.*

Banks, J.; Carson II, J.S.; Nelson, B.L.; Nicol, D.M. 2000. Discrete-Event System Simulation. *Prentice. Hall of India Private Ltd.*, New Delhi.

Dion, F., Karthik, S. and Ban, X. 2012. Evaluation of Traffic Simulation Model Use in the Development of Corridor System Management Plans (CSMPs). *California PATH Research Report*, UCB-ITS-PRR-2012-2.

Gowri, A. and Sivanandan, R. 2008. Evaluation of left turn channelization at a signalized intersection under heterogeneous traffic conditions. *Transport,* **23**(3): 221-229.

Institution of Transportation Engineers. 2010. *Manual of Transportation Engineering Studies*, ITE, 2nd edition, Washington, DC 20006 USA.

Keller, E.L. and Saklas, J.G. 1984. Passenger Car Equivalents from Network Simulation. *Journal of Transportation Engineering*, ASCE, **110**(4): 397- 411.

Mallikarjuna, Ch. and Rao, K.R. 2006. Area Occupancy Characteristics of Heterogeneous traffic, *Transportmetrica*, **2**(3): 223-236.

Mathew, V. 2017. Design Principles of Traffic Signal. *Lecture Notes in Transportation Systems Engineering,* IIT Bombay (tvm@civil.iitb.ac.in).

Naghawi, H. and Idewu, W. 2014. Analysing delay and queue length using microscopic simulation for the unconventional intersection design Superstreet. *Journal of the South African Institution of Civil Engineering,* **56**(1), 100–107.

Partha, S., Hossain, Q., Iqbal, H. and Islam, M. Passenger Car Equivalent (Pce) Of Through Vehicles At Signalized Intersections In Dhaka Metropolitan City, Bangladesh. *IATSS RESEARCH, **33**-2*

Park, B. and Qi, H. 2005. Development and Evaluation of a Procedure for the Calibration of Simulation Models. *Transportation Research Record: Journal of the Transportation Research Board, No. 1934,* Transportation Research Board of the National Academies, Washington, D.C., pp. 208–217.

PTV, V. 2012. *VISSIM 5.40 - User Manual*, PTV, Karlsruhe, Germany.

Radhakrishnan, P. and Mathew, T.V. 2011. Passenger car units and saturation flow models for highly heterogeneous traffic at urban signalised intersections. *Transportmetrica*, **7**(2): 141-162.

Sheela, A. and Kuncheria I. 2014. Traffic Simulation Model And Its Application For Estimating Saturation Flow At Signalised Intersection. *International Journal for Traffic and Transport Engineering*, **4**(3): 320 – 338.

Sunil K. and Ranjitha, J. 2013. Improvement of Traffic Operations in Congested Signalized Intersections - A Case Study In Bangalore City. *International Journal of Engineering Research & Technology (IJERT)*, **2**(7), ISSN: 2278-0181.

Sykes, P. 2010. Traffic Simulation with Paramics. *In J. Barcelo, ed. Fundamentals of Traffic Simulation.* Barcelona, Springer. pp.131-72.

Building Tomorrow's Society
Bâtir la Société de Demain

Fredericton, Canada
June 13 to 16, 2018 / *13 au 16 juin, 2018*

ESTIMATING THE USAGE OF CYCLING INFRASTRUCTURE IN CITIES WITH COLD CLIMATE

Shahriari, Siroos[1,3], Sadeghpour, Farnaz[2]
[1,2] University of Calgary, Canada
[3] siroos.shahriari@ucalgary.ca

Abstract: Cycling has gained increased attention among both researchers and public planners in the last two decades as a sustainable means of transportation. Compared to users of other modes of transportation, cyclists are more exposed to weather conditions. For example, harsh weather conditions such as cold temperatures or rain negatively affect an individual's decision to use a bicycle for transportation. As a result, cycling infrastructure usage changes throughout the year because of highly variable weather conditions. On the other hand, in recent years North American cities try to encourage cycling through the development of cycling infrastructure networks. For that, municipalities require estimates of infrastructure network usage throughout the entire year to make decisions about investments and maintenance costs. However, collecting data for all cycling infrastructure throughout the year is not economically feasible. The objective of this paper is to develop a model capable of estimating changes in cycling infrastructure usage for different months of the year based on observations from only one month. An estimation model will be developed using Generalized Estimating Equations (GEE). Changes in cycling frequency for each month will be estimated relative to a reference month. The usage change will be estimated based on weather variables, namely temperature, precipitation (i.e. rain, snow, hail), collected snow on the ground, wind speed and the number of sunlight hours in a day. As the model was developed, it was noticed that the effect of weather conditions on commuter cyclists and reactional cyclists is not the same. As a result, two separate sub-models will be developed. The first sub-model will be based on weekday usage data which will evaluate the impact of weather conditions on commuter cyclists. The second sub-model will be based on weekend usage to evaluate the impact of weather condition on utilitarian (non-commuter) cyclists. Additionally, it will be examined to what extent the above-mentioned weather variables are accountable for estimating cycling infrastructure usage. The developed model can help Cities and municipalities in decision making on investments, maintenance and modifications to the cycling infrastructure.

1 INTRODUCTION

In the last few decades, transportation planning efforts in North America has aimed to increase the share of active modes of transportation such as cycling and walking due to its benefits over the vehicular mode of transportation. While walking is limited to only close destinations, cycling offers a greater access range to users. Cycling is also an environmentally friendly mode of transportation, is cost-efficient, and increase in cycling will lead to higher social interaction and livelihood on streets. Since environment, social and economic are the three pillars of sustainability (Bell and Morse, 2008), increasing the share of cycling can help cities to move toward more sustainable transportation system (Buehler and Pucher 2012). To encourage cycling, it is important to improve infrastructure network (i.e. providing better conditions for cycling infrastructures) and expanding the infrastructure network (i.e. building new road sections).

Due to limited budget available within municipalities, planners and decision makers use infrastructure usage data as an evidence for identifying the optimal place to spend the money to improve the infrastructure network more efficiently. While using annual infrastructure usage is more reliable for decision making, currently municipalities use short count usage (partial data) which is collected for a limited period of the year (i.e. one month). Collecting infrastructure usage for an entire city throughout the year is not economically feasible. Since data collection requires either numerous bicycle counters installed on every road section or manpower assigned to each road section for the entire year. Hence, currently, short count usage data is used and it implement limitations for decision making since short count data does not provide any information about the usage variation throughout the year. Infrastructure usage variation throughout the year is needed especially in cities with cold climate such as most Canadian cities. There is a debate on whether keeping or removing cycling infrastructure during cold months of the year. It is assumed that cycling infrastructure will remain unused or will have a low usage during cold months.

The objective of this study is to develop a model with the goal of estimating cycling infrastructure usage through the year using observations from only one month. Literature suggests that many variables affect cycling. These variables can be categorized into five main group namely, cyclists' characteristics, attitudes toward cycling, cycling infrastructure, network connectivity and built environment and weather conditions. This study will focus on variables from the last group referred to as weather conditions. All the variables that affect infrastructure usage are constant for a specific infrastructure during a year except weather variables. Hence the variation in usage is attributed to weather variables. The weather variables examined in this study are temperature, precipitation (i.e. rain, snow, hail), snow on the ground, wind speed and sunlight hours in a day. This study first evaluates the impact of above-mentioned weather variables on cycling infrastructure usage and then estimates the change in usage during a year using weather variables. Generalized Estimated Equation (GEE) is used to develop the model since data used in this study was correlated over time. The model presented in this study estimates usage throughout the year and consequently the annual usage and can be used as a guide for decision making about funding allocation in municipalities.

2 LITERATURE REVIEW

In recent years, two studies in particular investigated cycling infrastructure usage trends and a number of studies developed estimation models to estimate infrastructure usage. Investigating cycling infrastructure usage trends helps us to better understand usage variation and the possible causes of the variation. A better understanding of cycling usage trends enables us to develop more accurate estimation models.

2.1 Infrastructure Usage Trends

Studies have shown that cycling infrastructure usage varies throughout the year and showed that cycling infrastructure usage during a year was affected by weather conditions (Lindsey et al., 2013; Miranda-Moreno et al. 2013). One study from Minneapolis found that traffic volumes for cyclists and pedestrians (non-motorized usage) varied over time. The usage varied depending on the month and day of the week. Usage variation trends were similar for different locations with different magnitudes. (Lindsey et al., 2012). Another research studied cycling patterns in five North-American cities namely, Montreal, Ottawa, Portland, San Francisco, and Vancouver. It was found that that monthly usage in cities with cold climates decreased more compared to usage in warmer cities. Cycling usage pattern was categorized into commuter, mixed commuter, recreational, and mixed recreational. The study showed that the impact of weather on recreational and commuter cyclists was not the same; infrastructure usage used by recreational cyclists decreased more compared to the usage of infrastructure used by commuter cyclists (Miranda-Moreno et al. 2013).

2.2 Weather Impact on Usage

Several studies have been published on the impact of weather conditions on cycling. These studies indicated that weather variables significantly impacted cycling infrastructure usage. Harsh weather conditions, such as days with extremely low or high temperatures and high precipitation, reduced the number of cyclists.

Temperature: Temperature was positively correlated with cycling frequency according to relevant studies. Different studies have confirmed that days with higher temperatures caused more people to cycle (Bergström and Magnusson, 2003; Buehler 2012; Flynn et al., 2012; Saneinejad et al., 2012; Brandenburg et al. 2007). A study surveyed employees of four companies in two Swedish cities showed a difference in mode choice between seasons. People cycled more during summer compared to winter. Instead of cycling, the majority of people used their cars in winter. Moreover, distance became a more important factor when choosing the mode of transportation during winter. For long distances, the number of cyclists remarkably dropped as distance increased in winter. The study showed that people almost never used bicycle for trips of more than 10 km in winter (Bergström and Magnusson, 2003). Similarly, Buehler (2012) showed that the likelihood of commuting to work by bicycle in summer was 73% more than the likelihood of using a bicycle to commute in winter. Another study from Vermont showed that people tended to cycle more on warmer days. The regression likelihood model presented in the study showed that an increase in one degree of Fahrenheit resulted in a 3% increase in the likelihood of cycling (Flynn et al., 2012). Similarly, another study from Toronto showed that cold temperatures negatively affected people's tendencies to cycle. In that study, it was found that using a bicycle was only sensitive to temperatures below 15° C (Saneinejad et al., 2012). Another study investigated the impact of weather conditions on different types of cyclists. With, using data from a suburban recreation area in Vienna, the study found that recreational cyclists were more affected by weather conditions compared to commuter cyclists. Consequently, it is possible that recreational cyclists checked the weather conditions beforehand, and as a result, they avoided cycling in harsh weather (Brandenburg et al. 2007).

Sunlight: Studies have shown that individuals tended to cycle during daylight hours rather than in the dark (Osberg et al., 1998; Rodgers, 1995; Spencer et al., 2013). A study that observed cyclists in Paris and Boston showed that the majority of cyclists, cycle during daylight hours. 65% and 76% of cyclists were observed during daylight hours in Paris and Boston, respectively (Osberg et al., 1998). Respondents in another study mentioned that during some parts of the year, they could not cycle safely due to insufficient daylight hours. Also, some respondents indicated that the effect of weather variables, such as rain, could compound the negative effects of cycling in the dark, which further prevented them from cycling (Spencer et al., 2013).

Wind: Wind speed was found to have a negative impact on a cycling infrastructure usage. Researchers believe that the main reason for the negative impact is that wind makes the temperature feel colder. A study from Toronto found that wind speed negatively affected cyclists and pedestrians. The magnitude of the impact of wind was twice on cyclists compared to pedestrians (Saneinejad et al., 2012). Another study surveyed 24 adult bicycle commuters from Vermont, and the results showed that wind could also have a positive impact on cycling when it was at a cyclist's back (Spencer et al., 2013).

Precipitation: According to relevant studies, precipitation is one of the preventing factors for cyclists; the number of cyclists drops on days with precipitation (Flynn et al., 2012; Nankervis, 1999; Saneinejad et al. 2012). A case study in Vermont found that people cycle less during rainy days. Among the weather variables considered in the study, rain was found to have the most negative effect on cycling. Days without rain had usage nearly doubled compared to rainy days. Each inch of snow resulted in a 10 % decrease in the likelihood of bicycle commuting (Flynn et al., 2012). Another study from Melbourne showed that rain was the most important preventing factor for cyclists as it caused more than half of the people surveyed (50.7%) to choose not to ride a bicycle on rainy days. Similarly, according to Saneinejad et al. (2012), precipitation in the form of rain and showers negatively impacted cyclists. However, rain was not always considered a negative factor for cyclists. Some respondents in Vermont described rain as refreshing and helpful (Spencer et al., 2013).

Snow on the ground: Snow on the ground also considered being one of the factors that prevented cyclists from cycling. According to Bergstrom and Magnusson (2003), road surfaces that were not cleared of snow was the most important road condition for mode choice according to the opinion of the surveyed cyclists.

2.3 Usage Estimation Models

Studies estimated cycling infrastructure usage by developing models using short count data (Nordback et al., 2013; El-Esawey, 2014; El-Esawey et al., 2013; El-Esawey and Mosa, 2017). These studies estimated cycling infrastructure usage by creating daily, monthly, and seasonal factors. Cycling infrastructure usage was estimated by multiplying the known short-term count by those factors.

Recent studies estimated cycling infrastructure usage using short count duration factors. The factors were calculated by dividing year usage to short-term count usage. However when estimating the usage for years other than the year in which factors were developed on, their estimation accuracy significantly drops since the variation in weather conditions is different in each year. While using short count factors is limited for only one year, this study estimates cycling infrastructure usage with considering the effect of weather variables on usage which estimates the usage with better accuracy and can be used for any year with known weather data.

3 METHODOLOGY

The objective of this study is to develop a model capable of estimating cycling infrastructure usage variation using short count data. To develop the model, potential variables that impact usage over a year were identified. Previous studies suggested that weather variables had a significant impact on cycling infrastructure usage variation (Lindsey et al., 2012; Miranda-Moreno et al. 2013). Investigating the association between temperature and one infrastructure section usage, confirmed the speculation about the effect of weather conditions on infrastructure usage. The magnitude of the impact of weather variables on cycling infrastructure usage was determined, and the usage was estimated based on those variables. To develop the model cycling infrastructure usage data from the city of Calgary was used. The Transportation Department at the City of Calgary provided daily cycling infrastructure usage throughout the year from 25 counters mainly located in downtown area. This source of data included 15750 data points; each data point represented specific road section usage for a specific day of the year. Weather data was provided by Canadian Government and (http://www.sunrise-and-sunset.com). Weather data included mean temperature, precipitation, cumulated snow on the ground, wind speed and sunlight hours in the day. Data was split into two sections. The majority of the collected data (90%) was used to develop the model, and a small portion (10%) was used as test data to validate the model and to determine the model accuracy for estimating cycling infrastructure usage.

3.1 Model Design

This study estimated the usage of cycling infrastructure compared to a reference month with known usage (the month with observation). The estimation was conducted with evaluating the change in usage as a result of a change in weather variables. The usage change contained five components. Each component represented the impact of one weather variable on usage. Total change in usage was computed by adding up the changes in usage resulted from the effect of each weather variables on usage. The equation can be written as follows where ΔF represents the usage change.

[1] $\Delta F = \beta_1 (\Delta \text{ Temp}) + \beta_2 (\Delta \text{ Prec}) + \beta_3 (\Delta \text{ Wind}) + \beta_4 (\Delta \text{ Snow}) + \beta_5 (\Delta \text{ Sunh}) + E$

Where β_1 to β_5 represent the coefficients of weather variables, weather variables (i.e. Temp, Prec, Wind, Snow, Sunh) represent the mean value of each weather variable for each month and *E* represents estimation error. Generalized Linear Model (GLM) is a general statistical approach that generally is used to estimate β. This study uses data that was collected with counters installed on cycling infrastructure sections. Each infrastructure section is a part of infrastructure with different characteristics from other parts of infrastructure which is separated from other sections by intersections. Cycling infrastructure usage was measured at successive times throughout the year. When data was collected from the same units at successive points over time, repeated observations are correlated. Since GEE is capable to estimate the coefficients when observations are correlated, Generalized Estimating Equations (GEE) was chosen over GLM analysis methods such as multiple regression, to analyze the data.

In GEE, quasi-likelihood estimates of *β's* are computed from the maximization of the normality-based log likelihood without assuming that the observations are normally distributed. In this study, the parameters β are coefficients for the weather variables, which are estimated by solving the following equation:

$$[2]\ \sum_{i=1}^{M} \left(\frac{\partial \mu_i}{\partial \beta}\right) V_i^{-1} (Y_i - \mu_i) = 0$$

where M is the number of infrastructure sections, Y_i represent the vector of observations for each infrastructure section throughout the time, μ_i called the mean vector of Y_i which is the estimated value of Y_i as a function of input variables (observations) and β parameters, and *V* is the covariance matrix of observations, which takes into the account how changes in one variable is associated with the changes in other variables. The covariance matrix includes three elements:

$$[3]\ V_i^{-1} = \phi\, A_i^{1/2} R_i A_i^{1/2}$$

where ϕ is an overdispersion parameter which takes into the account the difference in variability of dataset and model estimation; $A_i^{1/2}$ is an $n \times n$ diagonal matrix with the square root of variances of observation where *n* represents the number of observations; and R_i is an $n \times n$ correlation matrix. The correlation between observations is carried out by using a correlation matrix in GEE. Different types of correlation matrixes exist such as independent, exchangeable, auto-regressive and unstructured. Independent matrix is used when observations are independent. Exchangeable is used when all observations are equally correlated. Auto regressive is used when the correlation is known to decline through time and unstructured is used when there is no specific correlation between observations.

The GEE model is first fitted with computing initial estimates of β's, for example, using a naive linear regression. After computing initial β estimates, GEE estimates the dispersion parameter from residuals and computes the correlation matrix based on residuals. Then, the working covariance matrix is computed and β's are estimated again. These steps are continued until convergence of β's is obtained.

3.2 Model Development

The model estimated the change in usage of infrastructure for different months of the year relative to a reference month (a month with available usage count data), which in this study is September. A closer look at the daily usage revealed that there was a difference in usage during weekdays and weekends. Figure 1 shows daily usage of one cycling infrastructure section during weekdays and weekends. For example, as it can be seen in Figure 1, during three different periods of the year 2016 (early in the year, middle of the year and end of the year) weekend and weekday usages were different in values (A, B and C).

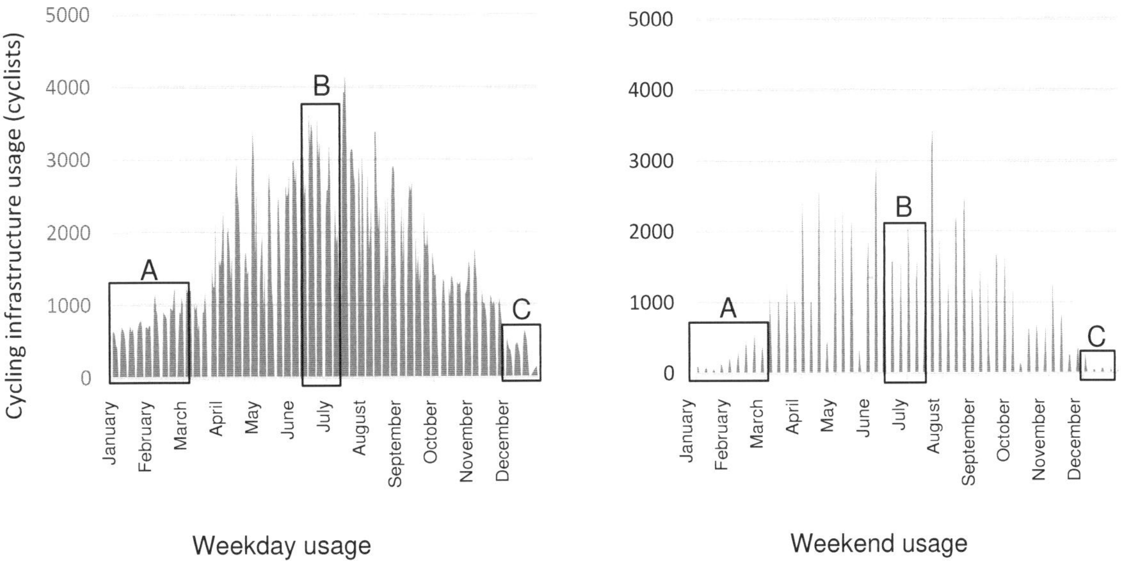

Figure 1: Daily infrastructure usage trends during weekdays and weekends

Due to differences in weekday and weekend usage, two separate models were developed. The first model (Commuter Model) was developed based on weekday usage, and the second model (Utilitarian Model) was developed based on weekend usage. Since cyclists who cycle on weekdays are generally commuter cyclists combined with utilitarian cyclists, commuter model evaluates the impact of weather conditions on a combination of commuters and utilitarian cyclists. The utilitarian model focuses on only utilitarian cyclists since in general, cyclists who cycle during weekends have only utilitarian purposes. They cycle for purposes such as shopping, going to restaurants, or for exercise, all of which fall into the category of utilitarian purposes. This study used auto regressive structure for both models since it showed the best fit on data compared to other correlation structures.

The model estimates usage based on only weather variable, hence other factors that affect cycling were removed to increase model accuracy. For example, national holidays were excluded from the database. To analyze data, the change in usage for each day of the week in each month was compared to the same day of the week in September to eliminate the impact of the day of the week. Also, since cycling infrastructure usages collected in this study had different standard deviations for different infrastructure sections, logarithmic data transformation was applied. Consequently, the logarithmic value of cycling infrastructure usage was used to develop the model.

Effect of weather variables on people's decision to cycle is not constant over a year. The effect of preventing factors, such as precipitation, are assumed to be greater in cold months due to the compounding effect of rain and cold temperatures. The variation in the magnitude of the effect of the variables makes it necessary to have separate models for different time periods. (Miranda-Moreno et al. 2013) found that hourly and weekly usage patterns were consistent in different cities with different characteristics while monthly usage patterns were different due to the impact of weather variables. Hence, monthly periods were selected for developing sub-models. Therefore, 11 sub-models were developed. Each sub-model evaluated the effect of weather conditions on changes in cycling frequency for each month of the year compared to September.

The commuter model was developed using weekday usage and weather data. Table 1 provides the results of the 11 sub-models of the commuter model. The coefficients represent the impact of each weather variable on the log value of infrastructure usage. Some of the coefficients were not significant nor rational. For example, Snow on the ground in winter can not positively impact cycling infrastructure usage. Therefore, those coefficients were removed from the model.

Table 1: Commuter model coefficients

Month	Temp	Total Precipitation	Snow on Ground	Wind Speed	Hours of Sunlight
January	0.021	-0.019		-0.003	0.097
February	0.020	-0.029		-0.005	0.026
March	0.017	-0.032	-0.008	-0.006	
April	0.025	-0.017	-0.012	-0.009	
May	0.020	-0.033		-0.006	
June	0.018	-0.011		-0.005	
July	0.027	-0.009		-0.006	
August	0.022	-0.013		-0.007	
September	-	-	-	-	-
October	0.015	-0.026	-0.052	-0.003	
November	0.014	-0.017	-0.021	-0.006	
December	0.017	-0.018	-0.035	-0.004	

All the coefficients presented are significant at p-value <0.05.

Table 2 provides the results obtained from the analysis on weekend usage for the 11 sub-models. The utilitarian model had fewer significant coefficients compared to the weekday model because it had a smaller sample size.

Table 2: Utilitarian model coefficients

Month	Temp	Total Precipitation	Snow on Ground	Wind Speed	Hours of Sunlight
January	0.034	-0.023		-0.006	
February	0.032	-0.014		-0.015	0.016
March	0.032	-0.046	-0.016		0.08
April	0.040	-0.015	-0.597	-0.009	
May	0.030	-0.023		-0.012	
June	0.023	-0.033		-0.002	
July	0.027	-0.028		-0.002	
August	0.028	-0.016		-0.01	
September	-	-	-	-	-
October	0.025	-0.015	-0.092	-0.001	
November	0.039	-0.044			
December	0.023	-0.023	-0.017	-0.012	

All the coefficients presented are significant at p-value <0.05.

3.3 Model applicability

To find which model had the best goodness of fit for estimating usage, both models were applied to estimate infrastructure usage during weekends and weekdays. To evaluate models accuracy, the usage for each month was estimated using September usage and weather variables. The mean value of percentage error (MAPE) was calculated as:

$$[4]\ \text{MAPE}=\frac{1}{N}\sum_{i=1}^{N}\left|\frac{Actal\ cycling\ frequency-Estimated\ cycling\ frequency}{Actal\ cycling\ frequency}\right|$$

Where *N* represents the number of observations used for validation which in this case is the number of months used for validation for each infrastructure section. The value of MAPE was calculated for both models. Months with an average usage of below 100 cyclists/month were excluded from the model. The small difference between the estimated and actual usage in those months led to a large MAPE value because of the small actual usage. This increased total MAPE value unrealistically. For example, for an infrastructure section with actual usage of 5 cyclists in a month and the estimated usage of 9 cyclists in month, the difference of model estimation and actual usage was only 4 cyclists while the estimation error was 80%.

Infrastructures usage during weekends was estimated with both models and the results were compared with actual usage. The utilitarian model showed a lower MAPE value compared to commuter model for estimating weekends usage for all infrastructure sections. Hence utilitarian model was selected to estimate infrastructure usage on weekends. Both models were applied to estimate weekday usage. By calculating MAPE values for two models it was noticed that for some infrastructure sections the commuter model had a better estimation result while for some other sections the utilitarian model better estimated the weekday usage.

In general, each infrastructure section has different types of users during weekdays and weekends. Weekend cyclists are mainly utilitarian cyclists, while weekday cyclists are a combination of commuter and utilitarian cyclists. The portion of commuter and utilitarian cyclists is different for different infrastructure sections. To discover more about the characteristics of infrastructure section users, a user type ratio (UTR) was defined as follows:

$$[5]\ \text{UTR}=\frac{Average\ number\ of\ cyclists\ on\ weekends}{Total\ number\ of\ cyclins\ t\ on\ weekdays\ and\ weekends}$$

A high UTR for an infrastructure section represents high usage on weekends compared to total usage. Usage on weekends is mainly attributed to utilitarian cyclists. Hence, infrastructure section with a high UTR contains a high portion of utilitarian cyclists among their users and mainly used for utilitarian purposes. By calculating UTR for different infrastructure sections, it was noted that sections with high UTR were mainly located at recreational locations. Also, looking at hourly trends of infrastructure sections usage, we noticed that usage trends of different sections were not the same. Sections with mainly commuter users (low UTR) had higher peak usage in the morning and afternoon and lower usage at midday during weekdays compared to infrastructure sections with utilitarian users (high UTR). While sections with utilitarian users had higher midday usage compared to sections with mainly commuter users during weekends. For the next step, we sorted infrastructure sections based on their UTR value. The result showed that for sections with UTR value more than 0.1,1 the utilitarian model gave a better fit while for sections with UTR value lower than 0.11 the commuter model gave a better estimation result. Figure 2 shows model selection procedure.

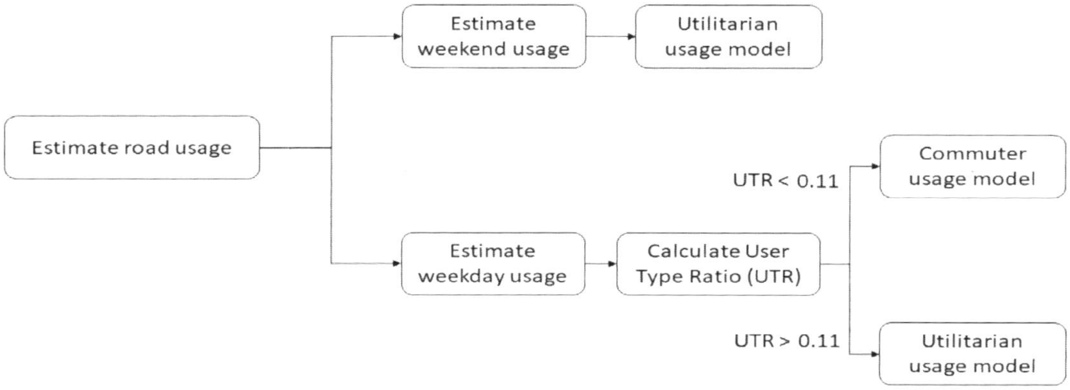

Figure 2: The model selection process

4 VALIDATION

To validate the model, the estimation model was used to estimate usage in 2017 for two infrastructure sections based on their September usage. Selected infrastructure sections had different infrastructure types and different locations that covered a variety of infrastructure sections with different characteristics. The results were compared with actual usage counted by bicycle counters. Table 3 provides details about selected infrastructure sections.

Table 3: Overview of selected infrastructure sections characteristics

Infrastructure section	Type of Infrastructure	Location	UTR	Infrastructure Neighbourhood
Parkdale	Off-road pathways	Outside downtown	0.1592	Recreational
5 Street North of 15 Ave	On-road Sep. physically	Downtown	0.1814	Commercial

Usage on weekends and weekdays were estimated separately based on September weekend and weekday usage. Since UTR values for selected infrastructure sections were higher than 0.11, the utilitarian model was used to estimate both weekend and weekday usage. The total monthly usage was calculated by adding the weekend and weekday usage together. Table 4 shows the observed and estimated annual usage and the estimation error. The estimated cycling frequency and actual cycling frequency are shown in Figure 3.

Table 4: Model result

Infrastructure section	Actual Annual Usage	Estimated Annual Usage	Annual Estimation Error	MAPE
Parkdale	25344	22894	0.096	0.192
5 St and 15 Ave SW	12265	12094	0.013	0.195

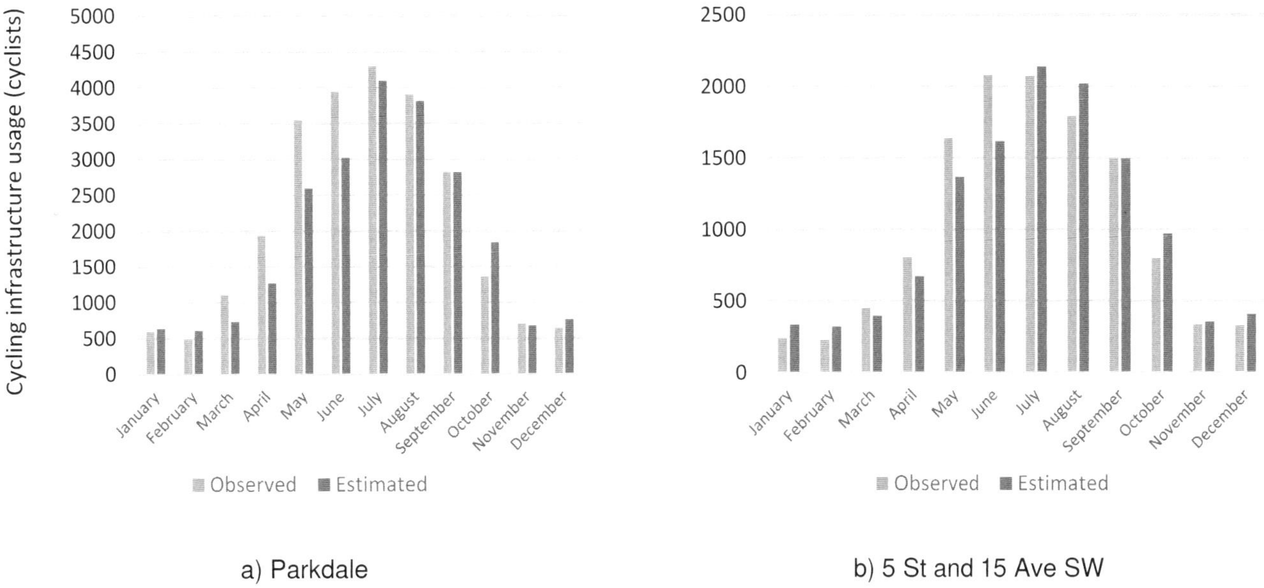

Figure 3: Comparison of the model result and observed usage

5 CONCLUSION

This study presented a framework for developing a model capable of estimating cycling infrastructure usage throughout the year. The results of this study showed the effect of different weather variables on cycling infrastructure usage. Among the weather variables, temperature and sunlight hours showed a positive impact on cycling infrastructure usage, while precipitation, collected snow on the ground, and wind speed showed a negative impact on cycling infrastructure usage. Also, the results indicated that the magnitude of the impact of these variables was not constant during the year. The impact of the variables with a negative impact, such as collected snow on the ground and precipitation, was generally higher during cold months of the year compared warm months. Temperature and precipitation were found as the two main weather variables that impact cycling as they showed higher impact on usage compared to other variables and their impact was significant for all months of the year.

The impact of the weather variables was evaluated separately for weekday and weekend usage because usage in weekends and weekdays were different in values. While in general weekday usage is attributed to commuter cyclists, usage on weekends is attributed to utilitarian cyclists. As the results of the model showed, the impact of weather variables was generally higher for weekend usage compared to weekday usage. The higher magnitude of the impact of weather variables on weekend usage indicated that utilitarian cyclists were more affected by weather conditions than commuter cyclists which is in line with (Miranda-Moreno et al. 2013; Brandenburg et al. 2007). The validation showed that when estimating usage for different months of the year, the model had an average error of 19.35%. This error represented the accuracy of the model in terms of estimating usage variation for different months of the year. When estimating the annual usage of infrastructure, the model had an average error value of 5%.

The developed estimation model in this study allows the usage throughout a year to be estimated based on the effect of weather conditions on cyclists. According to Köppen climate classification, climate can be classified into 5 groups namely, tropical, arid, temperate, cold, and polar. Calgary is a city with a cold climate. The effect of weather conditions on cyclists is not the same for people who live in different climate types. For example, people who live in cold cities are more resistant to cold weather compared to people who live in warmer cities; hence, the effect of temperature on cycling is not the same in different climate types. As a result, since the model presented in this study used data from a city with a cold climate, the model can be used to estimate cycling infrastructure usage for cities with a similar climate to Calgary.

References

Bell, S. and Morse, S. 2008. *Sustainability indicators: measuring the immeasurable?.*

Buehler, R. and Pucher, J. 2012. Cycling to work in 90 large American cities: New evidence on the role of bike paths and lanes. *Transportation*, ***39***(2): 409–432.

Brandenburg, C. and Matzarakis, A. and Arnberger, A. 2007. Weather and cycling – a first approach to the effects of weather conditions on cycling. *Meteorological applications,* **14**(1), 61–67.

Desert, B. W. A. and Steppe, S. and Tundra, E. T. P. Köppen climate classification.

Dill, J. and Carr, T. 2003. Bicycle Commuting and Facilities in Major U.S. Cities: If You Build Them, Commuters Will Use Them. *Transportation Research Record*, ***1828***(1): 116–123.

Esawey, M.El. and Lim, C. and Sayed, T. and Mosa, A.I. 2013. Development of Daily Adjustment Factors for Bicycle Traffic. *Journal of Transportation Engineering*, ***139***(8): 859–871.

Esawey, M.El. and Mosa, A.I. 2017. Determination and Application of Standard K Factors for Bicycle Traffic Determination and Application of Standard K Factors for Bicycle Traffic. *Transportation Research Record: Journal of the Transportation Research Board,* **2527** (2015): 58-68.

Flynn, B.S. and Dana, G.S. and Sears, J. and Aultman-hall, L. 2012. Weather factor impacts on commuting to work by bicycle. *Preventive Medicine*, ***54***(2): 122–124.

Lindsey, G. and Chen, J. and Hankey, S. 2013. Adjustment Factors for Estimating Miles Traveled by Non-Motorized Traffic. *Transportation Research Board 92nd Annual Meeting*, *6351*: 2–22.

Magnusson, R. 2003. Potential of transferring car trips to bicycle during winter. *Transportation Research Part A: Policy and Practice*, ***37***(8): 649–666.

Miranda-moreno, L.F. 2013. Classification of bicycle traffic patterns in five North American Cities. *Transportation Research Record: Journal of the Transportation Research Board,* (2339): 68-79.

Nankervis, M. 1999. The effect of weather and climate on bicycle commuting. Transportation Research Part A: Policy and Practice, **33**(6), 417–431.

Nordback, K. and Marshall, W. and Janson, B. and Stolz, E. 2013. Estimating annual average daily bicyclists: Error and accuracy. *Transportation Research Record: Journal of the Transportation Research Board*, (2339): 90-97.

Osberg, J.S. and Stiles, S.C. and Asare, O.K. 1998. Bicycle safety behavior in Paris and Boston. *Accident Analysis and Prevention*, ***30***(5): 679–687.

Rodgers, G. B. 1997. Factors Associated with the Crash Risk of Adult Bicyclists. *Journal of Safety Research*, **28**(4), 233–241.

Saneinejad, S. and Roorda, M.J. and Kennedy, C. 2012. Modelling the impact of weather conditions on active transportation travel behaviour. *Transportation Research Part D: Transport and Environment*, ***17***(2): 129–137.

Spencer, P. and Watts, R. and Vivanco, L. and Flynn, B. 2013. The effect of environmental factors on bicycle commuters in Vermont: Influences of a northern climate. *Journal of Transport Geography*, **31**, 11–17.

Building Tomorrow's Society
Bâtir la Société de Demain

Fredericton, Canada
June 13 to 16, 2018 / *13 au 16 juin, 2018*

PROPOSED MODEL OF ADOLESCENT RISKY DRIVING: INTEGRATING SOCIAL STRESSORS AND PHYSIOLOGICAL STRESS AS PREDICTORS OF DRIVER BEHAVIOUR

Sutherland, Jessica E.[1,3], Hassein, Udai[2], Easa, Said[2] and Day, David M.[1]
[1] Department of Psychology, Ryerson University, Toronto, Canada
[2] Department of Civil Engineering, Ryerson University, Toronto, Canada
[3] jessica.sutherland@psych.ryerson.ca

Abstract: Peer passengers increase the likelihood of fatal collisions among adolescent drivers. Previous research suggests that social acceptance is especially important to adolescents and that peer presence can induce physiological stress, compromising inhibitory control over risk-taking. Despite the evidence for this peer effect, an understanding of the specific mechanisms of how and when peer passengers can influence a young person's driving behaviour is lacking in the literature. Examining the effects of peer influence under different social conditions in a driving simulator will further our understanding of the relationship between young passengers and driving outcomes. Further, physiological stress has not been explored for its effects on driving in young drivers. This study tested the effects of a social stressor on driving behaviour in a driving simulator, including acceleration, speeding, and lateral displacement. Male and female participants between the ages of 16 and 25 were randomized to control and social stressor conditions (peer rejection or acceptance). Participants' cardiovascular stress physiology was measured while completing tasks either alone or partnered, and during a simulated drive. A significant interaction effect emerged with condition, gender, and physiological stress, as well as a significant main effect for condition on acceleration and lateral displacement. Assessing peer effects under different social circumstances while driving contributes to the growing literature on the connection between peer passengers and driving behaviour in young people.

1 Introduction

Motor vehicle collisions are a significant health concern during adolescence. The leading cause of death among adolescents and young adults in North America is unintentional injury, the majority being the result of motor vehicle collisions (Centers for Disease Control and Prevention, 2015; Statistics Canada, 2017). Peer passengers strongly affect young drivers' behaviour by increasing risky driving behaviours and motor vehicle injuries. However, there is currently limited research exploring the social mechanisms that precede risky driving behaviour when young people are driving with a passenger. By comparing different social circumstances in which risky driving might occur (i.e. feeling socially insecure or secure) and measuring individual differences in variables known to influence risky behavior (physiological stress), this study will test currently unexplored connections between peers and risky driving.

This paper presents preliminary analyses of peer passenger effects on driving outcomes and physiological stress as part of an ongoing study on adolescent risky driving behaviour. Following a literature review of peer effects on adolescent and young adult risk-taking behaviour, the experimental methodology, procedures, measures, and results are described. A discussion of key findings, implications, and plans for future research follows.

2 Literature Review

2.1 Peer Effects on Adolescent and Young Adult Risk Behaviour

Despite the fact that adolescents do not always engage in more risk-taking behavior than adults or children (Defoe, Dubas, Figner, & Aken, 2015), they are significantly more likely to do so when they are in the presence of peers compared to when they are alone (Chein, Albert, O'Brien, Uckert, & Steinberg, 2011; Gardner & Steinberg, 2005; O'Brien, Albert, Chein, & Steinberg, 2011). Peer presence can be a source of emotional arousal and reward, which can interfere with decision-making (Cowell, 2013; Zelazo & Carlson, 2012). Peers are believed to trigger affective arousal due to increasing neural sensitization to environmental social stimuli in adolescence, which can encourage risk-taking to achieve social goals, such as social dominance (Forbes & Dahl, 2010). Risk-taking activates neural reward pathways in adolescents, with peer presence enhancing this rewarding effect (Chein et al., 2011). Young men, in particular, engage in more risk-taking behaviour when there is a social gain to be had, such as popularity (Voroboyev, Kwon, Moe, Parkkola, & Hämäläinen, 2015) and when the peers explicitly promote risk-taking (Centifanti, Modecki, MacLellan, & Gowling, 2016).

2.2 Peers and Physiological Stress

Social evaluation is a stress-inducing experience in adolescence. After social stressors, adolescents experienced increased physiological stress, as measured by cortisol (Stroud et al., 2009; van den Bos, de Rooji, Miers, Bokhorst, & Michiel Westenberg, 2014) and skin conductance (Somerville, Jones, Ruberry, Dyke, Glover, & Casey, 2013), when compared to younger children. Neuroimaging research also finds decreased activation of neural structures involved in self-regulation and increased activation in affective structures (Sebastian, Viding, Williams, & Blakemore, 2011; Somerville et al., 2013) as well as stronger feelings of self-consciousness and embarrassment (Somerville et al., 2013) when adolescents believed they were being observed by a peer while completing a task. This suggests that social evaluation is physiologically stressful and emotionally arousing, potentially compromising executive functioning (Zelazo & Carlson, 2012).

Fear of social evaluation may be due to concerns about possible rejection. After experiencing social rejection, adolescents demonstrate an attentional bias for social cues that might help them achieve social acceptance, such as peer norms (Jones et al., 2014; Park & Baumeister, 2015; Peake, Dishion, Stromshak, Moore, & Pfeifer, 2013). A possible mechanism that might facilitate re-acceptance is compliance with norms (i.e. risky behaviour) communicated by the rejecting peer. Neuroimaging findings suggest that adolescents who are especially sensitive to rejection (as measured by self-reported rejection sensitivity) may be most susceptible to peer influence (Falk et al., 2014).

2.3 Peer Passengers and Driving Behaviour

Risky driving, or driving behaviours that increase the likelihood of motor vehicle accidents, are more common when young drivers have peer passengers. Driving with peers increases the likelihood of an accident due to distracted driving, excessive speed, and driver error (Ouimet, Pradhan, Brooks-Russell, Ehsani, Berbiche, & Simons-Morton, 2015; Curry, Mirman, Kallan, Winston, & Durbin, 2012; Simons-Morton et al., 2011; Simons-Morton, Lerner, & Singer, 2005). A peer passenger is also associated with reduced visual scanning of the road, diverting the driver's attention from changing road conditions and signage (Pradhan, Li, Bingham, Simons-Morton, Ouimet, & Shope, 2014).

A driver's gender also influences the effect of peer passengers. While adolescent males and females both engage in riskier driving with same-sex passengers compared to driving alone, adolescent males do this more often than do females (Simons-Morton et al., 2005). Moreover, collisions with adolescent female drivers are more likely to be preceded by an interior distractor (e.g. looking around at passengers) while collisions with male adolescent drivers are more likely to be preceded by exterior distractors, such as other traffic, even with passengers (Curry et al., 2012). Male adolescents tend to experience single-vehicle and head-on collisions more often, while female adolescents experience collisions on the left and right sides of the car (Bingham & Ehsani, 2012). These findings suggest that gender is a key predictor of different driving-related behaviors that increase injury risk and collisions.

3 Justification for Current Study and Research Questions.

While previous findings suggest that emotional or physiological arousal and social evaluation are possible mechanisms through which peers may influence behavior, no studies have experimentally manipulated these variables in the context of driving. Directly addressing this identified gap, this study integrates possible mechanisms of peer influence (i.e., social acceptance or rejection and physiological stress) into a proposed model and tests their effects on driving behaviours. The primary research questions include: (1) whether accepting or rejecting social conditions have different effects on risky driving outcomes, and (2) whether driving in a peer's presence produces a physiological stress. Gender is considered an exploratory variable in both questions.

4 Methods and Data Collection

4.1 Driving Scenario

The driving scenario was made using the STISIM driving simulation software with a rural two-lane highway design. The drive was 5 km and driven from south to north during good weather and visibility. Objects and road signs were placed at appropriate locations throughout the drive. The drive began at Point A (0 m) with traffic lights placed at Intersections B (600 m), C (1,100 m), and D (2,100 m), where a yellow light appeared for 4 s prior to a red light for 20 s. Between Point A and Intersection D, driving segments were straight drives with speed limits of 90 km per hour. Segment D-E (2,100 – 3,950 m) had one vertical curve length of 300 m at 70 km per hour. Intersection E occurred 3,950 into the drive and was a green light, requiring the participant to make a left-hand turn into oncoming traffic. The oncoming vehicles had gaps ranging from 0.5 - 2.5 s between them when the participant could choose to turn left or wait until opposing traffic had passed. Finally, segment E-F (3,950 – 5,000 m) had two horizontal curve radii of 300 m each at 70 km per hour. The first horizontal curve to the left began at 4,000 m and the second horizontal curve to the right began at 4,350 m. The scenario ended at Point 5 (5,000 m). The volume of opposing traffic represented approximately 350-400 vehicles per hour on the road. Passing was not permitted and no impeding vehicles were used during the simulation, because we were primarily interested in the influence of interior stimuli (i.e. passengers) on driver behaviour.

The primary outcome variables (driving behaviours) were measured via speeding, acceleration, and lateral displacement. Speed and acceleration were selected because they are frequently implicated as variables that precede collisions with young drivers (e.g. Curry et al., 2012; Ouimet et al., 2015). Lateral displacement was selected because no prior experimental research has measured the extent to which young drivers deviate from the midline of the road. Prior evidence suggests that peer passengers reduce visual scanning, which may distract drivers from attending to their vehicle's spatial location on the road (Pradhan et al., 2014). The drive was segmented into straight driving and curved sections between intersections (segments B-C, C-D, D-E, and E-F) and intersections (segments C, D, and E) for analyses. Segments A – B and Intersection B of the drive were used as trial drives so participants could adjust to the handling of the car and are excluded from analyses. See Tables 1, 2, and 3 for the means and standard deviations of speed, acceleration, and lateral displacement across experimental conditions, and Section 4.3 for a description of the experimental conditions (independent variables).

Table 1: Mean speed, acceleration, and lateral displacement in 90 km/hour drive segments

	Controls (N = 20) M(SD)	Acceptance (N = 15) M(SD)	Rejection (N = 19) M(SD)
Segment B-C			
Speed	19.41 (3.83)	15.84 (4.43)	17.44 (3.75)

Accel	.10 (0.52)	.55 (0.35)	.36 (0.42)
LD	2.28 (0.36)	2.26 (0.30)	2.25 (0.26)
Segment C-D			
Speed	22.03 (3.65)	21.30 (2.57)	20.68 (3.01)
Accel	0.03 (0.25)	-0.08 (0.14)	0.01 (0.16)
LD	2.27 (0.35)	2.35 (0.34)	2.25 (0.28)

Table 2: Mean speed, acceleration, and lateral displacement in 70 km/hour drive segments

	Controls (N = 20) M(SD)	Acceptance (N = 15) M(SD)	Rejection (N = 19) M(SD)
Segment D-E			
Speed	18.68 (2.48)	19.46 (2.45)	18.67 (1.60)
Accel	0.04 (0.14)	-0.02 (0.07)	-0.01 (0.05)
LD	2.25 (0.35)	2.31 (0.29)	2.23 (0.24)
Segment E-F			
Speed	17.81 (2.14)	17.65 (2.33)	17.66 (1.15)
Accel	0.22 (0.15)	0.19 (0.07)	0.16 (0.06)
LD	2.22 (0.24)	2.28 (0.22)	2.30 (0.16)

Table 3: Mean speed, acceleration, and lateral displacement in intersections

	Controls (N = 20) M(SD)	Acceptance (N = 15) M(SD)	Rejection (N = 19) M(SD)
Intersection C			
Speed	6.53 (6.57)	5.74 (4.54)	6.00 (5.94)
Accel	-0.31 (1.79)	-0.26 (0.71)	-0.43 (0.83)
LD	2.20 (0.30)	2.09 (0.30)	2.04 (0.35)
Intersection D			
Speed	4.36 (2.53)	3.22 (0.25)	3.10 (0.36)
Accel	-0.45 (1.74)	0.16 (0.18)	0.11 (0.16)
LD	1.83 (0.92)	2.11 (0.29)	2.13 (0.25)
Intersection E			
Speed	3.76 (1.63)	4.57 (3.41)	3.25 (0.29)
Accel	-0.23 (0.69)	0.03 (0.14)	0.05 (0.19)
LD	0.30 (1.16)	0.04 (0.98)	-0.43 (0.48)

4.2 Participants

A total of 54 participants were recruited from the community and an undergraduate psychology participant pool. Participants were eligible if they were between the ages of 16 and 25 (the minimum age to drive in Ontario and the maximum age to be considered an adolescent, respectively), reported no history of medical, psychiatric, or cognitive conditions that prevent them from driving or interacting with a passenger, and no history of motion sickness or nausea in three-dimension/virtual reality scenarios. The mean age was 19.4 [SD = 2.19]. The sample was 57.4% female and 33.3% Caucasian, 27.8% Asian, 14.8% Southeast Asian, and 24.1% other. A total of 36 participants (66.7%) reported a provisional full or full license status, an average of 2.35 (2.22) years of driving experience, and 7 (13.0%) reported a minor accident or traffic ticket in their driving history. See Table 4 for a breakdown of the sample's demographics and driving history across conditions.

4.3 Independent Variables

Using a 3x2 mixed factorial design, all participants were randomized into Control (N = 20), Social Rejection (hereafter referred to as *Rej*; N = 19) and Social Acceptance (hereafter referred to as *Acc*; N = 15) conditions. Controls completed the entire study protocol alone, while Rej and Acc participants completed the entire study protocol with a peer passenger. The peer passenger role was played by a confederate (a research assistant pretending to be a real participant) and matched to the participant's gender. Conditions were not significantly different from each other on gender, age, years of driving experience, and ethnicity. Acc and Rej conditions did not differ on their ratings of the confederate's social status or driving behaviour, but as expected, rejected participants felt significantly less accepted by their confederate. See Table 4 for a breakdown of participants' reporting of their perception of the confederate's social status, feelings of social acceptance by the confederate, and perception of the confederate's driving.

Table 4: Sample demographics, driving history, and confederate-related variables

	Controls (***N*** = 20) M(SD)	Acceptance (***N*** = 15) M(SD)	Rejection (***N*** = 19) M(SD)
Age	19.61 [SD = 2.43]	19.73 [SD = 2.19]	18.95 [SD = 1.99]
% Female	55.5%	53.3%	63.2%
% Caucasian	30.0%	40.0%	31.6%
Years of Driving Experience	2.45 (2.62)	2.71 (2.49)	1.97 (1.49)
Peer Status[1]	N/A	13.07 (3.73)	13.10 (2.13)
Social Acceptance[2]	N/A	12.73 (3.03)	8.58 (3.22)
Peer Driving[3]	N/A	7.33 (4.22)	7.21 (2.07)

[1]Participants rated confederates on their physical attractiveness, leadership, dominance, approachability, and desire to spend time with them on a scale of 1-5 after completing Task 1.

[2]Participants rated how well they worked together on tasks, desire to work together again, and how socially accepted they felt by the confederate on a scale of 1-5 after completing Task 2. Accepted and rejected participants significantly differed on their rating of how accepted they felt by the confederate (t[32] = 3.830, p = .001).

[3]Participants rated their perception of the confederate's willingness to engage in a range of risky driving behaviours (e.g. not use a seatbelt, drive while impaired, listen to loud music) on a scale of 1-5 after observing the confederate drive and completing their own drive.

4.4 Quasi-Experimental Variable: Physiological Stress

Physiological stress was measured via BIOPAC MP-150 (BIOPAC Systems, Inc., Goleta, CA) wearable biopotential recording systems. Heart rate and respiration were measured continuously via sensors on the torso and recorded via AcqKnowledge software (BIOPAC Systems Inc., Goleta, CA). Due to the high rate of non-responsive respiration data, only findings for mean heartbeats per minute are reported below. A lower heart rate has previously been shown to be associated with risk-taking and aggressive/antisocial behaviour broadly (Portnoy & Farrington, 2015). However, heart rate is also positively correlated with adolescents' and young adults' perceived stress during social stressor tasks (Evans, Greaves-Lord, Euser, Tuelen, Franken, & Huizink, 2012), which may in turn affect participants' physiological responsivity. These measures of physiological stress were selected due to the mobile nature of their data collection possibilities, since participants moved around a room during the study, and because other measures (i.e. finger sensors, blood pressure cuffs) can impede drivers' ability to manoeuver a steering wheel.

Each phase of the experiment (baseline, group tasks, stressor, confederate driving, participant driving) was segmented, allowing for temporal comparisons across the study. The driving phase of the experiment was further segmented by section of the drive (e.g. Segment B-C, Intersection D). See Table

5 for means and standard deviations of heart rate across experimental segments and driving segments between conditions.

Table 5: Mean heart rate in beats per minute (BPM) across experimental and driving segments

	Controls (N = 20) M(SD)	Acceptance (N = 15) M(SD)	Rejection (N = 19) M(SD)
Experimental			
Baseline	91.47 (17.10)	92.93 (15.10)	88.94 (11.10)
Task 1	87.97 (14.77)	92.77 (12.87)	88.85 (10.59)
Stressor	N/A	94.35 (13.86)	87.96 (9.88)
Task 2	N/A	93.55 (12.99)	90.47 (12.79)
Confederate Drive	N/A	100.18 (12.30)	93.77 (13.85)
Participant Drive	84.31 (15.11)	95.47 (11.52)	89.34 (14.11)
Return to Baseline	89.53 (15.04)	103.10 (11.56)	99.51 (17.98)
Driving			
B-C	93.36 (29.16)	92.95 (15.55)	89.22 (10.64)
C	92.75 (28.98)	92.46 (13.74)	89.56 (13.56)
C-D	93.70 (28.12)	93.79 (16.09)	91.70 (13.55)
D	91.26 (29.60)	92.69 (14.80)	88.63 (12.07)
D-E	86.56 (16.25)	92.85 (13.15)	88.08 (9.85)
E	90.50 (13.01)	93.63 (12.47)	88.64 (9.88)
E-F	92.30 (16.70)	92.47 (13.11)	88.91 (11.71)

4.5 Experimental Procedure

Control participants were told that the study is about driver stress and decision-making with or without a passenger. After obtaining consent, the BIOPAC sensors were placed on the participant's torso and they completed a brief driving history questionnaire, including driver's license status, years of driving experience, and any history of collisions while baseline physiological data was collected. They then spent five minutes on a road sign identification task (Task 1) before completing the driving course and then debriefed.

For social condition participants, the confederate was waiting in the lobby of the laboratory building when the participant arrived. Both were told the study is about driver stress and decision-making with or without a passenger and that the confederate has been randomized to be the passenger and the participant to be the driver, but that the confederate would also get to try out the simulator briefly. After obtaining consent, the BIOPAC sensors were placed on the participant and both the participant and confederate completed the driving history questionnaire while baseline physiological measurements were taken. Next, the participant and confederate completed Task 1 together. The confederate agreed more often (Acc condition) or appeared more skeptical (Rej condition) about most of the participant's suggestions. After Task 1, they completed the peer status questionnaire before being told that their next task was to complete a road rules test (Task 2) but that they could work together on it. As the stressor, the confederate then smiled and nodded (Acc condition) or interjected with, "*Uh, actually, could I do this one on my own?*" (Rej condition). After Task 2, the participant and confederate completed the social manipulation check and the confederate completed their simulated drive. The confederate followed a script and drove in a risk-averse (i.e., driving below the speed limit, commenting *"There are a lot of other cars, I should slow down")* or in a risk-accepting (i.e., speeding, commenting *"I better get through this light quickly"*) manner. When their trial was finished, the participant took the driver's seat to complete the driving course 2-3 times, depending on the time availability of the simulator. The participant and confederate then completed the peer driving manipulation check and were debriefed.

5 Results

5.1 Does a peer passenger influence driving outcomes?

Using repeated measures multivariate ANOVA (MANOVA), the effect of a passenger (Rej and Acc conditions) compared to controls was tested on acceleration, speed, and lateral displacement. Driving segments (B-C, C-D, D-E, and E-F) and intersections (C, D, and E) were considered in separate models as within-subjects dependent variables. Condition (control, Rej, and Acc) and gender (male or female) were used as the between-subjects factors. Age and years of driving experience were entered as covariates in both models.

For acceleration in driving segments, Box's *M* test was significant ($p < 0.001$), so Pillai's trace is reported. A Greenhouse-Geiser correction is used since Mauchsley's assumption of sphericity was violated. The condition x acceleration interaction [Pillai's trace = 0.404, $F(2.528, 54.358) = 9.173$, $p < 0.001$] was significant, with acceptance and rejection conditions differing from controls. Similarly, a condition x acceleration interaction in intersections approached significance ($p = .09$). For lateral displacement in driving segments, no significant main effects or interactions were found, but a significant condition x lateral displacement interaction was found between acceptance and rejection conditions and controls [Pillai's trace = 0.204, $F(2.374, 52.223) = 2.504$, $p = .048$] on intersections. No significant main or interaction effects were found for speed on driving segments or intersections.

5.2 Does a peer passenger influence physiological stress?

Again using MANOVA, the effect of a passenger (Rej and Acc conditions) compared to controls was tested on physiological stress during driving segments and intersections in separate models. Mean beats per minute were considered in separate models as within-subjects dependent variables. Condition (control, Rej, and Acc) and gender (male or female) were used as the between-subjects factors. Age and years of driving experience were entered as covariates in both models.

For physiological stress during driving segments and intersections, Box's *M* test was not significant and the assumption of sphericity was not violated, thus all findings are reported using Wilks' lambda and the assumption of sphericity. For driving segments, a significant condition x gender x heart rate interaction was found [Wilks' lambda = 0.733, $F(6, 132) = 3.424$, $p = .004$], such that female participants' heart rates remained significantly higher across all conditions compared to males, and was higher in the Acc condition (see Figure 1). For intersections, no significant main effects or interactions were found.

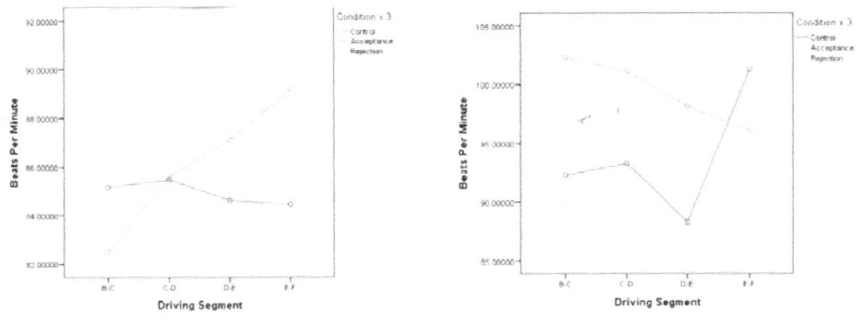

a) Male participants b) Female participants

Figure 1: Average beats per minute across conditions and driving segments for male and female participants

6 Discussion

This study is a preliminary analysis of the effects of a peer passenger on physiological stress and driving outcomes in young drivers. Our analyses revealed that acceleration and lateral displacement appear sensitive to effects of a peer passenger, and that physiological stress is influenced by interactions between social conditions and gender.

In both social conditions, participants accelerated and reached their top speed across driving segments more slowly than did controls. Though acceleration and speed are considered key risky driving behaviours influenced by peer passengers (Curry et al., 2012; Ouimet et al., 2015), it is also the case that peer passengers can distract drivers from noticing changing road conditions (e.g. the change of a red light to green, speed limit signage) as has been previously found (Pradhan et al., 2014). In this case, participants who drove with a peer passenger may have been slower to accelerate to their top speed due to their passenger serving as a distraction from road conditions.

Further, participants with a peer passenger also had significantly different lateral displacement from the midline of the road in intersections, and this effect appears to be particularly strong for Intersection E which required participants to make a left-hand turn into oncoming traffic. At Intersection E, participants who drove with a passenger were more likely to be closer to the midline (Acc condition) or over it (Rej condition) compared to controls while making a left turn. The ability of a driver to maintain their position in their lane while preparing to turn can influence the likelihood of a head-on collision with oncoming traffic. Thus, a peer passenger may again be serving a distraction, preventing the driver from maintaining a safe position to complete a left-hand turn.

The significant interaction among physiological stress, gender, and experimental condition also reveals that gender and physiological stress may affect driver behaviour. Males consistently had lower heart rates across all experimental and driving segments, regardless of condition, while female participants had significantly higher heart rates during the social stressor and driving segments of the study. Further, female participants also had significantly higher heart rates during both driving segments and intersections. Notably, this trend was unique to female participants in the Acc condition, suggesting that social acceptance increases heart rate, while social rejection does not. This is in contrast to what might be expected, given previous research that suggests social rejection is a significant stressor in adolescence and young adulthood (Jones et al., 2014; Peake et al., 2013).

However, given the gendered nature of this response to social acceptance, it may not be surprising. Previous research suggests that young men are more sensitive to social rewards in competitive environments, while young women tend to seek social rewards via group cohesion (Voroboyev et al., 2015). Thus, elevated heart rates after experiencing social acceptance among the female participants may have been in response to the rewarding nature of getting along with a peer. Under these socially rewarding conditions, it is possible that getting along well with the peer passenger serves as an additional element of distraction once inside the car, consistent with Curry et al.'s (2012) findings that young female drivers are more likely to be distracted by interior stimuli (i.e. passengers). Rejection, however, may remove the socially rewarding nature of interacting with a peer, reducing the peer passenger's power to distract.

7 Conclusions

Our preliminary findings indicate that physiological stress and peer passengers, both uniquely and in combination, affect driving behaviour (acceleration and lateral displacement). Though we did not find effects of stress or peer passengers on speeding as a specific outcome, peer passengers particularly affected acceleration and lateral displacement. It is possible that a peer passenger serves a distraction from changing road conditions, subsequently decreasing the driver's ability to modify their behaviour in accordance with road conditions. Gender and social acceptance, in particular, appear to be key variables influencing physiological stress while driving. At the moment, the current sample size is statistically underpowered to complete more complex modeling of physiological stress, driving outcomes, experimental conditions, and gender. However, our ongoing data collection will allow us to replicate these initial findings and utilize multi-level and structural equation models to predict driving outcomes and driving-related injury risk (i.e. collisions).

Acknowledgements

The financial support of the Social Sciences and Humanities Research Council, the Harry Rosen Institute for Stress Research, and the Ryerson University Career Boost Research Assistant program is acknowledged. Two Research Tools and Instruments grants from the Natural Sciences and Engineering Research Council of Canada have financially supported the driving simulator.

References

Bingham, C. R., & Ehsani, J. P. (2012). The relative odds of involvement in seven crash configurations by driver age and sex. *Journal of Adolescent Health, 51*(5), 484. doi:10.1016/j.jadohealth.2012.02.012

BIOPAC Systems, Inc. [Computer hardware/software]. Goleta, CA.

Centers for Disease Control and Prevention. (2015). *Teen drivers: get the facts.* Retrieved from https://www.cdc.gov/motorvehiclesafety/teen_drivers/teendrivers_factsheet.htmlTR

Centifanti, L. C. M., Modecki, K. L., MacLellan, S., & Gowling, H. (2016). Driving under the influence of risky peers: An experimental study of adolescent risk taking. *Journal of Research on Adolescence, 26*(1), 207-222. doi:10.1111/jora.12187

Chein, J., Albert, D., O'Brien, L., Uckert, K., & Steinberg, L. (2011). Peers increase adolescent risk taking by enhancing activity in the brain's reward circuitry. *Developmental Science, 14*(2), F1-F10. doi:10.1111/j.1467-7687.2010.01035.x

Cowell, R. A. (2013). *Do peers alter decision making processes in adolescence? An examination of peer influence on cool and hot executive function* (Unpublished doctoral dissertation). University of Minnesota.

Curry, A. E., Mirman, J. H., Kallan, M. J., Winston, F. K., & Durbin, D. R. (2012). Peer passengers: How do they affect teen crashes? *The Journal of Adolescent Health, 50*(6), 588-594. doi:10.1016/j.jadohealth.2011.10.016

Defoe, I. N., Dubas, J. S., Figner, B. C., & Aken, M. A. G. v. (2015). A meta-analysis on age differences in risky decision making: Adolescents versus children and adults. *Psychological Bulletin, 141*(1), 48-84. doi:10.1037/a0038088

Evans, B. E., Greaves-Lord, K., Euser, A. S., Tulen, J. H. M., Franken, I. H. A., & Huizink, A. C. (2012). Alcohol and tobacco use and heart rate reactivity to a psychosocial stressor in an adolescent population. *Drug and Alcohol Dependence, 126*(3), 296-303. doi:10.1016/j.drugalcdep.2012.05.031

Falk, E. B., Cascio, C. N., Brook O'Donnell, M., Carp, J., Tinney, F. J., Bingham, C. R., . . . Simons-Morton, B. G. (2014). Neural responses to exclusion predict susceptibility to social influence. *Journal of Adolescent Health, 54*(5), S22-S31. doi:10.1016/j.jadohealth.2013.12.035

Forbes, E. E., & Dahl, R. E. (2010). Pubertal development and behavior: Hormonal activation of social and motivational tendencies. *Brain and Cognition, 72*(1), 66-72. doi:10.1016/j.bandc.2009.10.007

Gardner, M., & Steinberg, L. (2005). Peer influence on risk taking, risk preference, and risk decision making in adolescence and adulthood: an experimental study. *Developmental Psychology, 41*(4), 625-635. doi:10.1037/a0026993

Jones, R. M., Somerville, L. H., Li, J., Ruberry, E. J., Powers, A., Mehta, N., . . . Casey, B. J. (2014). Adolescent-specific patterns of behavior and neural activity during social reinforcement learning. *Cognitive, Affective, & Behavioral Neuroscience, 14*(2), 683-697. doi:10.3758/s13415-014-0257-z

Ouimet, M. C., Pradhan, A. K., Brooks-Russell, A., Ehsani, J. P., Berbiche, D., & Simons-Morton, B. G. (2015). Young drivers and their passengers: A systematic review of epidemiological studies on crash risk. *Journal of Adolescent Health, 57*(1), S24-S35. doi:10.1016/j.jadohealth.2015.03.010

Ouimet, M. C., Pradhan, A. K., Simons-Morton, B. G., Divekar, G., Mehranian, H., & Fisher, D. L. (2013). The effect of male teenage passengers on male teenage drivers: Findings from a driving simulator study. *Accident; Analysis and Prevention, 58*, 132-139. doi:10.1016/j.aap.2013.04.024

O'Brien, L., Albert, D., Chein, J., & Steinberg, L. (2011). Adolescents prefer more immediate rewards when in the presence of their peers: peers and immediate rewards. *Journal of Research on Adolescence, 21*(4), 747-753. doi:10.1111/j.1532-7795.2011.00738.x

Park, J., & Baumeister, R. F. (2015). Social exclusion causes a shift toward prevention motivation. *Journal of Experimental Social Psychology, 56*, 153-159. doi:10.1016/j.jesp.2014.09.011

Peake, S. J., Dishion, T. J., Stormshak, E. A., Moore, W. E., & Pfeifer, J. H. (2013). Risk-taking and social exclusion in adolescence: Neural mechanisms underlying peer influences on decision-making. *Neuroimage, 82*, 23-34. doi:10.1016/j.neuroimage.2013.05.061

Portnoy, J., & Farrington, D. P. (2015). Resting heart rate and antisocial behavior: An updated systematic review and meta-analysis. *Aggression and Violent Behavior, 22*, 33-45. doi:10.1016/j.avb.2015.02.004

Pradhan, A. K., Li, K., Bingham, C. R., Simons-Morton, B. G., Ouimet, M. C., & Shope, J. T. (2014). Peer passenger influences on male adolescent drivers' visual scanning behavior during simulated driving. *Journal of Adolescent Health, 54*(5), S42-S49. doi:10.1016/j.jadohealth.2014.01.004

Sebastian, C., Viding, E., Williams, K. D., & Blakemore, S. (2011). Social brain development and the affective consequences of ostracism in adolescence. *Brain and Cognition, 72*(1), 134-145. doi:10.1016/j.bandc.2009.06.008

Simons-Morton, B.G, Lerner, N., & Singer, J. (2005). The observed effects of teenage passengers on the risky driving behavior of teenage drivers. *Accident Analysis and Prevention, 37*(6), 973-82. doi: 10.1016/j.aap.2005.04.014

Simons-Morton, B. G., Ouimet, M. C., Zhang, Z., Klauer, S. E., Lee, S. E., Wang, J., . . . Dingus, T. A. (2011). The effect of passengers and risk-taking friends on risky driving and Crashes/Near crashes among novice teenagers. *Journal of Adolescent Health, 49*(6), 587-593. doi:10.1016/j.jadohealth.2011.02.009

Simons-Morton, B. G., Bingham, C. R., Falk, E. B., Li, K., Pradhan, A. K., Ouimet, M. C., . . . Shope, J. T. (2014). Experimental effects of injunctive norms on simulated risky driving among teenage males. *Health Psychology, 33*(7), 616-627. doi:10.1037/a003483

Somerville, L. H., Jones, R. M., Ruberry, E. J., Dyke, J. P., Glover, G., & Casey, B. J. (2013). The medial prefrontal cortex and the emergence of self-conscious emotion in adolescence. *Psychological Science, 24*(8), 1554-1562. doi:10.1177/0956797613475633

Statistics Canada. (2017). *Leading causes of death, total population, by age group and sex, Canada, annual.* (CANSIM database table 102-0561). Retrieved from Statistics Canada website http://www5.statcan.gc.ca/cansim/a26?lang=eng&id=1020561

Stroud, L. R., Foster, E., Papandonatos, G. D., Handwerger, K., Granger, D. A., Kivlighan, K. T., & Niaura, R. (2009). Stress response and the adolescent transition: performance versus peer rejection stressors. *Development and Psychopathology, 21*(1), 47-68. doi:10.1017/S0954579409000042

van den Bos, E., Rooij, M., Miers, A. C., Bokhorst, C. L., & Westenberg, P. M. (2014). Adolescents' increasing stress response to social evaluation: Pubertal effects on cortisol and Alpha-Amylase during public speaking. *Child Development, 85*(1), 220-236. doi:10.1111/cdev.12118

Vorobyev, V., Kwon, M. S., Moe, D., Parkkola, R., & Hämäläinen, H. (2015). Risk-taking behavior in a computerized driving task: Brain activation correlates of decision-making, outcome, and peer influence in male adolescents. *PLoS One, 10*(6), 1-20. doi:10.1371/journal.pone.0129516

Zelazo, P. D., & Carlson, S. M. (2012). Hot and cool executive function in childhood and adolescence: Development and plasticity. *Child Development Perspectives, 6*(4), 354-360. doi:10.1111/j.1750-8606.2012.00246.x

Building Tomorrow's Society
Bâtir la Société de Demain

Fredericton, Canada
June 13 to 16, 2018 / *13 au 16 juin, 2018*

RELIABILITY OF STOPPING SIGHT DISTANCE DESIGN AT ROUNDABOUTS

Faizi, Jafar[1, 3] and Easa, Said[2]
[1] Ryerson University, Canada
[2] Ryerson University, Canada
[3] jafarfaizi89@gmail.com

Abstract: The existing stopping sight distance (SSD) design method for roundabouts is deterministic. This means that all of the design variables are predetermined, fixed values. This study presents a probabilistic (Reliability-based) method for the determination of SSD at roundabouts based on an equation recommended by "A Policy on Geometric Design of Highways and Streets" (6th Edition), named as AASHTO Green Book, for calculation of SSD. The reliability based method considers all design variables (design speed, perception-reaction time, and vehicle deceleration rate) as random variables. In this method, correlation of random variables is taken into consideration. Three types of SSD (SSD for approaches, SSD along the circulatory roadway, and SSD for exiting vehicles to the pedestrian crosswalk) are considered in this study. The safety margin is defined as SSD provided by the roundabout geometry minus SSD required based on vehicle and driver performance characteristics. The First-Order Second-Moment (FOSM) and Advanced First-Order Second-Moment (AFOSM) methods were used to model SSD. Once the required SSD are determined, lateral clearance design values were determined based on probabilistic SSD and geometry of roundabouts. The reliability model developed in this study was applied to a roundabout and comparison is made between the results of FOSM, AFOSM, and deterministic methods. The reliability-based SSD design provides designers with the option to choose the level of reliability of their design.

1 INTRODUCTION

Safety is one of the dominant parameters for the design of roundabouts. In order for roundabouts to be safe, the SSD requirements must be met in every single point of roundabouts. Currently, SSD is calculated using formula recommended by AASHTO, in which the design variables are predetermined, fixed, values. The design variables in SSD calculation are vehicle speed, perception-reaction time, and vehicle deceleration rate. In reality, design variables are not fixed values, rather random variables which are sometimes correlated with one another. In Probabilistic approach, safety margin is defined as provided design element dimension minus required design element dimension. Although in transportation engineering, a negative value for the safety margin does not mean that an accident would happen, but it would impose some restrictions and increase the likelihood of occurrence of collisions (Easa and Hussain 2016).

The reliability-based design method is applied in different branches of civil engineering. Some authors have applied probabilistic method to the geometric design of highways such as horizontal curve design (Himes and Donnell 2014), Sight distance at railroad crossings (Easa 1994), Evaluation of sight distance for traffic safety (Santos-Berbel et al. 2017), Intersection sight distance (Easa 2000), Calibration of road design guidelines (Hussein et al. 2014), sight distance for turning vehicles on traffic signals (Hussain and Easa 2016), and reliability of lateral clearance (Sarhan and Hassan 2011). The reliability-based method is used by some authors for the design of traffic signals such as pedestrian signal timing on traffic signals (Easa and Cheng 2013) and Intergreen interval design at traffic signals (Easa 1993). The results of studies of the application of reliability-based design method in transportation engineering indicate that deterministic design method sometimes underestimates or overestimates the design values. For example, the result of study on left turn lanes offset distance indicates that deterministic design method overestimates left-turn lanes offset distance which would increase the overall cost of the project (Hussain and Easa 2016). The results of studies of the application of probability in transportation engineering indicate that reliability-based design values provide a more accurate representation of the requirements of design values based on the randomness of the design variables.

There are many reliability analysis methods including First-Order Second-Moment Method, the Point-Estimate Method, and Exact Method. Because of its ease of application, the First-Order Second-Moment (FOSM) method is widely used in different areas of science and engineering. FOSM method expands objective function about mean values of random variables (Easa 2000). Although the FOSM reliability method is easy to use, it has some shortcomings. These shortcomings include the lack of availability of information about distribution of variables, errors because of shortening of Taylor series, and that the reliability index depends on the way the objective function is defined, meaning, when the objective function is written in two different ways, two different reliability indexes are obtained. To overcome this error, Advanced First-Order Second-Moment (AFOSM) method uses the invariant reliability index. AFOSM is an iterative method which resolves the error associated with FOSM method. In this study, both FOSM and AFOSM methods are used to calculate SSD at roundabouts.

2 DETERMINISTIC METHOD

Stopping sight distance is defined as the distance a vehicle travels from the time driver sees an object to the time brakes are applied, plus the distance travelled by the vehicle from the time brakes are applied to the time vehicle comes to a complete stop (A Policy on Geometric Design of Highways and Streets, 2011). SSD is checked at three locations at roundabouts: approaches, the circulatory roadway, and the exits. For the approaches of roundabouts, SSD is provided at the pedestrian crosswalk and at the yield line. Figure 1 illustrates stopping sight distances and driver's sightlines for each SSD for a single lane roundabout.

(Easa 2017) developed an analytical model for the determination of lateral clearance for symmetrical single-lane roundabouts. The author formulated lateral clearance for the roundabouts to satisfy SSD at the approaches, circulatory lane, and on the exits. SSD is considered to the pedestrian crosswalk and to the yield line for the approaches of roundabouts. Based on required SSD and the geometry of roundabouts, lateral clearance can be determined on every single point of the curves for the symmetrical single lane roundabout (Easa 2017).

Equation 1 is used for calculation of stopping sight distance at roundabouts (AASHTO 2011).

[1] $SSD = 0.278Vt + 0.039\frac{V^2}{a}$

where: V = vehicle speed (Km/h), t = perception-reaction time (s), and a = vehicle deceleration rate (m/s^2).

For calculation of stopping sight distance, a perception-reaction time (t) of 2.5 seconds and vehicle deceleration rate (a) of 3.4 m/s^2 is recommended (AASHTO 2011). For calculation of SSD using equation 1, design speed of the roundabout segment is considered as vehicle operating speed.

3 RELIABILITY-BASED METHOD

The First-Order Second-Moment and Advance First-Order Second-Moment methods are used to develop reliability-based models for the determination of SSD at roundabouts. Fundamentals of FOSM and AFOSM methods are briefly described, followed by the formulation of SSD with FOSM and AFOSM methods based on expression recommended by AASHTO.

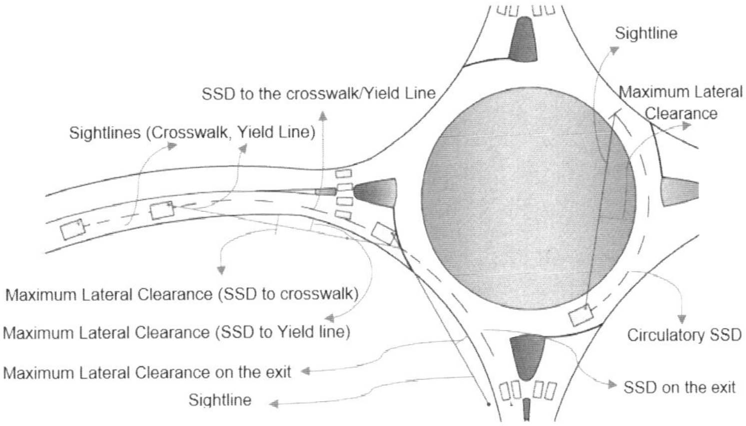

Figure 1: Stopping sight distances at roundabouts

3.1 First-Order Second-Moment (FOSM) Method

As the name of the First-Order Second-Moment (FOSM) implies, two moments (mean and variance) are required in the FOSM reliability method. The safety margin is expressed as:

[2] $F = f(X_1, X_2, X_3, \ldots\ldots, X_n)$

The Taylor series is used to expand and linearize the safety margin about the mean values of the variables $(\mu_{x1}, \mu_{x2}, \mu_{x3}, \ldots\ldots, \mu_{xn})$, and is expressed as:

[3] $F = f(\mu_{x1}, \mu_{x2}, \mu_{x3}, \ldots\ldots, \mu_{xn}) + \sum_{i=1}^{n} \frac{df(x_i)}{dX_i}(X_i - \mu_{xi}) + \frac{1}{2}\Sigma\Sigma \frac{\partial_f^2}{\partial X_i \partial X_j}(X_i - \mu_{xi})(X_i - \mu_{xj}) + \cdots$

In equation 3, the partial derivatives are calculated about the mean values of variables. The shortened form of Taylor series expansion is used to measure the approximate values of the mean and variance of the safety margin. The mean and variance of safety margin are expressed as:

[4] $E[F] \approx f(\mu_{x1}, \mu_{x2}, \mu_{x3}, \ldots\ldots, \mu_{xn})$

[5] $Var[F] = \sigma_F^2 = \Sigma\Sigma \frac{df(X_i)^2}{\partial X_i \partial X_j} Cov[x_i, x_j] + \sum_{i-1}^{n} (\frac{df(X_i)}{dX_i}\sigma_{xi})^2$

The partial derivatives in equation 5 are calculated using mean values and $Cov[x_i, x_j]$ is the covariance of x_i *and* x_j. Covariance describes how two variables are varying together. The covariance of x_i *and* x_j is expressed as:

[6] $Cov[x_i, x_j] = \rho_{x_i x_j} \sigma_{xi} \sigma_{xj}$

where: $\rho_{x_i x_j}$ = coefficient of correlation between x_i and x_j, σ_{xi} = standard deviation of random variable x_i, and σ_{xj} = standard deviation of random variable x_j. The coefficient of variation, CV, helps to illustrate how one random variable is dispersed compared to the mean of all other variables. The coefficient of variation (CV) is expressed as:

[7] $CV = \frac{\sigma_{xi}}{\mu_{xi}}$

where: σ_{xi} = standard deviation and μ_{xi} = mean value of variable x_i. Now, SSD at roundabouts is modeled with FOSM method based on formula recommended by (AASHTO 2011). Equation 1 is used for calculation of SSD at roundabouts. The design variables are denoted as (X) variables for ease of performance of mathematical operations and the required SSD, S_{Req}, is expressed as:

[8] $S_{Req} = 0.278X_1X_2 + 0.039\frac{X_1^2}{X_3}$

where: S_{Req} = required stopping sight distance (m), X_1 = vehicle speed (Km/h), X_2 = perception-reaction time (s), and X_3 = deceleration rate (m/s^2).

The safety margin (F) for SSD at Roundabouts is defined as SSD provided by the geometry of roundabouts minus SSD required based on vehicle and driver characteristics. Since the design elements in probability method are random variables, safety margin (F) is also a random variable. The safety margin is expressed as:

[9] $F = S_{sup} - S_{Req}$

where: S_{sup} = supplied SSD and S_{Req} = required SSD. The expected value of required SSD, $\mathrm{E}[S_{Req}]$, is given:

[10] $\mathrm{E}[S_{Req}] = 0.278\mu_{x1}\,\mu_{x2} + 0.039\frac{\mu_{x1}^2}{\mu_{x3}}$

where: μ_{x1} = mean value of vehicle speed $(\mathrm{Km/h})$, μ_{x2} = mean value of perception-reaction time (s), and μ_{x3} = mean value of vehicle deceleration rate $(\mathrm{m/s^2})$.

The mean and variance of safety margin (F) are given by equations 11 and 12, respectively.

[11] $E[\mathrm{F}] \approx S_{supply} - \mathrm{E}[S_{Req}]$

[12] $Var[F] = \sigma_F^2 = 2\frac{df(X)}{dX_1}\frac{df(X)}{dX_2}Cov(X_1, X_2) + 2\frac{df(X)}{dX_1}\frac{df(X)}{dX_3}Cov(X_1, X_3) + (\frac{df(X)}{dX_1}\sigma_{x1})^2 + \left(\frac{df(X)}{dX_2}\sigma_{x2}\right)^2 + (\frac{df(X)}{dX_3}\sigma_{x3})^2$

The derivatives of safety margin are given as:

[13] $\frac{df(X)}{dX_1} = 0.278X_2 + 0.078\frac{X_1}{X_3}$

[14] $\frac{df(X)}{dX_2} = 0.278X_1$

[15] $\frac{df(X)}{dX_3} = -\frac{0.039X_1^2}{X_3^{\,2}}$

The safety of proposed design can be expressed in term of reliability index (β). Reliability index is expressed as:

[16] $\beta = \frac{\mathrm{E}[\mathrm{F}]}{\sigma_\mathrm{F}}$

where: $\mathrm{E}[\mathrm{F}]$ and σ_F are given by Equations 4 and 5 (note that $\sigma_\mathrm{F} = \sqrt{Var[\mathrm{F}]}$).

The probability of failure is expressed as:

[17] $\mathrm{P_f} = \Phi(-\beta) = 1 - \Phi(\beta)$

where: $\Phi(-\beta)$ = sum of the area under the density curve of the safety margin from -∞ to $-\beta$. The table of standard normal variate or computer software's are used to obtain the value of $\Phi(-\beta)$. Based on assigned probability of failure (P_f), the corresponding value of β is used for the calculation of SSD at roundabouts.

Substituting value of E(F) from equation 16 to equation 11 and calculating for provided SSD (S_{prov}) will yield the following equation:

[18] $S_{sup} = \beta\sigma_F + E(S_{req})$

3.2 Advanced First-Order Second-Moment (AFOSM) Method

The advanced First-Order Second-Moment reliability method is developed by (Hasofer & Lind 1974). According to the author, the point chosen from failure boundary for linear approximation will provide the invariant reliability index. The safety margin (F) is already defined in equation 2. Generally, it is more favorable to work in terms of normalized variables. Normalized or standardized variables are dimensionless. The normalized variable (y_i) is given by:

[19] $y_i = \frac{x_i - \mu_{xi}}{\sigma_{xi}}$

where: σ_{xi} = standard deviation, μ_{xi} = mean value, and x_i = obtained value of a variable.

The mean value of normalized variable y_i is zero and its standard deviation is one. The safety margin is evolved with standardized variables, $y_1, y_2, y_3, \ldots\ldots, y_n$, and expressed as:

[20] $Z = h(y) = h(y_1, y_2, y_3, \ldots\ldots, y_n)$

The Taylor series, first-order approximation of Z, at the standard design values at which the approximation is taken, $y_i^* = h(y_1^*, y_2^*, y_3^*, \ldots\ldots, y_n^*)$, is expressed as:

[21] $Z = \sum_{i=1}^{n}(y_i - y_i^*)\, dh(y^*)/d(y_i^*)$

where: $dh(y^*)/d(y_i^*)$ = first derivatives of performance function with respect to y_i, calculated at design points, y_i^*. The mean, μ_z and standard deviation, σ_z, of Z are given by:

[22] $\mu_z = -\frac{\sum_{i=1}^{n}(y_i^*) dh(y^*)}{d(y_i^*)}$

[23] $\sigma_z = \sqrt{\sum_{i=1}^{n} dh(y^*)/d(y_i^*)^2}$

The solution in terms of normalized variables is given by:

[24] $y_i^* = -\frac{\frac{dh(y^*)}{d(y_i^*)}}{\sigma_z}(\beta + \frac{h(y)}{\sigma_z})$

where: σ_z = standard deviation, β = reliability index, and $h(y)$ = performance function.

The distance from the origin to y^*, which is the minimum distance, is the reliability index, β. The reliability index (β) is expressed as:

[25] $\beta = \sqrt{\sum_{i=1}^{n} y_i^2}$

The iterative algorithm for determination of reliability index (β) can be find on (Smith 1986).

When random variables, $X_1, X_2, X_3, \ldots\ldots, X_n$, are correlated with each other, a procedure is used to transform the correlated variables to noncorrelated variables, while allowing for correlation effects. After safety margin is evolved with transformed, uncorrelated, and reduced variables, the AFOSM procedure, described early, will be applied. The covariance matrix, denoted as CV_X, for correlated variables, $X_1, X_2, X_3, \ldots\ldots, X_n$, is expressed as:

$$[26]\ CV_X = \begin{bmatrix} \sigma_{x1}^2 & \cdots & COV(X_1, X_n) \\ \vdots & \ddots & \vdots \\ COV(X_n, X_1) & \cdots & \sigma_{xn}^2 \end{bmatrix}$$

The leading diagonal values of the covariance matrix, CV_X, are variances of the correlated variables and the off-diagonal values are the related covariances. The covariance matrix is used to find uncorrelated variables. Procedure for determination of uncorrelated variables is basically determination of eigenvalues and eigenvector of covariance matrix. The eigenvalue and eigenvector of covariance matrix can be obtained using Jacobi's method. The eigenvalues are variance of transformed variables and original variables, $X_1, X_2, X_3, \ldots\ldots, X_n$, multiplied by eigenvector will give the values of a set of transformed and uncorrelated random variables, $Y_1, Y_2, Y_3, \ldots\ldots, Y_n$. To obtain a set of reduced and uncorrelated variables, the uncorrelated variables, $Y_1, Y_2, Y_3, \ldots\ldots, Y_n$, are standardized. Once the reduced, uncorrelated variables, $y_1, y_2, y_3, \ldots\ldots, y_n$, are determined, $X_1, X_2, X_3, \ldots\ldots, X_n$ are calculated in terms of $y_1, y_2, y_3, \ldots\ldots, y_n$ and substituted into safety margin. Once the safety margin is expressed with transformed, reduced, and uncorrelated variables, the AFOSM method can be applied.

The safety margin (F) for the system is defined in equation 9. Before applying the AFOSM method to equation 9, the correlated variables need to be transformed into noncorrelated variables while allowing for the correlation effects. After the safety margin is evolved with transformed, uncorrelated, and reduced variables, y_1, y_2, y_3, the AFOSM method is applied. The iterative algorithm is applied for the determination of the invariant reliability index (Smith, 1986).

4 Design Values

Tables 1 and 2 present the lateral clearance design values for the approaches of roundabouts to satisfy SSD calculated with FOSM method to the crosswalk and the yield line, respectively. Tables 3 and 4 presents the lateral clearance design values for the approaches of roundabouts to satisfy SSD calculated with AFOSM method to the crosswalk and the yield line, respectively. The SSD used to calculate maximum lateral clearances for tables 1, 2, 3, and 4 are based on system probability of failure of 0.01% and coefficient of variation of 10% for all random variables.

Table 1: Design values of lateral clearance on the approaches (SSD to the crosswalk), FOSM Method

First Entry Curve Radius	Second Entry Curve Radius	Maximum Lateral Clearance (m)			
		Speed (Km/h) *			
		40	50	60	70
		Stopping Sight Distance (m)			
R_1 (m)	R_2 (m)	48	65	85	109
100	20	4.7	6.9		
	30	4.3	6.6		
	40	4.2	6.5		
500	20	3.3	3.6	3.9	4.8
	30	2.9	3.1	3.5	4.4
	40	2.6	2.9	3.2	4.2
1000	20	3.2	3.4	3.6	3.8
	30	2.8	3.0	3.1	3.3
	40	2.5	2.7	2.9	3.0

*For shaded cells, R_1 is less than R_{min} for the respective vehicle speed on the approaches and the superelevation of up to 0.06.

Table 2: Design values of lateral clearance on the approaches (SSD to Yield Line), FOSM Method

Entry Curve Radius R_1 (m)	Entry Curve Radius R_2 (m)	Maximum Lateral Clearance (m)			
		Speed (Km/h) *			
		40	50	60	70
		Stopping Sight Distance (m),			
		48	65	85	109
100	20	4.3	6.5		
	30	3.0	5.4		
	40	2.4	4.8		
500	20	3.1	3.7	4.3	5.1
	30	1.6	2.1	2.6	3.7
	40	0.8	1.3	1.9	3.0
1000	20	2.9	3.4	3.9	4.3
	30	1.4	1.8	2.2	2.5
	40	0.7	1.0	1.3	1.7

*For shaded cells, R_1 is less than R_{min} for the respective vehicle speed on the approaches and the superelevation of up to 0.06.

Table 3: Design values of lateral clearance on the approaches (SSD to the crosswalk), AFOSM Method

First Entry Curve Radius R_1 (m)	Second Entry Curve Radius R_2 (m)	Maximum Lateral Clearance (m)			
		Speed (Km/h) *			
		40	50	60	70
		Stopping Sight Distance (m),			
		51	70	95	130
100	20	5.0	7.6		
	30	4.7	7.4		
	40	4.5	7.3		
500	20	3.4	3.6	4.2	5.9
	30	3.0	3.2	3.8	5.5
	40	2.7	2.9	3.6	5.4
1000	20	3.2	3.5	3.7	4.1
	30	2.8	3.0	3.2	3.7
	40	2.5	2.7	3.0	3.5

*For shaded cells, R_1 is less than R_{min} for the respective vehicle speed on the approaches and the superelevation of up to 0.06.

The SSD calculated with AFOSM method is larger compared to FOSM method for the same probability of failure. The reason that FOSM method provides lower design value is that AFOSM removes the error associated with the FOSM reliability method. Since the AFOSM SSD design values are larger than FOSM SSD, the lateral clearance design values to satisfy AFOSM SSD is larger than lateral clearance values to satisfy FOSM SSD. The lateral clearance design values obtained to satisfy FOSM and AFOSM SSD are nearly equal when entry radii, R_1, R_2, are large and approach design speed is low and vice versa.

Table 4: Design values of lateral clearance for AFOSM, SSD on the approaches (SSD to Yield Line)

First Entry Curve Radius	Second Entry Curve Radius	Maximum Lateral Clearance (m)			
		Speed (Km/h) *			
		40	50	60	70
		Stopping Sight Distance (m),			
R_1 (m)	R_2 (m)	51	70	95	130
100	20	4.7	7.3		
	30	3.4	6.2		
	40	2.8	5.6		
500	20	3.2	3.9	4.6	6.1
	30	1.7	2.3	3.0	4.8
	40	0.9	1.4	2.3	4.2
1000	20	3.0	3.6	4.0	4.6
	30	1.5	1.9	2.3	2.9
	40	0.8	1.1	1.4	2.2

*For shaded cells, R_1 is less than R_{min} for the respective vehicle speed on the approaches and the superelevation of up to 0.06.

5 APPLICATION

The application involved the use of a symmetrical single-lane roundabout with two entry curves, curved in the same direction. The radius of first entry curve (R1) was 200 m and the radius of the second entry curve (R2) was 40 m. Figure 2 illustrates the geometry of roundabouts on the approach. The approach SSD was calculated based on probability of failure (P_f) of 0.01%. The reliability index (β) corresponding to $P_f = 0.01\%$ is 3.72. Since the mean values of design variables are required, mean values were calculated from the extreme values with the assumption that random variables are normally distributed. The expression relating mean and extreme value is given by:

[27] $\mu_X = \frac{X}{1+Z\,CV}$

where: X = extreme value, μ_X = mean value, CV = coefficient of variation, and Z = number of standard deviations of the normal distribution corresponding to a certain percentile. In this example, perception-reaction time is assumed to represents 95th percentile (Z=1.65), vehicle deceleration rate is assumed to represents 5th percentile (Z=-1.65), and vehicle speed is assumed to indicate 99th percentile speed (Z=2.32). The data used in reliability analysis, in both FOSM and AFOSM methods, are illustrated in table 5.

Table 5: Data used for probabilistic approach SSD design*

Variables	Extreme Value	Mean Value	Coefficient of Variation (%)
V (Km/h)	60	48.7	10
t (s)	2.5	2.15	10
a (m/s^2)	3.4	4.07	10

* $\rho_{x_1x_2} = +0.5 \quad \rho_{x_1x_3} = -0.5$

Based on design speed of 60 km/h and extreme values recommended for perception-reaction time (t = 2.5 s) and vehicle deceleration rate (a = 3.4 m/s^2), the required SSD is 83 meters (AASHTO, 2011).

When the FOSM method is applied, E(S_{req}) and σ_F were calculated as 51.83 and 9.28 using equations 10 and 12, respectively (note that $\sigma_F = \sqrt{Var[F]}$). Using the equation 18, the required (supplied) SSD was calculated as 86 meters.

The iterative algorithm presented in (Smith 1986) was used to produce stable reliability index. The required SSD was calculated with AFOSM method by trial as 95 meters. The calculated design parameters corresponding to the final iteration is illustrated in table 6.

Maximum lateral clearance to satisfy deterministic SSD was calculated as 5.5 m for to the pedestrian crosswalk and 4 m to the yield line, respectively. Maximum lateral clearance to satisfy FOSM SSD was calculated as 5.8 m to the crosswalk and 4.3 m to the yield line, respectively. Maximum lateral clearance to satisfy AFOSM SSD was 6.8 m to the crosswalk and 5.3 m to the yield line, respectively. Table 7 presents the results of the design elements calculated with the deterministic, FOSM, and AFOSM methods.

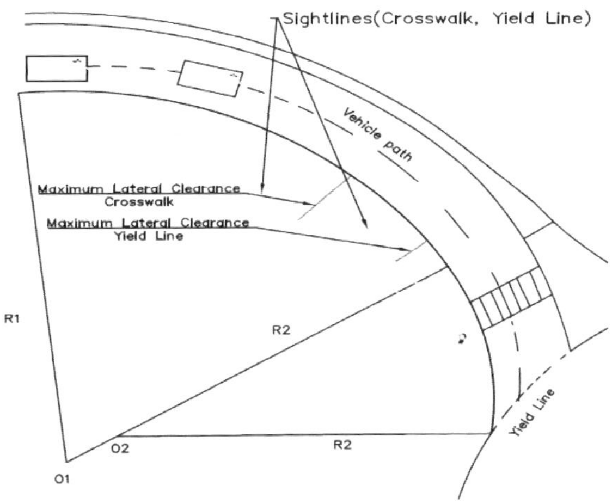

Figure 2: Stopping Sight Distance on the approach to the crosswalk

Table 6: Design parameters related to final iteration

Variables	y_i^*	X_i^*
V	0.77	65.70
T	1.05	2.08
a	-3.48	2.96

Table 7: Comparison of the results for design elements

Design Methods	Stopping Sight Distance (m)	Maximum Lateral Clearance (m) (SSD to crosswalk)	Maximum Lateral Clearance (m) (SSD to Yield line)
Deterministic	83	5.5	4.0
FOSM	86	5.8	4.3
AFOSM	95	6.8	5.3

The SSD design values of deterministic method was 3.5% lower than the SSD design value of the FOSM method and 12% lower than the SSD design value of the AFOSM method. The SSD calculated with AFOSM method was 9% larger than the SSD calculated with the FOSM method. The lateral clearance design value to satisfy deterministic SSD was 5% lower than the lateral clearance value to satisfy the FOSM SSD. Since the AFOSM provided larger SSD design value compared to the FOSM and deterministic methods, the lateral clearance value to satisfy AFOSM SSD was 19% larger than the lateral clearance value to satisfy deterministic SSD and 15% larger than lateral clearance value to satisfy FOSM SSD value.

6 CONCLUSION

The current method of calculating the SSD for roundabouts is deterministic (all of the design variables are predetermined, fixed values). The reliability-based design of SSD at roundabouts uses random variables as design variables. One of the major parameters in the design of SSD, in reliability-based design method, is accuracy of design variables. Since the random variables considered in the probabilistic methods vary, it is recommended that the vehicles operating speed, drivers perception-reaction time, and vehicles deceleration rate are measured on the field for the specific roundabout under study and the variability of these variables are used in the design.

A high level of reliability required larger SSD and vice versa. The literature suggests that the probability of failure for facilities are chosen based on the level of importance of the facilities. One of the major benefits of the probabilistic design method is that it provides designers insight into the reliability of their design. The SSD design value calculated with probabilistic method provide more precise lateral clearance design value based on the randomness of the design variables. The probability methods developed in this study would be useful for designers to select SSD design values based on their desired probability of failures. Design aids were developed for the lateral clearance requirements on the approaches to satisfy SSD. The largest of the lateral clearances design values obtained to satisfy SSD to the crosswalk and to the yield line should be provided for the entire length of the entry curves.

References

American Association of State Highway and Transportation Officials. 2011. "*A Policy on Geometric Design of Highways and Streets, 6th edition.", AASHTO*, Washington, DC, USA

Easa, S. 2000. Reliability Approach to Intersection Sight Distance Design. Transportation Research Record: Journal of the Transportation Research Board 1701:42–52. https://doi.org/10.3141/1701-06.

Easa, S. 2017. Lateral Clearance Needs for Stopping Sight Distance at Single-Lane Symmetrical Roundabouts. Transportation Research Board, Washington, DC, United States, **17-06289**, 1–18.

Easa, S. 1993. Reliability-Based Design of Intergreen Interval at Traffic Signals. Journal of Transportation Engineering, 119(2): 255-271.

Easa, S. 1994. Reliability-Based Design of Sight Distance at Railroad Grade Crossings. Transportation Research Part A: Policy and Practice, 28A (1):1–15.

Easa, S, and Cheng, J. 2013. Reliability Analysis of Minimum Pedestrian Green Interval for Traffic Signals. Journal of Transportation Engineering, 139(7):651–659. https://doi.org/10.1061/(ASCE)TE.1943-5436.0000549.

Hasofer, A, M, and Lind, N, C. 1974. Exact and invariant second-moment code format. J. Eng. Div. ASCE. 100(1): 111-121

Hussain, A, and Easa, S. 2016. Reliability Analysis of Left-Turn Sight Distance at Signalized Intersections. J. Transp. Eng., Part A: Systems, 142(3).

Hussein, M, Sayed, T, Ismail, K, Espen, A, V. 2014. "Calibrating Road Design Guides Using Risk-Based Reliability Analysis." Journal of Transportation Engineering 140 (9):04014041-1–04014041-6. https://doi.org/10.1061/(ASCE)TE.1943-5436.0000694.

Robinson, B, W, Troutbeck, R, B, Werner, B, L, Courage, K, K, Michael, M, J, FLannery, A, M, Edward, B, J, and Jacquemart, G. 2010. Roundabouts: An Informational Guide. Transportation Research Board, Washington, DC, USA. https://doi.org/10.17226/22914.

Santos-Berbel, C, Essa, M, Sayed, T, and Castro, M. 2017. Reliability-Based Analysis of Sight Distance Modelling for Traffic Safety. Journal of Advanced Transportation 2017:1–12. https://doi.org/10.1155/2017/5612849.

Sarhan, M, and Hassan, Y. 2011. Reliability-based three-dimenstion design of horizontal lateral clearance. Can. J. Civ.Eng. 38: 900-908

Scott, C, and Donnell, T. 2014. Reliability Approach to Horizontal Curve Design. Transportation Research

Record: Journal of the Transportation Research Board 2436 (2436):51–59. https://doi.org/10.3141/2436-06.
Smith, G, N. 1986. *Probability and statistics in civil engineering,* Nichols Publishing Company, New York.

Building Tomorrow's Society
Bâtir la Société de Demain

Fredericton, Canada
June 13 to 16, 2018 / *13 au 16 juin, 2018*

THE DETERIORATION OF PAVEMENT MARKING RETROREFLECTIVITY: ARE WE READY TO ADOPT MINIMUM STANDARDS?

Carly MacEacheron[1], Eric Hildebrand[2,3], and Trevor Hanson[2]
[1] WSP Canada Group Ltd., Canada
[2] University of New Brunswick, Canada
[3] edh@unb.ca

Abstract: Pavement markings with retroreflective properties provide delineation and safety benefits for drivers during nighttime conditions. Many road authorities do not have a history of retroreflectivity performance for pavement markings throughout their life-cycles. This makes it difficult to assess the impacts of potential minimum retroreflectivity standards such as those proposed by the Federal Highway Administration (FHWA) or to adopt performance-based specifications. A year-long study of pavement markings on 24 sampled road sections was performed to better understand the state of pavement markings in New Brunswick and the causes/rates of their deterioration. Analyses found the factors that have a statistically significant effect on pavement marking retroreflectivity include age, traffic volume, road class, season when the marking was applied, and paint colour. Analyses of the sampled retroreflectivity values found that they generally deteriorate over time, and that white markings consistently produce higher readings than yellow markings through their life-cycle even though white markings deteriorate more quickly. Overall, 27% and 38% of sampled pavement markings met the 2014 or 2017 proposed FHWA minimum retroreflectivity standards at the time of their replacement, respectively. These findings suggest that the current pavement marking maintenance schedule used in New Brunswick (which is consistent with many other jurisdictions) would be inadequate if minimum standards consistent with the FHWA were to be adopted. It is recommended that road authorities develop pavement marking policies that will improve compliance with an adopted set of minimum retroreflectivity standards or guidelines.

1 INTRODUCTION

Longitudinal pavement markings are among the most important traffic control devices for their ability to delineate roadways and provide safety benefits to drivers. To provide the same benefits during nighttime and inclement weather conditions, pavement markings must have retroreflective properties to be adequately visible. Retroreflectivity, defined as "a measure of an object's ability to reflect light back towards a light source along the same axis from which it strikes the object" (FHWA 2007), is usually achieved in pavement markings by applying glass retroreflective beads onto the marking material's surface. Resulting retroreflective properties allows vehicles' headlights to by reflected approximately back towards the driver's eyes.

Retroreflectivity of pavement markings is most commonly measured using the Coefficient of Retroreflected Luminance (R_L). It is the ratio of luminance (the total amount of light reflected back to the driver) and normal illuminance (the illuminance of the headlights on the marking) and is measured in millicandelas per lux square metre (mcd/m^2/lx). Measurements are taken using a retroreflectometer, which emits a light at an angle considered to represent standard geometry, as defined by ASTM Standard E1710-11, and measures the amount of light reflected back. This simulates what an average driver would see during nighttime and inclement weather conditions.

The United States Manual on Uniform Traffic Control Devices (US-MUTCD) requires that pavement markings be retroreflective but does not currently specify any minimum retroreflectivity standards. In reaction to this, the U.S. Federal Highway Administration (FHWA) proposed minimum pavement marking retroreflectivity standards in 2014 to be published in the Manual on Uniform Traffic Control Devices (MUTCD), outlined in Table 1.

Table 1: Recommended MUTCD standards for minimum retroreflectivity levels of longitudinal pavement markings (FHWA 2014)

	Posted Speed (mph)		
	≤ 30	35 - 50	≥ 55
Two-lane roads with centerline markings only	n/a	100	250
All other roads	n/a	50	100

More recently, the FHWA (2017) issued for comment a simpler, much more lenient proposed standard whereby a minimum retroreflectivity of 50 mcd/m^2/lux is required if posted speeds exceed 35 mph, and 100 mcd/m^2/lux where posted speeds exceed 70 mph. The new proposed standard only applies to road with an AADT of 6,000 or above. It is unknown whether this supplemental 2017 proposal will ultimately be adopted, but this study includes consideration for the potential adoption of either standard.

The adoption of minimum standards by U.S. jurisdictions may lead to similar standards or guidelines in Canada, but little if any research has explored prospective impacts on Canadian jurisdictions that adopt minimum retroreflectivity standards. The New Brunswick Department of Transportation and Infrastructure (NBDTI) operates a pavement marking program that is based on traffic volume-based repainting thresholds detailed in its pavement marking specifications, which resulted in the striping of 7,100 km of the Province's 18,000 route-km of paved roads in 2015-2016 (GNB 2016). The only retroreflectivity standard specification employed by NBDTI concerns material type (e.g. markings embedded with glass beads) and minimum retroreflectivity levels for new markings as a quality control measure. The introduction of retroreflectivity-based thresholds for in-service markings has the potential to significantly impact re-striping operations, though the degree of this impact is unclear. Like many road authorities, NBDTI does not have historical retroreflectivity data that describes pavement marking deterioration over their life-cycles in practice and, as such, it would be difficult for NBDTI to make any informed decision regarding adopting minimum retroreflectivity standards without further research.

A year-long study was undertaken to monitor the retroreflectivity of pavement markings on 24 sections of various classes of New Brunswick roads to develop an understanding of the rates and causes of pavement marking deterioration in the province and to assess compliance with FHWA's proposed minimum standards. The results of this study will be of interest to all Canadian highway agencies in advance of the possible adoption of minimum retroreflectivity standards or for the development of performance-based specifications.

2 BACKGROUND

2.1 Factors Affecting the Deterioration of Pavement Marking Retroreflectivity

Factors identified in the literature that are known to cause pavement marking retroreflectivity deterioration include: marking material, marking colour, traffic volume and heavy vehicle percentage, application methods, lateral line location, directionality, winter maintenance, and road surface type and condition. The causes and magnitude of the major contributing factors are described below.

Pavement marking materials are generally grouped in either a durable or non-durable category. Non-durable materials include traffic paints that have shorter service lives than more durable materials such as thermoplastics and preformed tapes. Paints have the shortest service lives of all pavement marking materials and have much lower costs, resulting in many road authorities including New Brunswick to use paints despite their lack of durability. NBDTI primarily uses waterborne paints for pavement marking application since oil-based paints can only be used between mid-October and May in Canada due to volatile organic compound (VOC) emission standards. The two traffic paints have been found to have comparable service lives and retroreflectivity levels throughout their life-cycle. Additionally, white pavement markings have been shown to produce significantly higher retroreflectivity readings than yellow markings (Migletz *et al.* 1999). White markings have been found to degrade more rapidly than yellow markings, but white markings' higher initial values usually offset their faster degradation rate and yellow markings remain at lower R_L values throughout their life-cycle (Bahar *et al.* 2006).

It is generally accepted that traffic volume and composition affect the deterioration of pavement marking retroreflectivity. This is because tires passing over the lines lift and scatter the retroreflective beads or abrade them enough for them to eventually lose their retroreflective properties. As such, pavement markings on roads with higher AADTs have generally been found to have higher rates of pavement marking retroreflectivity deterioration (Migletz *et al.* 1994). Marking location and road geometry are also determining factors of how significant the effect of traffic volume is on retroreflectivity degradation. Centreline and lane line retroreflectivity has been found to degrade more quickly than edge lines due to higher numbers of traffic crossovers at the centre of a road compared to the edge (Craig *et al.* 2007). Road geometry characteristics such as lane width, horizontal curvature, and the presence of driveways have also been found to cause pavement marking degradation due to vehicle crossovers (Migletz *et al.* 1994; Bowman & Abboud 2001).

Winter maintenance activities are generally considered to have the largest effect on pavement marking retroreflectivity compared to any other factor, creating a substantial issue for jurisdictions with snowfall (Scheuer *et al.* 1997). Deterioration from winter maintenance is primarily caused by snowplow blades scraping the markings, which lifts and abrades the beads as well as the marking itself. Similar effects can be found with studded tires, which are prevalent during winter seasons. Salts and other de-icing agents can cause retroreflectivity deterioration through chemical disintegration of the marking material and/or loosening of the bond between the marking and pavement. Sand can also act as an abrasive that degrades the translucent quality of the retroreflective beads (Bowman & Abboud 2001). Studies have found that yellow and white traffic paint markings' retroreflectivity readings can deteriorate up to 21% and 62%, respectively, over a winter season (Lu & Barter 1998). It appears to be the consensus in literature that paint pavement markings do not exhibit adequate durability for use in regions with significant amounts of snowfall such as New Brunswick.

The more texture a road surface has, the poorer the retroreflectivity of its pavement markings. This is because irregularities in the road surface may allow marking materials or retroreflective beads to sink into voids and aggregates to cast a shadow on them, making them less visible to drivers. This applies to chip seal surfaces with their large proportion of surface voids, and cracked or uneven asphalt surfaces. Retroreflectivity readings have been found to be 23% and 38% lower for white and yellow thermoplastic markings, respectively, on chip seal compared to hot-mix asphalt surfaces (Gates *et al.* 2003).

2.2 Service Life and Degradation of Pavement Markings

The performance of pavement markings is usually described in terms of "service life", the time expended between application and the point at which retroreflectivity readings are below a pre-determined minimum level. Literature indicates that service lives for paint pavement markings are highly variable and range from six months and two years.

Pavement marking retroreflectivity deterioration is usually described as the relationship between retroreflectivity and age, but a consensus has not yet been reached on the graphical form of this relationship. This is likely due to the highly variable nature of the rate of deterioration and its dependence on various factors. There is therefore no defined method of predicting pavement marking retroreflectivity levels over time and there is great value in individual road authorities having historical retroreflectivity data available to determine re-striping frequency.

2.3 NBDTI Re-Striping Practice

NBDTI employs a traffic volume-based warrant to guide its re-striping program. NBDTI's pavement marking specifications stipulate that roads with volumes above 1,000 AADT are re-striped yearly, and those with volumes between 500 and 1,000 AADT are only required to be re-striped every two years; roads with volumes below 500 AADT are not striped.

3 METHODOLOGY

3.1 Data Collection

A sample of 24 local, collector and arterial NBDTI highways were chosen to represent selected categories of traffic volumes and road surface type (asphalt and chip seal). All sample locations were approximately within a 30-minute drive of Fredericton, New Brunswick based on the assumption that they were representative of climate conditions, winter maintenance and pavement marking application methods throughout the province. Sample location selection was primarily based on capturing full pavement marking life-cycles over a year-long data collection period. Since the sampling period was only a year, samples representing roads re-striped biennially were selected with some in their first year and some in their second year since the last pavement marking application piecing together a two-year cycle.

Data from the 24 sample locations were collected monthly between December 2014 and November 2015, with the exception of the three month period between January and March of 2015 when constant salt, sand and/or ice on the roads prevented the dry, clean surface required for accurate retroreflectivity readings. A Mirolux Plus 30-metre handheld retroreflectometer was used as per the requirements of ASTM E1710-11. Three retroreflectivity readings were taken longitudinally at each of the edge lines, centrelines and/or lane lines, depending on the marking configuration at that sample location, providing an average reading for each pavement marking.

3.2 Data Analyses

Analyses of sampled data included graphical analysis examining the deterioration of the sampled pavement markings' retroreflectivity over time and their compliance with FHWA's proposed minimum standards, followed by a modelling effort to identify factors contributing to the deterioration of the pavement markings' retroreflective properties.

Scatter plots of pavement marking retroreflectivity readings versus age were plotted for the sampled data over their one or two-year life-cycles. Data sets were fit with linear trend lines representing the deterioration curve over time. Different plots were made according to marking configuration and speed limit to compare to different proposed minimum values in the standards. The sample analysis then included a multiple linear regression analysis to test the effects of independent variables including age, season, road class, surface material, traffic volume, and paint type on the dependent variable, retroreflectivity.

4 RESULTS

4.1 Graphical Analysis

The analysis included scatter plots of retroreflectivity values over time at individual sample locations to develop an understanding of the general rate of pavement marking deterioration over a full installation cycle. The readings generally drop over time, but a significant amount of variability was seen with some locations even having slightly higher retroreflectivity readings at times before resuming in a downward trend. This anomaly may be due to inherent variability in using a handheld retroreflectometer, or because of sampling variability associated with meter placement (despite best efforts to remain consistent and the use of averages for readings to alleviate such issues). Similar variability in pavement marking retroreflectivity data has been identified in several other studies (Scheuer *et al.* 1997; Kopf 2004). Note, any 'negative' retroreflective values can be attributed to retroreflectometer calibration variability.

All data points from the 24 test sections were combined into Figure 1 to develop an overall trend of retroreflectivity over time. Linear trend lines were fit to all data points, and specifically to white (edge line) and yellow (centreline) markings separately. Trend lines helped to develop a general understanding and comparison of deterioration rates over time. Sampled white pavement markings consistently had higher retroreflectivity readings than yellow, but white markings appeared to deteriorate more quickly than yellow as shown by the model slopes. White readings were high enough compared to yellow upon initial application that the higher deterioration rate still led to the white markings being higher through most of their life-cycle.

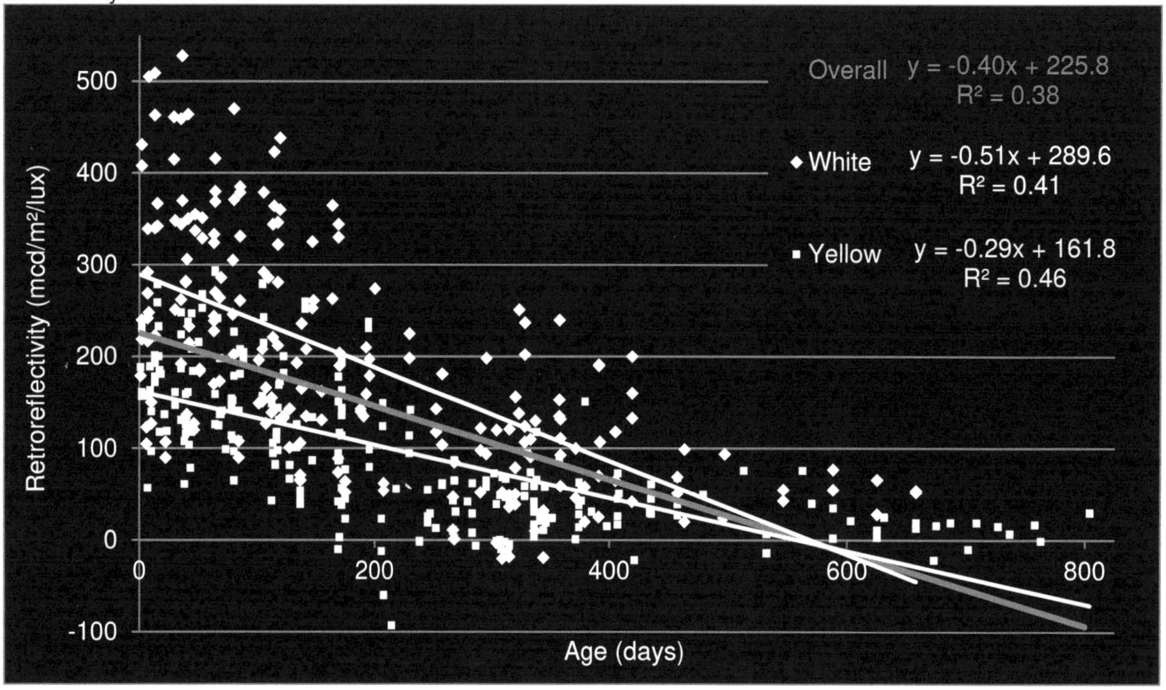

Figure 1: Retroreflectivity deterioration of white, yellow (and overall) sampled markings over time

4.1.1 Comparison to Proposed FHWA Standards

The data were then compared to FHWA's 2014 and 2017 proposed minimum retroreflectivity standards by grouping the sampled pavement markings by speed limit and configuration (consistent with the standards) and plotting them over time. The plots include a horizontal line representing the proposed minimum retroreflectivity level for that given speed zone and marking configuration to compare the data to the FHWA standards.

Retroreflectivity data from samples with centrelines only and a posted speed limit between 35 and 50 mph (56 to 80 km/h) are plotted against age in Figure 2. The red and blue horizontal lines in the graph represent the 2014 and 2017 minimum standards of 100 and 50 mcd/m^2/lux, respectively. A linear trend line was fit to the data, showing a negative slope that falls below the 2014 minimum standard just beyond 100 days following application, and below the 2017 standard just beyond a year. It is noteworthy that apart from the fitted regression line, many of the sampled sites show retroreflectivity levels below either proposed standard beginning at approximately 150 days after installation. This is well short of their current intended life-cycles of either 1 or 2 years.

Sampled roads with centrelines only and speed limits between 55 and 70 mph (90-110 km/h) are plotted in Figure 3 comparing their compliance with the 2014 and 2017 minimum retroreflectivity standards of 250 and 50 mcd/m^2/lux, respectively. There was only one sample road section included in the study that fit this category because road sections with higher speed limits typically also have edge lines in New Brunswick.

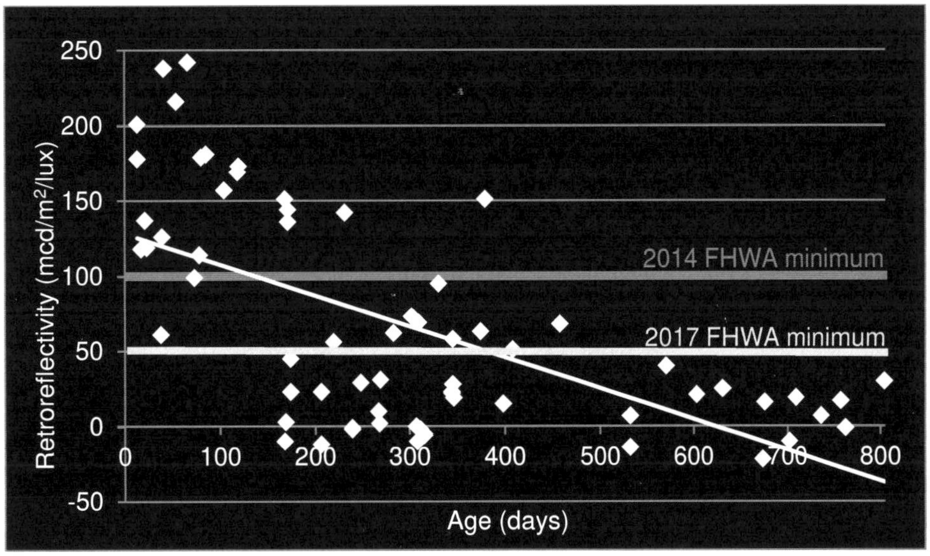

Figure 2: Compliance to FHWA's minimum retroreflectivity standards (centreline markings only and speed limits 35-50 mph)

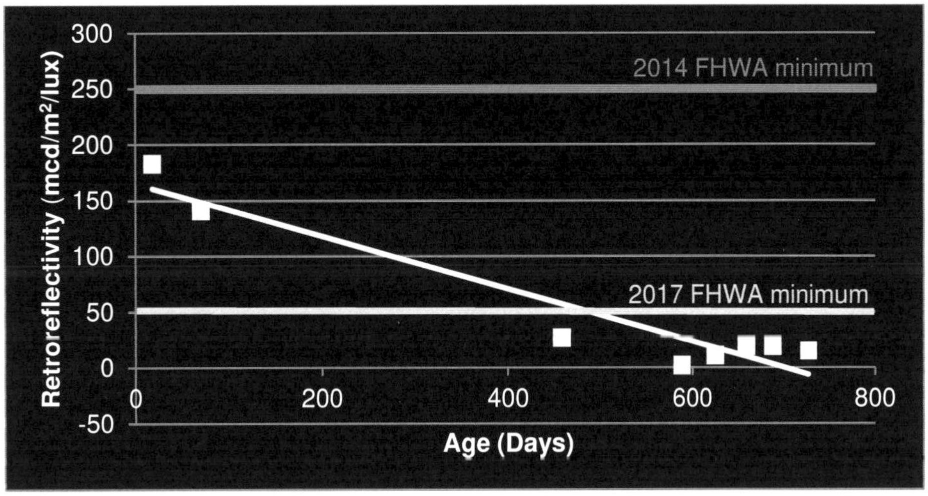

Figure 3: Compliance to FHWA's minimum retroreflectivity standards (centreline markings only and speed limits 55-70 mph)

Observed retroreflectivity values at this sample section are well below the minimum retroreflectivity standard of 250 mcd/m^2/lux throughout its entire life-cycle but fall below the 50 mcd/m^2/lux threshold more than a year following re-striping. NBDTI's quality control standards require that yellow pavement markings are at a minimum of only 200 mcd/m^2/lux two to four weeks after installation, making it unlikely that markings in this category meet the FHWA standard of 250 mcd/m^2/lux even at the beginning of their life-cycle, let alone throughout one or two-year cycles. If the 2014 standard were put into place, a modification of retroreflectivity requirements at the time of application or a change in marking configuration for sections in this category would need to be implemented by NBDTI.

Sample sections categorized as "all other roads" by the FHWA, or, in the case of this study roads with edge lines as well as centrelines, and having speed limits between 35 and 50 mph (56-80 km/h) are plotted and compared to their minimum standard of 50 mcd/m^2/lux (proposed in both 2014 and 2017) in Figure 4. The sample dataset complies well with minimum standards, with the yellow trend line reaching below the minimum standard at approximately a year and the white at close to 500 days. The compliance with the standard is likely because it is a low standard compared to the initial retroreflectivity readings of 200 and 250 mcd/m²/lux required by NBDTI's quality control standards for yellow and white lines, respectively.

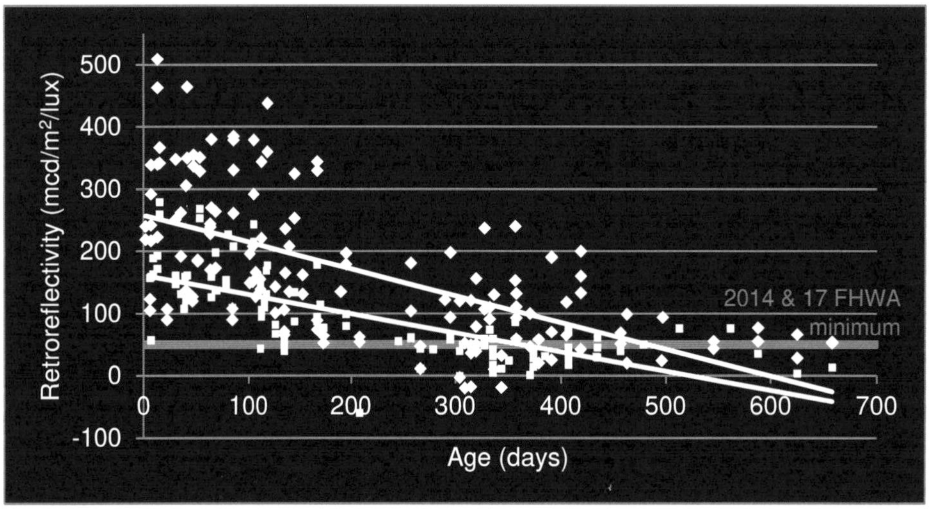

Figure 4: Compliance to FHWA's minimum retroreflectivity standards (centreline and edge markings and speed limits 35-50 mph)

Yellow centrelines and white edge lines from sample sections in the all other roads with speed limits between 55 and 70 mph are plotted with their 2014 and 2017 minimum standards of 100 and 50 mcd/m²/lux, respectively, in Figure 5. White markings comply well with both sets of standards, while the yellow trend line reaches both the minimum standards within a year.

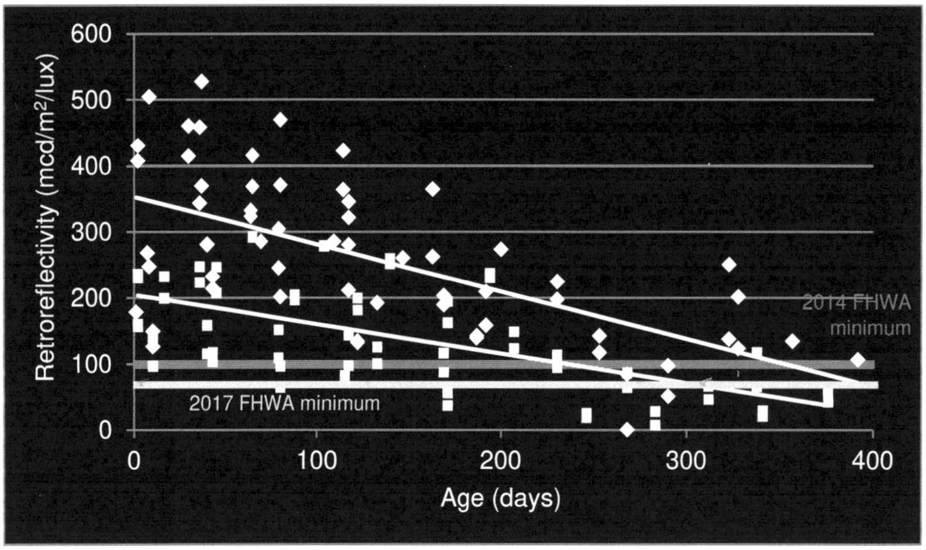

Figure 5: Compliance to FHWA's minimum retroreflectivity standards (centreline and edge line markings and speed limits 55-70 mph)

Results of the graphical analysis were summarized according to the percentage of all sample sections complying with 2014 FHWA's proposed minimum retroreflectivity standards over time in order to develop an understanding of the overall compliance to the proposed standards and to provide service life estimates. Cumulative percentages of sampled markings complying with the standard over three, six, and twelve month intervals were recorded for each marking configuration/speed category in Table 2.

Table 2: Sampled pavement markings passing FHWA's 2014 proposed minimum retroreflectivity standards

Marking configuration and speed (mph)	Paint Colour	Sample size	% Passing at 3 months	% Passing at 6 months	% Passing at 1 year
Centreline only (35-50)	Yellow	7	71%	71%	0%
Centreline only (≥ 55)	Yellow	1	0%	0%	0%
Centreline and edge line (35-50)	White	20	100%	100%	65%
	Yellow	13	100%	77%	23%
Centreline and edge line (≥ 55)	White	7	100%	100%	43%
	Yellow	8	75%	50%	13%
Overall	W&Y	56	91%	82%	36%

Overall compliance of the sample sections were then compared to the 2017 FHWA proposed minimum standards. A minimum retroreflectivity level of 50 mcd/m^2/lux was applied to all sections, since all samples were taken on roads with speed limits between 35 and 70 mph. Sampled roads with AADT values below 6,000 were included in the analysis despite being excluded from the new proposed standards because it is considered important to maintain some level of retroreflectivity on the Province's roads. Most New Brunswick roads do not have AADTs that exceed 6,000 and, given the more lenient standards, it would be achievable maintain the standards on roads currently included in the Province's re-striping schedule. The analysis found improved compliance to the new proposed standard, with all markings conforming at three months, 82% at six months and 51% at a year from their re-striping date.

Observing the proportion of markings passing both sets of minimum standards over a year, it appears that compliance within a 1-year cycle is low even with the more lenient 2017 standards. Only 38% of sampled markings met the 2017 standard at the time of their replacement (compared to 27% compliance to the 2014 standards). Compliance with both sets of standards appeared to drop off between six months and one year (46% and 31% of samples based on 2014 and 2017 standards, respectively), often because the winter season occurred during that period. The 2014 standard indicates that yellow lines are less compliant than white, with only up to 23% of yellow markings passing after a year, and yellow markings on road sections with edge lines also last longer than roads with centreline(s) only. Comparing individual sampled markings' compliance with overall compliance described by the trend lines in Figures 2 through 5, it appears that better compliance is exhibited when analyzing data grouped by category compared to estimated service lives of individual markings. For example, the trend line for yellow markings in Figure 4 shows a service life of approximately one year, but individual analysis of markings in the same category found only 23% complied within a year. This suggests that if a pavement marking management plan were to be implemented at a network level, individual sections of road would have markings below minimum standards. Road agencies may consider implementing a management plan that considers more specific road characteristics, such as AADT or road class, to maximize compliance.

Findings from the graphical analysis suggest that the current pavement marking maintenance schedule would be inadequate if either of the proposed FHWA minimum retroreflectivity standards were put into place in New Brunswick. Assuming the 7,100 route-kilometres re-striped in the 2015-2016 NBDTI re-striping program is the typical annual value, increasing the frequency of pavement marking maintenance at this magnitude to meet the standards would likely represent a substantial increase in resource requirements.

4.2 Modelling Analysis

A multiple linear regression analysis was performed on the data to evaluate which of the independent variables (age, volume, road class, striping season, and surface type) have a statistically significant effect on the dependent variable, retroreflectivity. Several models including various groupings of data based on pavement marking colour were created with the goal of creating an explanatory model for retroreflectivity. Such an exercise should be undertaken by any road authority considering implementing a pavement marking management plan in order to meet minimum standards or guidelines.

All variables were found to be significant except the asphalt surface type (compared to chip seal). Age and traffic volume were found to have negative effects on retroreflectivity as expected. White pavement markings were found to produce significantly higher readings than yellow pavement markings, and arterial and local functional road classes both model higher effects on retroreflectivity than collector roads. Retroreflectivity readings were shown to be higher the earlier in the year they were re-striped (beginning in May), which reflects the increased deterioration that occurs during the harsh winter conditions.

Results of the multiple regression model generally output the relationships between independent variables and retroreflectivity as expected based on the literature review, with the exception of the road class and road surface type variables. It was expected that collector roads would produce higher readings than local roads because they are generally re-striped more often, but the results were that arterial and local roads produce higher retroreflectivity readings than collector roads. It is suspected that correlation between traffic volume and road class led to this result. It was also expected based on the literature review that pavement surface would have a significant effect on retroreflectivity values, but it is predicted that this result was not shown in this model because of correlation between road class and pavement surface material, since most chip seal surfaces are found on local roads.

5 CONCLUSIONS AND RECOMMENDATIONS

A sample of pavement marking retroreflectivity in New Brunswick over a year-long period revealed information on rates, causes, and characteristics of deterioration over time that will be beneficial to many jurisdictions considering implementing minimum standards (as proposed in the U.S.).

Scatter plots of sampled data found that pavement marking retroreflectivity generally decreased over time. White markings produce retroreflectivity readings over 120 mcd/m^2/lux higher than yellow markings upon their initial installation and generally continue to produce higher readings through their life-cycle, though white markings deteriorate more quickly than yellow markings. Reading variability/uncertainty is an issue given current retroreflectometer technology that should be considered with any plan implementation.

Graphical analysis comparing sampled retroreflectivity data to FHWA's proposed minimum standards found that only 27% of the pavement markings sampled complied with the 2014 standard at the time of their replacement. This value increased to 38% when compared with the more lenient 2017 proposed standards. With many markings having a service life of six months to one year, NBDTI's current maintenance schedule of re-striping roads every one to two years would therefore be inadequate if FHWA's retroreflectivity standards were to be adopted.

Modelling analyses of the sampled data found factors that have a statistically significant effect on pavement marking retroreflectivity to be age, traffic volume, road class, season in which the marking was applied, and paint colour. Age and traffic volume were found to have a negative effect on retroreflectivity readings. Functional road classification has an impact as does the season of marking application.

Based on the sampled pavement markings' overall noncompliance, major modifications to NBDTI's current pavement marking management plan would be required to meet any minimum standards. It is likely than many other road authorities would be in a similar position. Any management plan should incorporate road characteristics found to be statistically significant in the regression model in order to improve compliance over a network-wide plan. Paint pavement markings should be re-striped at least once a year on all roads, and the use of more durable pavement marking materials should be considered if more infrequent maintenance is desired. Other efforts that may be taken to maximize the service life of pavement markings include re-striping pavement markings on road sections that are more prone to deterioration, such as high-traffic roads, as early in the season as possible to offset the effects of the winter season. Management plans may be amended pending publication of an approved set of standards, but in the meantime, these recommendations may be carried forward by NBDTI and other road authorities in a staged approach to create pavement marking management plans and policies that may improve compliance with FHWA's proposed minimum standards, ultimately improving the safety performance of the road network.

Acknowledgements

The participation and support of the Traffic and Engineering Branch of NBDTI toward the design and completion of this research study is greatly appreciated. The study would not have been possible without funding provided by the Natural Sciences and Engineering Research Council of Canada.

References

American Society for Testing Materials (ASTM). 2011. *ASTM Standard 1710-11: Standard Test Method for Measurement of Retroreflective Pavement Marking Materials with CEN-Prescribed Geometry using a Portable Retroreflectometer.* ASTM International, West Conshohocken, PA.

Bahar, G., Masliah, M., Erwin, T., Tan, E., & Hauer, E. 2006. *Pavement Marking Materials and Markers: Real-world Relationship between Retroreflectivity and Safety over Time.* National Cooperative Highway Research Program Project 17-28. Transportation Research Board, Washington, D.C.

Bowman, B. & Abboud, N. 2001. *Estimating the Effective Life of Pavement Markings based on Crash History.* Auburn University Department of Civil Engineering, Auburn, AL. Available from http://www.eng.auburn.edu/files/centers/hrc/IR-01-02.pdf [accessed 4 June 2015].

Craig, W.N., Sitzabee, W.E., Rasdorf, W.J. & Hummer, J.E. 2007. Statistical Validation of the Effect of Lateral Line Location on Pavement Marking Retroreflectivity Degradation. *Public Works Management & Policy*, **12**(2): 431-450. doi: https://doi.org/10.1177/1087724X07308773.

Federal Highway Administration (FHWA). 2007. *Updates to Research on Recommended Minimum Levels for Pavement Marking Retroreflectivity to meet Driver Night Visibility Needs.* U.S. Department of Transportation, McLean, VA.

Federal Highway Administration (FHWA). 2014. *Summary of the MUTCD Pavement Marking Retroreflectivity Standard.* U.S. Department of Transportation, McLean, VA.

Federal Highway Administration (FHWA). 2017. *Pavement Markings Regulations / Standards, Supplemental Notice of Proposed Amendment (SNPA).* U.S. Department of Transportation, McLean, VA. Available from https://safety.fhwa.dot.gov/roadway_dept/night_visib/pavementreg.cfm [accessed 29 Jan 29 2018].

Gates, T.J., Hawkins, H.G. & Rose, E.R. 2003. *Effective Pavement Marking Practices for Sealcoat and Hot-Mix Asphalt Pavements.* Texas Transportation Institute, College Station, TX. Available from http://d2dtl5nnlpfr0r.cloudfront.net/tti.tamu.edu/documents/0-4150-4.pdf [accessed 24 June 2015].

Government of New Brunswick, 2016. *Annual Report 2015-2016* Transportation and Infrastructure, Fredericton, NB. Available from http://www2.gnb.ca/content/dam/gnb/Departments/trans/pdf/en/Publications/2016AnnualReport.pdf [accessed 23 Jan 2018].

Kopf, J. 2004. *Reflectivity of Pavement Markings: Analysis of Retroreflectivity Degradation Curves.* Washington State Transportation Center, Seattle, WA. Available from http://www.wsdot.wa.gov/research/reports/fullreports/592.1.pdf [accessed 21 May 2015].

Lu, J.J. & Barter, T. 1998. Evaluation of Traffic Markings in Cold Regions. *Journal of Transportation Engineering*, **124**(1): 42-51.

Migletz, J., Fish. J.K. & Graham, J.L. 1994. *Roadway Delineation Practices Handbook.* U.S. Department of Transportation, Washington, D.C.

Migletz, J., Graham, J.L., Bauer, K.M. & Harwood, D.W. 1999. Field Surveys of Pavement-Marking Retroreflectivity. *Transportation Research Record*, **1657**: 71-78. doi: https://doi.org/10.3141/1657-10.

Scheuer, M., Maleck, T.L., & Lighthizer, D.R. 1997. Paint-Line Retroreflectivity over Time. *Transportation Research Record*, **1585**: 53-63. doi: https://doi.org/10.3141/1585-08.

Building Tomorrow's Society
Bâtir la Société de Demain

Fredericton, Canada
June 13 to 16, 2018 / *13 au 16 juin, 2018*

THE IMPACT OF PAVEMENT CONDITION ON ROAD SAFETY IN NEW BRUNSWICK

Arash Ghanbari [1], Xiomara Sanchez[1], Eric Hildebrand[1,2], and James Christie[1]
[1] University of New Brunswick, Canada
[2] edh@unb.ca

Abstract: This study developed an understanding of the relationship between pavement condition and road safety on rural arterial and collector roads in the province of New Brunswick. The intent was to use this relationship for consideration as a decision-making criterion in asset management programs. Two crash prediction models were developed for single-vehicle and multiple-vehicle collisions using a negative binomial regression that relates pavement condition indicators to accident frequency. International Roughness Index was found to be significant in both models which indicate that IRI is a contributing factor to single-vehicle and multiple-vehicle accident involvements. It is noteworthy that the model coefficient for IRI was determined to be negative in both models which reflects that a higher number of collisions is expected to occur on road sections with smoother pavement surfaces. Crash modification functions were developed for both single-vehicle and multiple-vehicle collisions. The results indicate that a unit of increase in IRI in m/km is expected to decrease single and multiple-vehicle accidents by 7.7% and 14.8%, respectively. It was also found that the percentage of change in multiple-vehicle collisions due to pavement roughness changes, while all other variables in the models remain constant, is approximately two times greater than the percentage of change in single-vehicle accidents. According to the crash modification functions (CMF) associated with pavement age, the results indicated that collector roads are slightly more impacted by pavement roughness than arterial roads.

1 INTRODUCTION

There were 1,669 fatal and 116,735 injury motor vehicle accidents in Canada in 2015 where more than 1,800 lives were lost and nearly 162,000 people were injured (Transport Canada 2015). Statistics Canada (2018) information shows accidents were the fourth or fifth leading cause of deaths in Canada from 2009 to 2014. These results indicate that, despite many recent studies undertaken with the goal of reducing collisions, safety remains a significant issue necessitating further investigations to identify more efficient approaches to reduce accidents.

Most of the studies on traffic safety have evaluated the impact of human factors, road geometrics, roadway environmental factors, and vehicle attributes on collisions. Few studies have investigated the impact of pavement condition on road safety due to a lack of a quantitative pavement data collection technology before the 1990s (Chan *et al.* 2009).

Pavement condition is assessed using pavement condition indicators which are used to translate agency objectives into specific, quantifiable measures required for pavement management (TAC 2013). Exploring the relationship between pavement condition indicators and collision experience can help transportation agencies to quantitatively evaluate their pavement condition in terms of safety to provide baseline data for asset management systems.

2 IMPACT OF PAVEMENT CONDITION ON COLLISION OCCURRENCE, TYPE, AND SEVERITY

Several studies have quantitatively evaluated the impact of pavement condition on accidents using various pavement condition indicators including the impact on accident occurrence, collisions types, and accident severity. It is noted that only international roughness index (IRI) and rut depth (RD) were used in this study. Pavement roughness is defined as the deviation of a surface from a true planar surface with characteristic dimensions that affect vehicle dynamics, ride quality, dynamic loads and drainage, for example, longitudinal profile, transverse profile, and cross slope. International roughness index is used to quantify pavement roughness. The profile of the road is measured by road profilers which produce a series of offsets related to the "true profile" of the road allowing IRI to be calculated in m/km or in/mi.

Start *et al.* (2007) quantified the impact of rutting on collision rates on rural undivided highways where he found that rut depth levels greater than 7.6 mm significantly affect road safety. Anastasopoulos and Mannering (2009) found that international roughness index (IRI) is an excellent binary variable significantly influencing collision occurrences on low volume rural interstate highways. They also indicated that a unit increase in the maximum IRI (1 in/mi = 0.016 m/km), results in an average of 0.8 percent increase in the number of annual collisions.

Chan *et al.* (2009) revealed that IRI is a significant variable to predict collision frequency on urban interstate highways with moderate to high traffic volumes, while RD is not a contributing factor. According to their analysis, a 100 in/mi increase in IRI would increase collision frequency by 1.6 times. Li and Huang (2014) found that impact of poor pavement condition on accidents is more significant on urban collectors, and high-speed roads (speed limits of greater than 104.6 km/hr).

Al-Masaeid (1997) found that the increase in the IRI level would reduce the single-vehicle accident rate while increasing the multiple-vehicle accident rate on two-lane undivided roads in Jordan. He also indicated that IRI has no statistical effect on the total accident rate. He suggested that it would be beneficial from a safety perspective to keep the IRI below 5 m/km on rural two-lane roads.

Li *et al.* (2013) quantified the impact of pavement condition on crash severity and found the severity is more significant on multi-lane, non-freeway arterials with relatively high-speed limits, dry pavement surface, and daylight conditions. They determined that fatal or incapacitating crash severity level is more impacted by IRI than other crash severity levels. Zeng *et al.* (2014) conducted a before-after study to quantitatively evaluate the safety effectiveness of good pavement conditions versus deficient pavement conditions. They found that good pavement condition can reduce fatal and injury crashes by 25% on rural two-lane undivided primary highway segments with a traffic volume of from 76 to 29,142 vehicles per day. Lee *et al.* (2015) investigated the impact of poor pavement condition on crash severity levels for low (≤56.3 km/hr), medium (64.4 km/hr -72.4 km/hr), and high (≥80.5 km/hr) speed roads. They indicated that pavement condition increases the severity of multiple-vehicle crashes on all three speed-level roads.

3 RESEARCH GOAL AND OBJECTIVES

The overall goal of this study was to develop a better understanding of the relationship between pavement condition and safety in the province of New Brunswick for consideration as a decision-making criterion in the New Brunswick Department of Transportation and Infrastructure's (NBDTI) asset management system.

The specific objectives set out to meet the study goal were to:

1. Develop crash prediction models for single-vehicle and multiple-vehicle accidents to relate collision frequency to the pavement condition indicators.

2. Develop crash modification factors (CMF) for single-vehicle and multiple-vehicle accidents to estimate the effectiveness of pavement condition improvements on collisions.

4 METHODOLOGY

The targeted facility types were segments (control sections) of rural two-lane, two-way, undivided collector and arterial roads throughout the province of New Brunswick. For these facility types, nearly 3,100 kilometres of roads in the province were found to be appropriate for this study which included about 950 kilometres of arterial roads and over 2,100 kilometres of collector roads.

The following variables were gleaned from the selected control sections and included for analysis:

1. HORIZONTAL CURVATURE (>4 degrees): including the number of sharp horizontal curves per kilometre (NSHCPK) and weighted average degree of sharp horizontal curves (ADHC)
2. GRADIENT (>5%): including number of sharp grades per kilometre (NSGPK), and weighted average of sharp grades (AVSG)
3. INTERSECTION DENSITY: number of intersections per kilometre (IPK)
4. POSTED SPEED LIMIT (km/hr): weighted average posted speed (APSL)
5. TRAFFIC VOLUME: average annual daily traffic (AADT) of the selected control sections were obtained for use as potential predictors in the crash prediction models
6. PAVEMENT CONDITION INDICATORS: data were used to calculate the average IRI and RD on the control sections available on arterial roads
7. ACCIDENT FREQUENCY: To reduce the number of control sections with zero accident frequency and consequently improve the accuracy of the models, three-year total accident frequencies (single and multiple) from NBDTI datasets were used in this study.

5 STATISTICAL ANALYSIS

Since collision frequencies are count data which refer to observations that have only non-negative integer values ranging from zero to some greater undetermined values such as accident frequency, Poisson and Negative Binomial (NB) distributions can be used for developing the crash prediction models (Abdel-Aty and Radwan 2000). If collision data are equi-dispersed then Poisson regression is an appropriate approach. Given that the variance of the collision frequencies for the control sections were found to exceed the means, the Negative Binomial (NB) regression was a more appropriate modelling approach to employ for this study.

The NB model is a two-parameter model-with mean (μ) and dispersion parameter (α). The traditional parameterization of the negative binomial is known as the NB2 negative binomial model which was used in this study. Different methods have been proposed for estimating the coefficients of the NB models with the method of maximum likelihood estimation (MLE) being the most widely used. With respect to the dispersion parameter of the NB model, the traditional negative binomial model (TNB) which assumes that the dispersion parameter of the negative binomial distribution is fixed for all locations was selected for use in this study.

The minimum sample size requirements recommended by Lord (2006) as a function of the sample mean values were used in this study resulting in minimum required sample size for single-vehicle and multiple-vehicle collision models estimated to be 84 and 218, respectively which were satisfied given the number of available control sections (243).

5.1 Model Specification

The mathematical form of the crash prediction models should meet the following two criteria: it must yield logical results, meaning that it should not lead to the prediction of a negative number of accidents and it should ensure a prediction of zero accident frequency for zero values of the exposure and length variables.

There must exist a known link function that can linearize this form for the purpose of coefficient estimation (Sawalha and Sayed 2001). The following model used in this study satisfies these criteria.

$$[1]\ AF = a_0 * L^{a_1} * AADT^{a_2} * \exp\left(\sum_{j=1}^{m} b_j * x_j\right)$$

where, AF: Predicted accident frequency

L: Segment length (km)

AADT: Average annual daily traffic

x_j: Any of m variables additional to length and AADT

a_0, a_1, a_2 and b_j: Model parameters to be estimated

Explanatory variables for the final models were selected in a stepped approach beginning with AADT and length. Correlation and multicollinearity among variables were tested to ensure independence. Outlier data were identified with a procedure known as Cooks distance (Cook 1977, SAS/STAT 2015).

5.2 Assessment of Fit

While there are several measures that can be used for estimating how well a model fits the base data, in this study two approaches were used. The Scaled Deviance and Pearson χ^2 are asymptotically χ^2-distributed with n-p degree of freedom where n is the number of observations used in the model and p is the number model parameters including intercept and predictors included in the model. A model is valid if its scaled deviance and Pearson χ^2 are less than ${\chi^2}_{0.05,n-p}$ where ${\chi^2}_{0.05,n-p}$ is critical χ^2 value at the 95% confidence level which can be obtained from the chi-square distribution table.

The second criterion used in this study to assess goodness-of-fit of crash prediction models was the plot of the observed crash frequency versus predicted crash frequency. For a perfect model, the linear trendline fitted to the data, should have the equation y=x, while R^2 is equal to 1 or close to 1. In this study, regression analysis was performed to evaluate the intercept and slope of the linear trendline fitted to the data.

5.3 Pavement Deterioration Equations

Transportation agencies use pavement deterioration curves to predict how pavement condition will change over time in their road network. NBDTI uses different pavement deterioration curves in terms of IRI for the road classification types in New Brunswick. In order to simplify the application of these pavement deterioration curves, pavement deterioration equations were obtained through fitting the best trend line (highest R^2) to the data that the curves were based on using Microsoft Excel. Therefore, the following local pavement deterioration models were obtained for arterial and collector roads separately:

Arterial Roads:

[2] $IRI_{\_i}$ =0.8803*Exp(0.0442*i)

Collector Roads:

[3] $IRI_{\_i}$ =1.18*Exp(0.0395*i)

where:

$IRI_{\_i}$: IRI in the ith year

i: Pavement age.

5.4 Crash Modification Factors (CMF)

According to FHWA (2010), A CMF is simply defined as a multiplicative factor used to reflect the expected change in safety performance associated with the corresponding change in highway design and/or the traffic control feature. CMF for a countermeasure i (CMF_i) is calculated as follows:

$$[4]\ CMF_i = \frac{\text{Expected number of crashes if change I is made}}{\text{Expected number of crashes if change I is not made}}$$

There are several methods for developing CMFs including before-after with a comparison group, full Bayes, cross sectional, case control, etc. (FHWA, 2010). Since the method used in this study was a cross-sectional approach, crash modification functions (CMFunctions) associated with IRI for single-vehicle accidents and multiple-vehicle accidents were developed assuming that only IRI changes, while all other predictors in the models remain constant. For practical purposes, pavement deterioration equations were used to translate CMFunctions to pavement age.

6 ANALYSIS AND RESULTS

6.1 Single-Vehicle Collision Model

For the single-vehicle accident model, average annual daily traffic (AADT), length (L), number of intersections per kilometre (IPK), average vertical grades (AVG), and international roughness index (IRI) were identified to be significant. The highest value in the correlation matrix was determined to be -0.51 (between AADT and IRI) which shows that there is no major correlation between the pairs of variables included in the model. All the VIFs were calculated to be below two which is an indication that, there is no significant multicollinearity among the variables. Indeed, five observations were detected as outliers and removed from the model. As a result, the final model was developed using 238 observations (243 total -5 outliers). The result of the maximum likelihood parameter estimation analysis is presented in Table 1. The mathematical form of the single-vehicle accident model can be written as follows:

$$[5]\ SAF = 0.0085 * AADT^{0.59} * L^{1.04} * Exp(0.2 * IPK + 0.03 * AVG - 0.08 * IRI)$$

where: SAF: Single-vehicle accident frequency (accidents/3 years)

AADT: Average annual daily traffic (vehicles/day)

L: Length (km)

IPK: Intersection density per kilometre

AVG: Weighted average gradient (%)

IRI: International roughness index (m/km)

Table 1: Analysis of maximum likelihood parameter estimates (single-vehicle collision model)

Parameter	DF	Estimate	Standard Error	Wald 95% Confidence Limits		Wald Chi-Square	p-value
Intercept	1	-4.77	0.51	-5.76	-3.77	88.18	<.0001
AADT	1	0.59	0.05	0.5	0.69	147.62	<.0001
Length	1	1.04	0.11	0.83	1.24	96.74	<.0001
IPK	1	0.2	0.07	0.07	0.33	9.43	0.0021

IRI	1	-0.08	0.04	-0.15	-0.01	4.89	0.027
AVG	1	0.03	0.01	0.01	0.06	8.36	0.0038
Dispersion Parameter	1	0.1017	0.02	0.07	0.14	31.92	<.0001

The degrees of freedom of the single-vehicle accident model were determined to be 232; therefore the critical chi-squared value for this model at the 95% confidence level was estimated to be 268.53. The Scaled deviance (SD) and Pearson Chi-squared (Pearson χ^2) of this model were obtained to be 251.06 and 235.64 respectively which are both below $\chi^2_{0.05,232}$ (268.54). This is an indication that according to the first criteria, the single-vehicle accident model is valid. The plot of the actual accident frequency versus predicted accident frequency can be found in Figure 1.

Performing regression analysis, the intercept of the linear trendline fitted to the data was estimated to be 0.8 which was found not to be significantly different than 0. The slope was estimated to be 0.93 which is close to 1 and was identified to be significant at the 95% confidence level as its p-value is below 0.05. Finally, the R-squared was determined to be 0.58 which means that the model explains about 60% of the variability of response data.

These results indicate that the model is valid as the equation of the trend line is not significantly different from the ideal equation (y=x), although there is an indication of missing some important variables for predicting single-vehicle accident occurrence such as variables related to roadside condition.

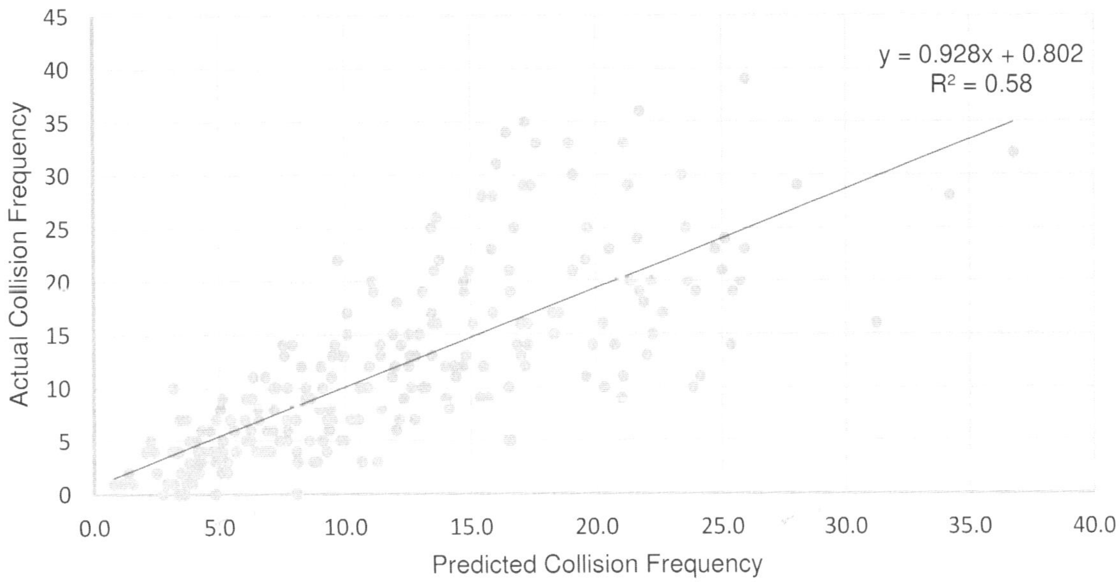

Figure 1: Actual collision frequency versus predicted frequency (single-vehicle)

6.2 Multiple-Vehicle Collision Model

For the multiple-vehicle collision model, average annual daily traffic (AADT), length (L), number of intersections per kilometre (IPK), average posted speed limit (APSL), and international roughness index (IRI) were identified to be significant. The highest value in the correlation matrix was determined to be -0.51 (between AADT and IRI) which shows that there is no strong correlation between the pairs of variables included in the model. All the VIFs were calculated to be below two which is an indication that, there is no significant multicollinearity among the variables. Indeed, only one observation was detected as an outlier and removed from the model. As a result, the final model was developed using 242 observations

(243 total -1 outlier). The result of the maximum likelihood parameter estimation analysis is presented in Table 2.

Table 2: Analysis of maximum likelihood parameter estimates (multiple-vehicle collision model)

Parameter	DF	Estimate	Standard Error	Wald 95% Confidence Limits		Wald Chi-Square	P-value
Intercept	1	-4.43	0.92	-6.23	-2.63	23.26	<.0001
LOGA	1	1.01	0.08	0.86	1.16	170.53	<.0001
LOGL	1	0.89	0.15	0.58	1.19	32.97	<.0001
IPK	1	0.49	0.1	0.3	0.68	25.16	<.0001
APSL	1	-0.05	0.01	-0.06	-0.04	67.77	<.0001
IRI	1	-0.16	0.06	-0.27	-0.04	6.91	0.0086
Dispersion Parameter	1	0.14	0.04	0.09	0.23	16.83	<.0001

The mathematical form of the multiple-vehicle accident model can be written as follows:

[6] $\mathrm{MAF} = 0.0119 * \mathrm{AADT}^{1.01} * \mathrm{L}^{0.89} * \mathrm{Exp}(0.49 * \mathrm{IPK} - 0.05 * \mathrm{APSL} - 0.16 * \mathrm{IRI})$

where: MAF: Multiple-vehicle accident frequency (accident/3 years)

AADT: Average annual daily traffic (vehicles/day)

L: Length (km)

IPK: Intersection density per kilometre

APSL: Average posted speed limit (km/h)

IRI: International roughness index (m/km)

The degrees of freedom of the multiple-vehicle accident model was determined to be 236, therefore the critical chi-squared value for this model at the 95% confidence level was estimated to be 272.84 ($\chi^2_{0.05,236}$). The Scaled deviance (SD) and Pearson Chi-squared (Pearson χ^2) of this model were obtained to be 262.76 and 254 respectively which are both below $\chi^2_{0.05,236}$ (272.84). This is an indication that according to the first criteria, the multiple-vehicle accident model is valid.

The plot of the actual accident frequency versus predicted accident frequency is very similar to that shown in Figure 1. Performing regression analysis, the intercept was estimated to be 0.16 which is statistically no different than 0. The slope was estimated to be 0.96 which is statistically no different than 1. These results indicate that the equation obtained from the plot is very close to the y=x line which is the ideal relationship. Finally, the R-squared was determined to be 0.67 which means that the model explains about 70% of the variability of the response data. In summary, the model is valid and fits the data better than the single-vehicle accident model.

Rut depth (RD) was not found to be significant in both models which is an indication that rut depth is not a contributing factor in accident occurrence on rural two-lane undivided arterial and collector roads. One of the reasons might be the range of RD used in this study which was between 0.07mm and 12.7mm, while according to TxDOT, RD values below 12.7 mm are classified in "no rutting" category.

The coefficient of IRI was estimated to be negative in both models, which means that as pavement deteriorates, while all other significant variables in the models remain constant, single-vehicle accidents and multiple-vehicle accidents are predicted to be decreased. One of the main reasons might be that people are more cautious on rougher roads and tend to slow down on poor pavements.

6.3 Crash Modification Functions

Crash modification functions (CMFunctions) were developed for single-vehicle and multiple-vehicle accidents using equation 4 as follows:

[7] $CMF_s = \text{Exp}(-0.08(IRI_2 - IRI_1))$

[8] $CMF_M = \text{Exp}(-0.16(IRI_2 - IRI_1))$

where: CMF_s: CMFunction for the single-vehicle accidents

CMF_M: CMFunction for the multi-vehicle accidents

IRI_1: The initial IRI (m/km)

IRI_2: The IRI value at the time of safety evaluation (m/km)

CMFunctions indicate that one-unit increase in IRI in m/km, while all other variables in the models remains constant, is predicted to reduce single-vehicle and multiple-vehicle accident frequencies by 7.7% and 14.8%, respectively.

In order to make the CMFunctions more practical, CMFunctions associated with pavement age were developed using available pavement deterioration equations. Since deterioration rates on arterial and collector roads are different, four CMFunctions were obtained which can be found in Table 3.

In order to compare the safety performance of new pavements with deteriorated pavements, CMFs associated with PA_1 of 0 and PA_2 of from 0 to 15, were calculated and are depicted in Figures 2 and 3. These figures indicate that, as pavement deteriorates, single-vehicle and multiple-vehicle accident frequencies are expected to decrease gradually on both road classifications types. The percentage reduction in multiple-vehicle accident frequency is approximately two times higher than single-vehicle accident frequency. Collector roads are slightly more impacted by road roughness than arterial roads.

Table 3: CMFunctions Associated with Pavement Age

Road Classification	Single-Vehicle Accidents	Multiple-Vehicle Accidents
Arterial Roads	$\text{Exp}(-0.0704(1.0452^{PA_2} - 1.0452^{PA_1}))$	$\text{Exp}(-0.1408(1.0452^{PA_2} - 1.0452^{PA_1}))$
Collector Roads	$\text{Exp}(-0.0944(1.0403^{PA_2} - 1.0403^{PA_1}))$	$\text{Exp}(-0.1888(1.0403^{PA_2} - 1.0403^{PA_1}))$

where: PA_1: Initial age of the pavement section.

PA_2: Pavement age in which safety evaluation is performed.

With respect to pavement treatment, the results indicate that applying pavement treatments is predicted to (counter-intuitively) increase accidents. The main reason may be that drivers tend to increase their speed on pavements with excellent condition which can result in higher collision experience. For example, if it is assumed that pavements are reconstructed after 15 years, single-vehicle and multiple-vehicle accidents are predicted to increase by 7% and 14% respectively, on collector roads.

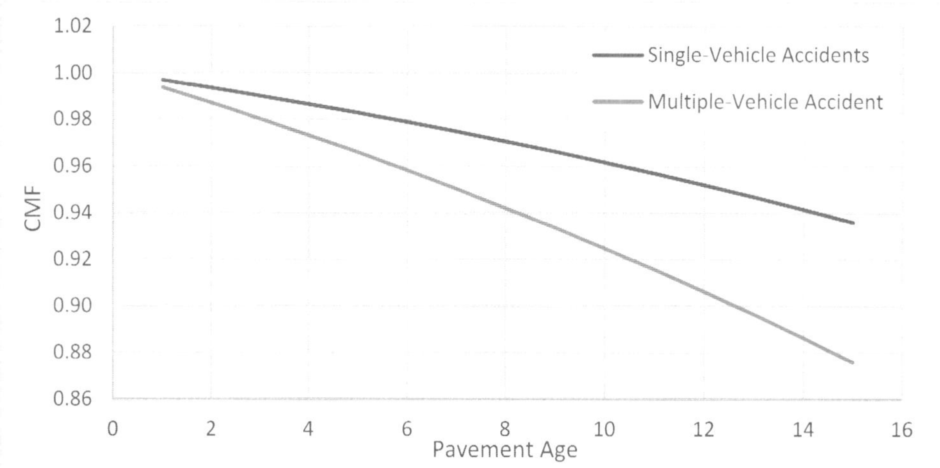

Figure 2: CMF Associated with Pavement Deterioration on Arterial Roads

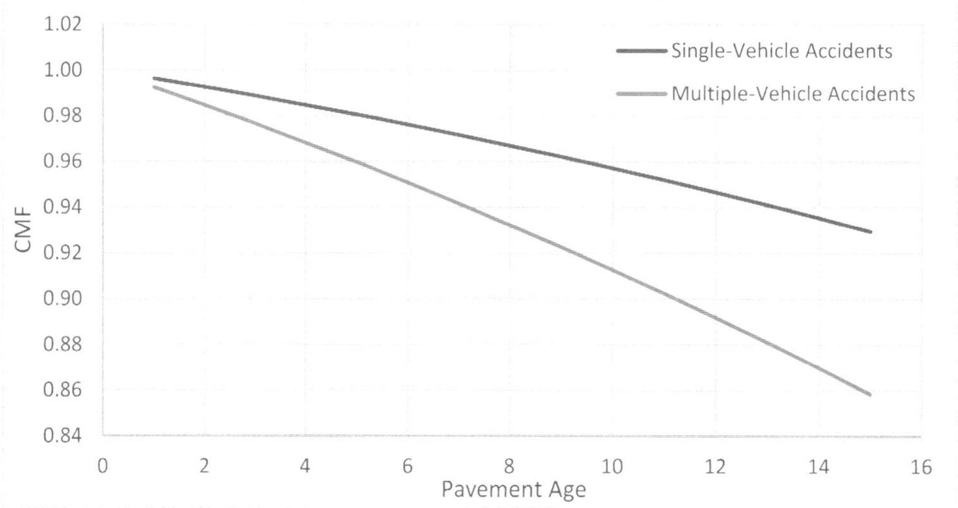

Figure 3: CMF Associated with Pavement Deterioration on Collector Roads

7 CONCLUSIONS

Two crash prediction models were developed separately for single and multiple-vehicle collisions. International roughness index (IRI) was found to be significant in both models which indicates that road roughness contributes to collision experience on rural two-lane undivided arterial and collector roads. An inverse relationship was found between IRI and collision experience (collisions tend to decrease with rougher roads). Rut depth (RD) was not found to be significant in both models which is an indication that rut depth is not a statistically significant explanatory variable for collision occurrence on rural roads.

Crash modification functions were developed for single-vehicle and multiple-vehicle collisions using a cross-sectional approach recommended by FHWA (2010). The results indicate that the impact of road roughness on multiple-vehicle accidents is twice the impact of road roughness on single-vehicle accidents (a unit of increase in IRI in m/km is expected to decrease multiple and single-vehicle accidents by 14.8% and 7.7%, respectively). Developing crash modification functions associated with pavement age, it was found that pavement deterioration decreases accidents gradually.The results also indicate that the impact of pavement roughness is slightly more significant on collector roads than on arterial roads.

8 RECOMMENDATIONS

The crash prediction models and crash modification functions developed in this study are applicable only for rural-two-lane undivided arterial and collector roads and are calibrated for the province of New Brunswick. Transferability of these models to other jurisdictions may require reformulation/calibration.

AADTs on the roads used in this study are between 60 and 8,000 vehicles/day. It is highly recommended that new models be developed for application to roads with AADTs outside of this volume range.

It is recommended that higher values of rut depth (RD) (0.07mm to 12.7mm used in this study) be investigated to identify if greater values have a predictive impact on collision occurrence.

Acknowledgements

The authors wish to acknowledge the financial support of the Natural Sciences and Engineering Research Council of Canada and generous in-kind contributions of the New Brunswick Department of Transportation and Infrastructure.

References

Abdel-Aty, M. A. and Radwan, A. E. 2000. Modeling traffic accident occurrence and involvement. *Accident Analysis and Prevention*, **32**(5): 633–642.

Al-Masaeid, H. R. 1997. Impact of pavement condition on rural road accidents. *Canadian Journal of Civil Engineering*, **24**(4): 523–531.

Al-Masaeid, H. R., Hamed, M., Aboul-Ela, M. and Ghannam, A. G. 1995. Consistency of Horizontal Alignment for Different Vehicle Classes', *Transportation research record 1500*: 178–183.

Anastasopoulos, P. C. and Mannering, F. L. 2009. A note on modeling vehicle accident frequencies with random-parameters count models. *Accident Analysis and Prevention*, **41**(1): 153–159.

Chan, C. Y., B. Huang, X. Yan, and S. Richards. (2008). Effects of Asphalt pavement conditions on traffic accidents in Tennessee utilizing management system (PMS). *Journal of advanced transportation*, Vol. 44, No. 3: 150-161.

Chan, C. Y., Huang, B., Yan, X. and Richards, S. 2009. Relationship Between Highway Pavement Condition, Crash Frequency, and Crash Type. *Journal of Transportation Safety & Security*, 1(4): 268–281.

Chen, D. H., Bilyeu, J., Walker, D. and Murphy, M. 2001. Study of Rut-Depth Measurements. *Transportation Research Record: Journal of the Transportation Research Board,* ***1764***(1): 78–88.

Cook, R. D. 1977. Detection of Influential Observation in Linear Regression. *Technometrics*, **19**(1): 15–18.

Federal Highway Administration (FHWA). 2010. A Guide to Developing Quality Crash Modification Factors. Federal Highway Administration, Washington, DC.

Lamm, R., Psaranos, B., Mailaender, T. 1999. Highway Design and Traffic Safety Handbook, McGraw-Hill.

Lee, J., Nam, B. and Abdel-aty, M. 2015. Effects of Pavement Surface Conditions on Traffic Crash Severity. *Journal of Transportation Engineering*, **11**(1996): 1–11.

Li, Y. and Huang, J. 2014. Safety Impact of Pavement Conditions. *Transportation Research Board 93rd Annual Meeting*, Washington, D.C.

Li, Y., Liu, C. and Ding, L. 2013. Impact of pavement conditions on crash severity. *Accident Analysis and Prevention.* Elsevier Ltd, 59: 399–406.

Lord, D. 2006. Modeling motor vehicle crashes using Poisson-gamma models: Examining the effects of low sample mean values and small sample size on the estimation of the fixed dispersion parameter. *Accident Analysis and Prevention*, **38**(4): 751–766.

SAS/STAT 14.1 user's guide. [accessed January 22, 2018]. https://support.sas.com/documentation/cdl

Start, M., Kim, J. and Berg, W. 2007. Potential Safety Cost-Effectiveness of Treating Rutted Pavements. *Transportation Research Record: Journal of the Transportation Research Board 98*: 208–213.

Statistics Canada, CANSIM table 102-0561, [accessed January 22, 2018]. http://www.statcan.gc.ca/tables-tableaux/sum-som/l01/cst01/hlth36a-eng.htm.

Transport Canada. 2015. Canadian Motor Vehicle Traffic Collision Statistics, [accessed January 22, 2018]. https://www.tc.gc.ca/eng/motorvehiclesafety/tp-tp3322-2015-1487.html.

Transportation Association of Canada (TAC). 2013. *Pavement Asset Design and Management Guide.* Ottawa, ON.

Zeng, H., Fontaine D, M. and Smith L, B. 2014. Estimation of the Safety Effect of Pavement Condition on Rural, Two-Lane Highways. *Transportation Research Record: Journal of the Transportation Research Board 2435*: 45–52.

Building Tomorrow's Society
Bâtir la Société de Demain

Fredericton, Canada
June 13 to 16, 2018 / *13 au 16 juin, 2018*

THROUGHPUT EVALUATION OF WORKZONES ON HIGH VOLUME HIGHWAYS OF ONTARIO, CANADA

Nahidi, Seyedata[1,4], Rahman, Sonia[2] and Tighe, Susan[3]
[1] University of Waterloo, Canada
[2] Stantec, Canada
[3] University of Waterloo, Canada
[4] snahidi@uwaterloo.ca

Abstract: Highways as the primary type of transportation infrastructure, allow for haulage of huge amounts of goods and services in North America. An aging infrastructure combined with several other parameters such as high volume of heavy truck traffic and long harsh winters could lead to faster pavement deterioration. To improve safety, traffic planning, geometry, and structural performance of the highways, transportation agencies should have a detailed program to maintain, preserve, and reconstruct these infrastructures. Workzones interrupt the regular traffic flow on the highways which could potentially lead to an increase in the number of collisions. To prevent this, agencies are encouraged to present a comprehensive strategy to minimize the queuing in the workzones. The Ministry of Transportation Ontario (MTO) have identified that accurate prediction of workzone throughput could significantly reduce the user delay costs, and increase the safety of the high volume highways. This paper involves the evaluation of the workzone's throughput performance. To that end, various strategies such as Multiple Linear Regression, Negative Binomial Regression, and Truncated Regression models were developed to identify the parameters affecting workzone throughputs. An innovative concept of random parameter modeling was also used to account for unobserved heterogeneity issue. Results show that along with traditional factors, further parameters such as number of closed lanes, length of the workzone, presence of police, time of day, and percentage of heavy trucks have a significant effect on the throughput of workzones on high volume highways.

1 Introduction

Maintenance, rehabilitation, and preservation of the major high volume highways within the province of Ontario is one the most important responsibilities of the Ministry of Transportation of Ontario (MTO). Based on the definition on the King's Highways official website, 400-series highways could be defined as follows[1],

"At the top of the classification ladder are the 400-Series King's Highways. These roads are almost exclusively controlled access freeways, and are relatively few in number (15 in total, counting the Queen Elizabeth Way & 407 Express Toll Route). These important routes form the backbone of Ontario's road infrastructure, and are being expanded constantly to meet the needs of Ontario's economy."

[1] http://www.thekingshighway.ca/intro.html

The 400-series highways inside and around the Grand Toronto Area (GTA) are responsible for transporting vast numbers of people, but also goods and sources (Statistics Canada 2016). Thus, setting up a workzone on these highways should be utterly well organized to eliminate any possible traffic flow.

High volume highways are generally confronting issues such as rapid pavement deterioration, and need of more frequent workzones in order to maintain their serviceability. Setting up workzones in high volume highways would considerably increase the user delay cost and accident frequency. These factors would significantly affect MTO's strategies regarding the policies related to the workzones.

In order to solve the issues inside and around the workzones, several advanced and traditional methods were examined by several private and public agencies all around the world. Installing portable barriers, prohibition of heavy vehicles, reducing the speed limit, and strict traffic enforcement were some of the methods that were used to control the safety and traffic of the workzones. Presence of the Police at the beginning of the workzones also aimed to reduce the violations in these vulnerable areas. Moreover, in some workzones in order to capture the influence of the trucks and heavy vehicles on possible lane blockage and secondary crashes in the workzones, entrance of these vehicle types was prohibited (Chen and Tarko 2012).

Generally, government agencies prefer to schedule the workzone activities for high volume highways at night because of lower traffic volume and less disruptions for motorists in comparison to the day time peak hours. This arrangement also avoids excessive traffic and queuing on major highways, and inconvenience for motorists (Shepard and Cottrell 1986).

This study aimed to investigate whether the workzone strategies and guidelines are appropriate for current workzone throughputs, and if not, which additional parameters should be accounted for in order to increase the performance of the workzones. To that end, statistical models was used to predict the possible range of throughput in these workzones at queuing time. Then, by analyzing the results derived from these models the most important countermeasures were identified in order to improve the performance of the workzones.

2 Background

Workzone could be defined as a section of a roadway with construction, maintenance, or utility-work activities (Tuner 1999). Previously, the MTO conducted several studies from 2007 to 2011 to evaluate the workzone throughput at queuing times in high volume 400-series highways. Since 2007 several of these workzones were visited for evaluation and data collection in order to enrich the existing dataset. The dataset was then analyzed and several multiple linear regression models were developed to predict the possible future throughput. These generic models had the benefit of providing a broader insight about the factors which can affect workzone throughput in queuing condition. These studies did not focus on the safety of the workzones (Mushtaq 2011; Hicks 2006).

Although there are significant improvements in highway design, vehicle safety, and workzone guidelines in recent decades, the number of accidents which resulted in fatalities and injuries are still unacceptably high. Canada Transport (6) reported that around 150,000 of accidents occurred in 2014 led to serious injuries for drivers, passengers, and pedestrians. Based on this report around 2000 of these accidents resulted in fatalities. Report did not focused on the workzones separately, however workzones hypothetically could be considered as high-risk locations. Factors such as low visibility, more impaired drivers, and higher speed could increase the likelihood of experiencing severe injuries and fatalities in the workzones during the night.

Traffic Management in highway workzones could be considered as one of the most challenging tasks of transportation agencies. Reducing the motorists' delay and the risk of accidents are the major concerns that should be addressed under a proper traffic management system. This problem could be predominantly captured on the segments prior to the workzones in which one or more lanes are dropped. Gipps (1986) conducted a study to capture the possible influential parameters on drivers' behavior during lane changing. He claimed that parameters such as lane selection, urgency of lane change, entry and departure of non-transit vehicles into and from transit lanes, and effect of heavy vehicles can all significantly affect the drivers' lane-changing behavior. Moreover, these sections encounter higher traffic turbulence, resulting in higher likelihood of accidents and longer delays (Tarko et al. 1998).

Generally, the likelihood of experiencing conflicts, accidents, and delays is much higher in the entry section of the workzones. Lane closures are leading to capacity reduction and subsequent aggressive lane changes. These sudden maneuvers at the merging point creates extremely hazardous environment for both motorists and workers.

Recently, several studies were conducted to investigate the factors that can significantly affect the throughput of the workzones. Some of these studies used data from high volume 400-series highways located in Ontario, Canada (Mushtaq 2011; Hicks 2006). Findings of these studies captured that the presence of police, number of closed lanes, and types of barriers are the most significant factors for throughput prediction. These studies only focused on fixed parameter multiple linear regression modelling approach which doesn't account for unobserved heterogeneity issue.

Although, statistical models are highly data dependent and coefficient magnitudes could vary from one model to another, they are still expected to capture similar behavior. For example, in most of the studies queuing will increase by closing more than two lanes.

3 Data Collection

The main idea of estimating the throughput in workzones was initiated in 2006. After the first phase of data collection, more data was collected in 2011 due to the expansion of the comprehensive database (Mushtaq 2011; Hicks 2006). Finally, in 2016 data collection continued to provide an appropriate context to develop more reliable statistical models. The data collection methodology was consistent over the three data collection periods. In 2016 data collection, it was assumed that when the traffic speed in construction zone is 10 km/hr or less, queuing occurs.

A site characteristics form, which was designed and used in early phases, was used in 2016 data collection process as well. Factors that were documented on the form included:

- Date and time of site visit
- Location
- Weather
- Speed limit (posted and temporary in km/h)
- Length of work zone
- Duration of closure
- Etc.

During the data collection, vehicle counts were split into three categories: passenger vehicles, medium heavy vehicles, and heavy vehicles. On fifteen-minute intervals, the number of vehicles passing through the workzones was recorded. The hourly throughput frequency was calculated for the sites with queuing based on the collected data. In addition to the throughput records, videos and images were recorded to capture any anomalies. These videos allowed researchers to conduct further analysis by reviewing them if necessary.

The MTO Central Region Rehabilitation Program provided the schedule and characteristics of the potential workzone projects on 400-series highways. The provided program included various information such as the location, length, and type of project. This list also presented the possible operational duration of each workzone.

Although thirty-six construction sites were identified by the Ministry as the potential sites for data collection, only 9 of the them were visited. The rest of the sites could not be visited due to:

- No queuing reported by the Construction Administrators;
- No lane closures;
- Cancellation of construction work due to weather etc.

3.1 Descriptive Statistics

Based on the available dataset, it was captured that police officials were not present in 91% of the closures. 83% of the dataset was collected during the weekdays. Around 88% of the closures occurred at night time which was expected. In 94 % of the workzones barrel were used for closure, and in only 6% of the construction areas concrete barrier were installed. Right and left lanes were closed almost equally in the visited workzones. In addition, in 73% of the construction sites 2 or more lanes were closed to conduct the required maintenance, rehabilitation, and reconstruction activities. Among all visited construction sites 401 and Queen Elizabeth Way (QEW) were visited the most.

4 Methodology

4.1 Multiple Linear Regression

Multiple linear regression could be defined as a model which includes at least two independent variables. These models are common for continuous dependent variables (Washington et al. 2011). Most regression applications generally try to detect a list of explanatory variables that are significantly influence the dependent variable.

In general, explanatory models are based on data achieved from well-controlled experimentations in the laboratory, analytical models are based on data obtained from observational studies such as data collected from workzones. It has been investigated that any changes on dependent variable due to changes of explanatory variables is depend on numerous factors other than independent variables. To observe the effect of the independent variable, some assumptions should be considered (Washington et al. 2011). When any of the requirements is not satisfied, corrective actions should be considered, and in some cases, alternative modeling approaches should be tried.

The main goal of the regression is to state the relationship between dependent variable Y with at least one independent variable. Generally, these regression models aim to present some realistic information and properties of the population by only reviewing properties of the sample.

The matrix expression of the linear regression model can be presented as follows,

[1] $Y = X\beta + \varepsilon$

$$\varepsilon = Y - \hat{Y} = Y - X\hat{\beta}$$

where X is properties matrix, β is the factor's coefficient matrix, Y is dependent variable, $\hat{Y}$ is the prediction matrix, and ε is the difference between actual and predicted values.

One of the most common ways to solve the above problem is Least Squares Method. This method is also known as "Ordinary least squares" or "OLS". In this method, the goal is to estimate the model parameters by using the sample data. To that end, the following formula could be used to estimate these parameters,

[2] $\beta = (X^T X)^{-1} X^T Y$

Equation 2 presents the estimation of model parameters.

4.2 Truncated Regression

The truncated regression model is common for the no limit observations in the tobit formulation. This model is not a linear regression model and could be considered as non-linear regression (LIMDEP 2012). For general explanation of this model in which the range of the dependent variable is truncated in both tails, the conditional mean function is,

$$[3]\ E[Y_i|x_i, L_i \leq Y_i \leq U_i] = \acute{\beta} x_i + \sigma \frac{\phi_L - \phi_U}{\Phi_U - \Phi_L}$$

Where $x \sim N[\mu, \sigma^2]$ and σ is a constant. $\phi(.)$ and $\Phi(.)$ are standard normal pdf and cdf, respectively. L represents lower limit; however, U is upper limits representative.

The specification of the both lower and upper limits are same as the tobit model,

$$[4]\ \alpha_{limit} = [limit - \acute{\beta}x]/\sigma\ ], limit = upper\ or\ lower$$

$$\Phi_{limit} = \Phi(\alpha_{limit})$$

$$\phi_{limit} = \phi(\alpha_{limit})$$

The least squares analysis is inconsistent and inaccurate in solving this type of problems. To that end, maximum likelihood method was chosen to estimate the coefficients.

The log likelihood function of the truncated regression could be maximized by using Olsen's transformation parameters. After transformation, this function could be written as,

$$[5]\ \sum_i logL_i = \sum_i log\theta - \frac{1}{2}\log 2\pi - \frac{1}{2}\varepsilon_i^2 - \log[\Phi(\theta U_i - \acute{\gamma} x_i) - \Phi(\theta L_i - \acute{\gamma} x_i)]$$

where $\theta = 1/\sigma$ and $\gamma = \beta/\sigma$.

Since the truncated regression is a nonlinear regression, coefficients are not the marginal effects. Differentiation of conditional mean function with respect to the x_i produces the vector of slopes,

$$[6]\ \frac{\partial E[Y_i|x_i, L_i \leq Y_i \leq U_i]}{\partial x_i} = \beta \left[1 - \frac{\alpha_L \phi_L - \alpha_U \phi_U}{\Phi_U - \Phi_L} - \left(\frac{\phi_L - \phi_U}{\Phi_U - \Phi_L}\right)^2\right]$$

The term in the bracket is the scale factor of the marginal effect.

4.3 Poisson and Negative Binomial Regression Models

One of the goals of this study is to determine the number of vehicles passing the work zone during the queuing time. Equation 7 presents the Poisson Regression model, where the probability of work zone i having y_i throughput per hour.

$$[7]\ P(y_i) = \frac{EXP(-\lambda_i)\lambda_i^{y_i}}{y_i!}$$

Where $P(y_i)$ is the probability of workzone i having y_i throughput per hour and λ_i is the Poisson parameter of workzone i, which is equal to workzone i's expected number of throughput per hour, $E[y_i]$.

The relationship between independent (explanatory) variables with Poisson parameter can be presented as Equation 8.

$$[8]\ \lambda_i = EXP(\beta X_i)\ \text{or, equivalently}\ \ Ln\ N\ (\lambda_i) = \beta X_i$$

X_is are the vector of the statistically significant explanatory variables directly derived from the collected dataset; βs are vector of the estimated parameters.

Sometimes the data do not meet the requirement of equality of mean and variance to permit the use of Poisson Regression. Negative Binomial Regression is one of the most popular models that can account for this inequality (Washington et al. 2011).

The Negative Binomial model has a similar formulation to Poisson model and it only counts for error term (ε) (Washington et al. 2011):

[9] $\lambda_i = \text{EXP}(\beta X_i + \varepsilon_i)$ or, equivalently $\text{Ln N}(\lambda_i) = \beta X_i + \varepsilon_i$

Where EXP(ε_i) is Gamma-distributed disturbance term with the mean 1 and variance α.

One of the methods to estimate the accuracy of the Poisson and negative binomial models is McFadden ρ^2. This value is equivalent to R^2 in linear regression and can be calculated as Equation 10.

[10] $\rho^2 = 1 - \frac{LL(\beta)}{LL(0)}$

where LL(β) is the log-likelihood at convergence with parameter vector β, and LL(0) is the log-likelihood at zero. A perfect model has ρ^2 equal to 1, however, the model with ρ^2 equal to zero is considered flawed (Washington et al. 2011).

4.4 Random parameters models

The modeling approaches that were discussed earlier consider the coefficients constant across the observations. However, this fixed-parameter assumption might not be corrected in all cases. This fluctuating behavior of some parameters is known as unobserved heterogeneity. Random parameter models account for this issue and led the coefficients change across the observations. This is a significant consideration, as if unobserved heterogeneity is not addressed, the parameter estimates can be incorrect, the inferences unreliable, and the predictions inaccurate (Washington et al. 2011; Anastasopoulos 2009; Nahidi et al. 2017).

In random parameters modeling, the coefficients become:

[11] $\beta_n = \beta + \omega_n$

where β_n is a vector of estimable parameters, and ω_n is a randomly distributed term (for example, a normally distributed term with mean zero and variance σ^2). Based on these assumptions, β_n is the vector of the coefficients for an individual parameter across the observations. The variation of this factor could be described by a density function - mixing distribution - q($\beta|\omega$), where ω presents the parameters of the density distribution (LIMDEP 2012). A simulation-based maximum likelihood estimation method was used to simplify the additional complexity due to presence of random parameters in the calculation of outcome probabilities. To achieve simulated probabilities numerous iterations of the same procedure should be conducted (Washington et al. 2011).

One of the major concerns in random parameter models is the sampling technique of β values from the selected distribution. In order to solve this issue, the technique of Halton draws (Halton 1960) is chosen. This method can provide an appropriate context to efficiently approximate probabilities by limiting the number of draws to the smallest possible extent. Moreover, the non-random selection of draws provides more robust estimates, as compared to the use of merely random draws (Bhat 2003; Train 2003) for further details on the simulation approach. Based on the literature, selecting 200 Halton draws can possibly yield to have stable and accurate estimation of parameters (Bhat 2003; Milton et al. 2008). For the functional form of the parameter density function several distributions, such as normal, Weibull, lognormal, uniform, and triangular distributions, will be considered and results will be compared in order to choose the best candidate in terms of statistical fit. A parameter will be considered as random when the mean and standard deviation of the parameter density function is statistically significant with 95% confidence level.

5 Discussion of results

Tables 1 and 3 show the statistically significant parameters of the various modeling strategies used for this study in order to estimate the best throughput estimation for the workzones in high volume highways in Ontario, Canada. A concise argument about each significant factor in the developed models will be presented in the following section.

5.1 Model results

Table 1 presents the estimated fixed-parameter models. The constant in these models is a representation of the average capacity of the highways from where the data was collected. Queuing usually occurs when departure rate at the termination area is lower than the arrival rate at the transition area. Developed models show that throughput decreases when two or more lanes with a length of more than 3km are closed. In addition, multiple linear regression and truncated regression models, captured that when only one lane is closed (in comparison with 2 or more closed lanes) more vehicles are passing through the workzones, which was expected. 401 highway is responsible for transportation of vast amount of people and trade goods, and is one of the busiest highways of the region. Accounting for this fact guarantees the justification of the significancy factor attributed to this highway in the models. Highway 401 was identified as a highway which has more queuing than the other highways due to the amount of traffic. On the other hand, 400, 409, and 410 highways behaved in an opposite way and models show that less queuing was experienced in the workzones located in these highways. The developed models also show that an increase in the percentage of trucks would lead to a significant increase in queuing, which is in-line with previous studies (Mushtaq 2011; Hicks 2006). In order to capture a more detailed effect of the heavy vehicles, this factor was divided in two major categories:1- more than 20% trucks, 2- less than 6% trucks. when the truck percentage is higher than 20%, throughput decreases, and queuing increases which was expected. Moreover, when this factor is less than 6%, throughput increases. This factor can emphasize that low percentage of trucks balances the speed of the vehicles in highways.

Surprisingly, it was captured that the presence of Police in workzone areas could significantly reduce the number of vehicles passing through the workzones. It highlights the fact that drivers act more cautiously by observing police cars prior to the workzones and aim to reduce their speed more rapidly. In addition, if two or more lanes are closed throughput frequency decreases. The RMSE for each of these models were calculated and this measure is presented for each model in Table 2. Lower RMSE shows better fit. Among all these models, truncated regression model performs better than the other two. Also, this model predicts more correct data points within 200 vcl/h/l, which is around 87%.

Table 3 shows the model results from random parameter modeling process. One the major differences of the random parameter models is that the constant values in these models are much closer to the design capacity, and are therefore expected to be more accurate. Similar to the fixed parameter multiple linear regression model, presence of police shows a significant reduction on throughput in workzones at the queuing time. Most of the factors in random parameter models are similar to the fixed parameter models. One of the parameters which was significant in the truncated random parameter model is left lane closure. This model captured that when the left lane is closed more vehicles are passing the workzones. This could be because of higher speed in left lanes and hence a quicker merging process. Based on the standard deviation and mean value of the night factor in the negative binomial model, when lane closure occurs during the night time, throughput increases which means less queuing. This was expected but note that the magnitude of this factor changes across the observations.

Another interesting finding of this study was that installation of the barrels instead of concrete barriers could reduce the throughput. Type of barriers could possibly affect the behavior of drivers and motorists while they are passing the workzones. Because of this fact, this factor was significant as random parameter, which means the effect of this factor changes across the observations. Finally, based on the Truncated regression model, having barrels as barriers in the workzones increase the queuing in 97% of the occasions. However, in 3% of the workzones concrete barriers increased the queuing. This fact slightly changes under the Negative Binomial model. In this model, in 93% of the closures with barrels queuing increases.

Table 4 shows the percentage of the correct predictions of random parameter models within 25, 50, 75, 100, 200, and 300 vcl/h/l. This table also presents the RMSE value for three different random parameter models for sake of comparison. Based on prediction performance and RMSE values Truncated Regression model performs better than the other two models.

6 Summary and Conclusion

This article aimed to investigate a practical analysis of workzone characteristics, configurations, and conditions on the throughput of workzones on high volume highways in Ontario, Canada. The proposed statistical modeling approach provided an appropriate context in order to identify the parameters that affect the throughput of the workzones. Also, this study accounted for unobserved heterogeneity by benefiting from random parameter concept in the modeling procedure. Modelling strategies such as Multiple Linear regression, Negative Binomial regression and Truncated regression were used in this paper. The results from statistical analysis presented a broad perspective about factors which can play a significant role in predicting the throughput of the workzones. Lengthier workzones, and high percentage of trucks were the two major factors that increased queuing in these regions. Therefore, by setting up shorter workzones and also limiting trucks' access to the highways within active construction zones could help reduce queuing. In addition, based on the estimated models, it is recommended to have only one lane closure in high volume highways instead of two or more lanes closures.

Table 1: Fixed Parameter Models' Summary

Fixed Parameter Models	MLR*		TR**		NBNR***	
Factors	Coef.	t-value	Coef.	t-value	Coef.	t-value
Constant	1397.450	28.170	1049.090	13.610	7.138	193.160
Presence of Police (1 if present, 0 otherwise)	-229.571	-3.940				
WZ**** Configuration (1 if only one lane closed and it is day time, 0 otherwise)	499.296	7.330	765.145	6.190		
WZ Configuration (1 if two or more lanes closed and length of the workzone is greater than 3 km)	-158.116	-3.780	-212.311	-5.100	-0.144	-3.600
WZ Configuration (1 if it's weekend night and only one lane closed, 0 otherwise)					0.284	3.590
WZ Condition (1 if 2 or more lanes closed, 0 otherwise)					-0.323	-4.210
WZ Location (1 if 401 Highway, 0 otherwise)	-76.099	-2.100			-0.053	-1.670
WZ Location (1 if 400, 409, and 410 Highways, 0 otherwise)			87.965	2.000		
WZ Condition (1 if speed limit is 80 km/h and it' is night time, 0 otherwise)			-273.785	-4.600	-0.226	-3.780
WZ Condition (1 if it's night, 0 otherwise)			164.958	2.150		
Natural Logarithm of Truck Percentage	-70.050	-3.400				

Truck Percentage (1 if it's higher than 20%, 0 otherwise)			-112.874	-2.440	-0.077	-1.870
Truck Percentage (1 if it's less than 6%, 0 otherwise)			79.702	1.990	0.088	2.070
Sigma			143.440	11.530		
Dispersion Parameter					0.014	4.890
R-Squared or Rho-Square	0.61		0.69		0.68	
Adjusted R-Squared or Rho-Square	0.58		0.66		0.66	

* Multiple Linear Regression ** Truncated Regression *** Negative Binomial Regression **** Workzone

Table 2: Fixed Parameter Models' RMSE and Prediction Performance

Models	Within 100 vcl/h/l	Within 200 vcl/h/l	Within 300 vcl/h/l	RMSE
TR	57%	87%	98%	134.369
NBNR	55%	82%	98%	141.795
MLR	59%	79%	98%	146.946

Table 3: Random Parameter Models' Summary

Random Parameter Model	MLR		TR		NBNR	
Factors	Coef.	t-value	Coef.	t-value	Coef.	t-value
Constant	1853.730	49.580	1670.230	59.320	7.012	271.98
Presence of Police (1 if present, 0 otherwise)	-195.473	-7.130				
WZ Configuration (1 if two or more lanes closed and length of the workzone is greater than 3 km)	-114.172	-5.680			-0.152	-9.410
WZ Configuration (1 if two or more lanes closed, 0 otherwise)	-118.377	-6.220	-102.688	-6.720	-0.461	-14.190
WZ Configuration (1 if left lane closure, 0 otherwise)			33.516	2.480		
WZ Configuration (1 if it's weekend night and only one lane closed, 0 otherwise)					0.424	12.410
WZ Condition (1 if night, 0 otherwise)	-120.459	-4.230	-111.753	-4.780	0.104	4.820
Standard deviation of parameter distribution					*0.131*	*20.080*
WZ Condition (1 if speed limit is 80 km/h and it's night, 0 otherwise)			-242.796	-10.900	-0.209	-9.650
WZ Location (1 if 401 Highway, 0 otherwise)	-64.968	-3.680				
WZ Location (1 if 401 or QEW Highways, 0 otherwise)			-67.483	-3.950		

Natural Logarithm of Truck Percentage	-66.029	-7.000				
Truck Percentage (1 if it's higher than 20%, 0 otherwise)	-0.088				-0.088	-5.120
Truck Percentage (1 if it's higher than 18%, 0 otherwise)			-103.014	-6.200		
Truck Percentage (1 if it's less than 6%, 0 otherwise)					0.060	4.130
Type of Barrier (1: Barrel, 0: concrete)	-279.830	-7.890	-215.003	-6.460		
Standard deviation of parameter distribution	*152.883*	*18.570*	*150.927*	*23.010*		
Sigma			56.428	12.330		
Dispersion Parameter					524.384	4.320

Table 4: Random Parameter Models' RMSE and Prediction Performance

Models	Within 25 vcl/h/l	Within 50 vcl/h/l	Within 75 vcl/h/l	Within 100 vcl/h/l	Within 200 vcl/h/l	Within 300 vcl/h/l	RMSE
TR	74%	94%	98%	100%	100%	100%	24.668
NGBN	67%	91%	99%	99%	100%	100%	30.886
MLR	59%	88%	98%	100%	100%	100%	33.271

References

Anastasopoulos, P., Mannering, F. 2009. A Note on Modeling Vehicle Accident Frequencies with Random-parameter Count Models. *Journal of Accident Analysis and Prevention*, 41**(1):** 153-159.

Bhat, C. 2003. Simulation Estimation of Mixed Discrete Choice Models Using Randomized and Scrambled Halton Sequences. *Journal of Transportation Research Part B: Methodological*, 2003**(9)**: 837-855. DOI: 10.1016/S0191-2615(02)00090-5

Chen, E., and Tarko, A. P. 2012. Best Practices for INDOT-Funded Work Zone Police Patrols. Publication FHWA/IN/JTRP-2012/36. *Joint Transportation Research Program*, Indiana Department of Transportation and Purdue University, West Lafayette, Indiana.

Gipps, P. A. 1986. Model for the Structure of Lane-changing Decisions. *Transportation Research Part B: Methodological*, 20: 403-414.

Halton, J. H. 1960. On the Efficiency of Certain Quasi-Random Sequences of Points in Evaluating Multi-Dimensional Integrals. *Journal of Numerische Mathematik*, 2**(1)**: 84-90. DOI: 10.1007/BF01386213

Hicks, C. 2006. Measuring Highway Work Zone Throughput." MA thesis. U of Toronto.

LIMDEP Econometric Modeling Guide. 2012. LIMDEP, Plainview, NY. 1162–1167.

Milton, J., Shankar, V. and Mannering, F. 2008. Highway Accident Severities and the Mixed Logit Model: An Exploratory Empirical Analysis. *Journal of Accident Analysis and Prevention*, 40: 260-266.

Ministry of Transportation of Ontario, Book 7: Ontario Traffic Manual for Temporary Conditions. Queen's Printer, Ontario, 2014. ISBN: 978-1-4606-3505-6

Mushtaq, A. 2011. Measuring Work Zone Throughput and User Delays. MA thesis. U of Waterloo.

Nahidi, S., Fountas, G., Sarvani SP, Sarwar, M. T., Anastasopoulos, P.C. 2017. Project Discrepancies in Roadway Construction and Preservation: A Statistical Analysis of Public–Private Partnership Contract Types in the US. *Journal of Frontier Built Environment*, 3. DOI=10.3389/fbuil.2017.00015

Shepard, F. D., Cottrell, B.H. Jr., 1986. Benefits and Safety Impacts of Night Work-Zone Activities. Transportation Research Record: *Journal of the Transportation Research Board*, TRR, 1083: 31-36.

Statistics Canada. Population by Year, by Province and Territory. 2016. Government of Canada, CANSIM (database). http://www.statcan.gc.ca/tables-tableaux/sum-som/l01/cst01/demo02a-eng.htm.

Tarko, A., Kanipakapatnam, S., Wasson, J. 1998. Modeling and Optimization of the Indiana Lane Merge Control System on Approaches to Freeway Workzones, Part I. Publication FHWA/IN/JTRP-97/12. *Joint Transportation Research Program*, Indiana Department of Transportation and Purdue University, West Lafayette, Indiana.

Train, K. 2003. Discrete Choice Methods with Simulation, Cambridge University Press, Cambridge, UK .

Transports Canada. Transportation in Canada. Government of Canada. https://www.tc.gc.ca/media/documents/policy/2014_TC_Annual_Report_Overview-EN.pdf. ISSN: 1920-0846. Accessed 2015.

Turner, J.N. 1999. What's a Work Zone? Journal of Public Road, 62**(6)**.

Washington, S., Karliftis, M., Mannering, F. 2011. Statistical and Econometric Methods for Transportation Data Analysis (2nd ed.), Chapman and Hall/CRC, Boca Raton, FL.

Building Tomorrow's Society
Bâtir la Société de Demain

Fredericton, Canada
June 13 to 16, 2018 / *13 au 16 juin, 2018*

USING NATURALISTIC DRIVING DATA TO QUANTIFY DRIVER FOLLOWING BEHAVIOR DURING BRAKING

Tawfeek, Mostafa H.[1,3], and El-Basyouny, Karim[2]
[1] Ph.D. Candidate, University of Alberta, Canada
[2] Associate Professor and City of Edmonton Urban Traffic Safety Research Chair, University of Alberta, Canada
[3] mostafa.h.tawfeek@ualberta.ca

Abstract: This study aims at quantifying and comparing drivers' following behavior during braking at intersections and midblock road segments (i.e., non-intersection) using Naturalistic Driving Data (NDS) data. The quantification of driver behavior was based on exploring the Probability Density Functions (PDF) of two parameters (i.e., minimum following distance and time-to-collision) which were extracted from NDS data. To classify drivers' following behavior at intersection and segments, all events were mapped onto the road network using ArcGIS. The intersection events were defined based on the driver's stopping sight distance to the stop line of an intersection. The results showed that there was a significant statistical difference between the PDFs of the parameters at intersection and segment events. Also, the results showed that drivers tend to keep less following distances and high deceleration rates (i.e., more aggressive) during intersection events when compared to segment events. Generally, the results of this study highlight the importance of considering the driver location before judging or classifying his/her behavior.

INTRODUCTION

Driver assistance or in-vehicle warning systems (e.g., Red Light Violation Warning, Forward Collision Warning (FCW), etc.) are expected to have a significant role in crash avoidance. The expected benefits from such systems will be mainly realized by enhancing the driver's awareness regarding their surrounding environment (Farah et al., 2012; Mohammed et al., 2016; Olia et al., 2015). These systems can enhance the awareness of drivers by alerting them of safety-critical events through the use of in-vehicle messaging (e.g., audible, visual, etc.). For instance, FCW provides a warning to the driver when a rear-end collision is expected with a leading vehicle in the same lane and the same direction of travel (Harding et al. 2014). In other words, such a warning is warranted when the actual following distance is equal or less than a minimum following distance which is predefined by the warning system.

One of the major challenges facing these systems is to accommodate a wide range of driving behaviors (e.g., from cautious through neutral to aggressive driving situations) and satisfy various drivers' needs (e.g., reliable warnings). For a cautious driving situation (i.e., when the driver usually keeps a larger following distance to a leading vehicle), an FCW might be considered as a late warning in this situation. On the other hand, for an aggressive driving situation (i.e., when the driver usually keeps a lesser following distance to a leading vehicle) an FCW, in this situation, might be considered as an early warning. The evaluation of warning systems showed that too early warnings, reduce users' acceptance and decrease the impact of the warning on driver behavior (van Driel et al. 2007, Staubach et al. 2014). Conversely, in the case of late warnings, drivers could start a maneuver before the warning is relayed, and hence, drivers will develop a mistrust due to the conflict between their expectations and the warning system performance (Abe and Richardson 2005). Therefore, it is essential to consider the situations/contexts when the warning will be relayed to the driver and drivers' needs.

Several studies investigated and classified drivers based on their behavior to different categories as a preliminary step of the development of an efficient and safe performing driving assistance systems (Wang et al. 2010, Jensen et al. 2011, Martinez et al. 2017). In these studies, driver behavior was analyzed regardless the location of the driver (i.e., at intersections or on midblock segments). However, it is expected that the driver behavior will vary based on several factors such as the roadway geometry (Ben-Bassat and Shinar 2011), driver mental workload (Messer 1980), in-vehicle advisory messages (Mohammed et al. 2016), etc. Therefore, this study aims at, first, investigating the driver following behavior during braking at different road network locations (i.e., at intersections and on midblock segments). This investigation is carried out by quantifying two behavioral measures (i.e., minimum following distance and time-to-collision) which are extracted from naturalistic driving data. These events are then mapped onto the road network using ArcGIS to differentiate between intersection and non-intersection events. Second, a comparison between these measures at intersections and on midblock segments is conducted. The probability distributions of the behavioral measures are generated for each location for three speed groups (i.e., low, medium, and high). To further explore the differences in driver behavior, the probability distributions are compared using statistical testing.

1 LITERATURE REVIEW

The knowledge of unassisted driving behavior is essential for the development of an efficient warning system (Berndt et al. 2007). Therefore, several studies investigated drivers' nominal behavior to develop and/or enhance warning systems. To achieve this goal, drivers' behavior characteristics in terms of TTC were analyzed in different classes such as age, gender groups, and speed ranges (Montgomery et al. 2014). The purpose of this study was to characterize the differences between age and genders groups during car-following while braking situations (Montgomery et al. 2014). In a different study, an approach for driving behavior classification was proposed using several measurable parameters such as time headway, the inverse of TTC, and the time between accelerator pedal release to brake activation. The drivers were classified according to their aggression, stability, conflict proneness, and skillfulness (Wang et al. 2010).

Feng et al. (2017) assessed the longitudinal vehicle jerk as a measure to classify aggressive drivers using Naturalistic Driving Study (NDS) data. Two jerk-based measures were proposed, namely, the frequent utilization of large positive or negative jerk. First, from the NDS data, the drivers recognized to be aggressive according to their excessive speeding, tailgating, or crash/near-crash frequency in the data. Based on that, three groups of aggressive drivers were generated, and the proposed jerk-based measures were validated using these three groups. The results showed that the proposed measures could successfully identify aggressive drivers in the three groups of aggressive drivers. However, the results indicated that the large negative jerk was more efficient when compared to the large positive jerk in detecting aggressive drivers (Feng et al. 2017).

Using speed and lateral and longitudinal acceleration, Gonzalez et al. (2014) developed a linear filter based on Gaussian Mixture Model to detect driving aggressiveness. This model was validated by monitoring the driving behavioral measures of 10 drivers on a customized test route. This test route, which was part of a private road network, included five different road features such as a roundabout, curves, uphill and downhill segments, and sharp bend. The drivers conducted several driving rounds on the route including only one aggressive round for each driver. It is worth to mention that the drivers were not given any instruction/definition of the aggressive driving (Gonzalez et al. 2014).

Zheng et al. (2017) defined the aggressive driving based on a questioner given to 15 drivers before and after local road tests (i.e., within University of Florida campus area). This study aimed to assess the relation between the driver type and driver behavior in a high vehicle-pedestrian interaction low-speed environment. Such an environment was within campus area with speed limit between 20 mph and 35 mph. The driving aggressiveness, obtained from the questioner, was compared to behavioral measures, namely, driving desired speed and yield to pedestrian behavior. The results revealed that the two measures could efficiently identify the driver aggressiveness (Zheng et al. 2017).

Based on smartphone accelerometer data (i.e., the acceleration in X, Y, and Z axes of the smartphone), Osafune et al. (2017) classified the drivers into safe and risky drivers. This classification was validated based on drivers' crash records in 20 years. The data was collected for 15 months covering almost all areas of Japan. Three thresholds were selected to identify risky drivers including acceleration, deceleration, and left acceleration (Osafune et al. 2017).

Sundbom et al. (2013) defined aggressive and normal driving based on drivers' steering behavior in terms of curve lateral acceleration, yaw rate, and lateral movement in the lane. Only two drivers participated in this study on a test track where each driver had a mixture of aggressive and normal driving. A model was developed to online classify the two driving styles (i.e., aggressive and normal). It is worth to mention that the test track had no traffic and consisted only of curved sections and straight sections. Also, the straight segments were removed when estimating the model performance because of drivers' lack of excitation in these segments which generated classification errors (Sundbom et al. 2013). In a simulation environment, Casucci et al. (2010) differentiate between the prudent and aggressive drivers based on braking behavior and lateral performance. It was found that aggressive drivers completed an overtaking task faster than prudent drivers. In addition, the cruising speed (i.e., the final speed after completing the overtaking maneuver) of aggressive drivers is higher than the prudent drivers (Casucci et al. 2010).

In summary, driver behavior classification is an approach to investigate and understand nominal driver behavior which is used to enhance the development of different warning systems. A wide range of measures were used for driver behavior classification such as drivers' speed (Casucci et al. 2010, Rodriguez Gonzalez et al. 2014, Zheng et al. 2017), TTC (Montgomery et al. 2014), jerk (Feng et al. 2017), and acceleration (Osafune et al., 2017; Gonzalez et al., 2014). In almost all the mentioned studies, these measures were used as an aggregate measure (e.g., averaged over a certain period of time or specific maneuver) regardless the location of the driver. For instance, Gonzalez et al. (2014) classified the driver behavior while driving on a part of a private road network which consisted of several road features (i.e., roundabout, curves, etc.). However, the behavioral measures were aggregated on the entire route without considering the change in the driver behavior from one road feature (e.g., roundabout) to another (e.g., curve). In addition, Osafune et al. (2017) and Feng et al. (2017) extracted behavioral measures from smartphone data and NDS (both are high-frequency data) without referring to the location of the events of the interest.

Therefore, this paper is investigating whether the driver will be relatively more aggressive in a specific location on the network when compared to other locations. In other words, the hypothesis in this paper is that there is a difference in driver following behavior when driving on midblock segments and around intersections. Moreover, this paper will characterize driver behavior during braking on these two locations. The results of this paper could be used as the core of a context-aware application. Such an application adopts automatically to the driver behavior variation due to the change in driver's surrounding environment (i.e., other vehicles, pedestrians, infrastructure, weather, etc.) over time.

2 METHODOLOGY

To characterize driver behavior during braking, the NDS data was used to identify car following events as well as extract various driver behavioral measures (e.g., following distance and TTC). The candidate car following events during braking were split into three groups based on the host vehicle speed and plotted on a map using ArcGIS. The detailed procedure is discussed in the below subsections.

2.1 Data Description

Big data has a wide range of definitions in the literature. For instance, big data is defined as the type of data which is hard to store, manage, process, and visualize (Cox and Ellsworth 1997, Manyika et al. 2011). In addition, big data could be defined as the methods and technologies which enable unveiling hidden values from large, complex, and diverse datasets (Hashem et al. 2015). Moreover, big data could be described by high volume (i.e., large storage), variety (i.e., data format), high velocity (i.e., change in data over time), veracity (i.e., data credibility and certainty), and value (i.e., data usefulness) (Elragal 2014). Big data has a significant role in the automotive and transportation industry transformation. For instance, big data is used to develop intelligent vehicles applications such as Self-Driving cars, Autonomous Vehicles, and Connected Vehicles (Luckow et al. 2015). In addition, big data, in the form of NDS, was used for understanding driver behavior to develop robust intelligent vehicles applications (Wang et al. 2017). NDS data, which is considered as big data, is a rich source of detailed real-world driving behavioral and performance data. NDS was carried out to investigate driver behavior and performance by collecting detailed data on volunteer drivers, vehicles, and the surrounding environment during normal every day driving.

The NDS, which is used in this paper, was collected as part of the Safety Pilot Model Deployment (SPMD). SPMD was a data collection effort under real-life conditions with about 3,000 vehicles equipped with V2V communication devices in Ann Arbor, Michigan, US (Henclewood and Rajiwade 2015). SPMD was part of the "Connected Vehicle (CV) Safety Pilot Program" research initiative which was aimed at evaluating the safety benefits of CV technologies. The dataset comes from the Data Acquisition System (DAS) which was developed by Virginia Tech Transportation Institute. This dataset is available in the ITS Public Data Hub which is operated by the US DOT (USDOT 2018). The dataset was collected from 64 host vehicles equipped with Integrated Safety Devices (ISD) for two months (October 2012 and April 2013). For each second, ISD unit recorded data including position and GPS details (e.g., latitude, longitude, elevation, speed and GPS-based data fidelity measures) and driver behavior data (e.g., braking status, speed, yaw rate, and acceleration). In addition, the dataset contains radar data collected from the ISD unit. The radar data provided information pertaining to the detected vehicle type (i.e. light vehicle, heavy vehicle, bike, pedestrian, or unclassified), status (i.e. moving and in path), the distance from the host vehicle, and the relative speed to the object in front of the host vehicle (Henclewood and Rajiwade 2015). The data was stored in two *.csv* files with a total file size of around 40.7 GB. The first file housed the information related to the vehicle operation and position of the host vehicle. The second file stored the radar data of the detected vehicles around the host vehicle.

2.2 Identifying Drivers' Following Events During Braking

One of the car following identification algorithms in naturalistic data was developed by Kusano et al. (2014) which aimed at identifying host vehicle braking events while following another vehicle. This algorithm was validated and successfully identified approximately 92% of the car following events when compared to manual inspection of video data in a100-car NDS. This algorithm was used to investigate various aspects of driving following behavior during braking. For instance, this algorithm was used to identify the situations during normal car-following situations with brake application to compare between the TTC and the Enhanced TTC as thresholds for triggering FCW (Chen et al. 2016). In addition, the same algorithm was utilized to compare the TTC for different speed bins or different drivers' demographic groups (i.e., age and gender) in car-following scenarios while braking (Montgomery et al. 2014, Kusano et al. 2015).

Consequently, this algorithm will be adapted to serve the purpose of this study. The following steps, which were coded in MATLAB, summarize the NDS data processing and events identification:

1. Preparing and filtering the data:
    1. Remove all object types except the object type which was recognized as a light vehicle (i.e., passenger car).
    2. Remove all instances which were marked with invalid GPS data.
2. Identifying car following events during braking using Kusano et al. algorithm (Kusano et al. 2014). For more details about the algorithm, readers are directed to (Kusano et al. 2014, Tawfeek and El-Basyouny 2018)This algorithm could be summarized in the following steps:

1. Identify all braking events for the host vehicle only when the vehicle's speed is above 10 mph (4.47 m/s) for at least 1 second of the braking time.
2. Identify candidate leading vehicles within 3 seconds headway in front of the host vehicle during the braking events. These candidate leading vehicles should be within 2 meters laterally from the path of the host vehicle.
3. Remove fixed objects (i.e., parked vehicle and slow vehicles in adjacent lanes). This step depends on the speed of the leading vehicle, the gap distance between the leading and the following vehicle, and the azimuth change rate. The azimuth here is defined as the angle between the host vehicle path and the candidate leading vehicle.
4. Select leading vehicle, when multiple candidate vehicles were selected from the previous step, based on the lateral distance between the host and the leading vehicle and proximity to the host vehicle.

3. Selecting incidents of interest:
   1. Scan each identified following event for the minimum following distance.
   2. Extract and calculate driver behavior measures (i.e., TTC) occurring at the minimum following distance while braking.

A total of 43,738 car-following events during braking were identified as candidate events. The incidents, when the following distance is the minimum during the event, were prepared to be plotted on a map to differentiate between intersections and segments events as discussed in the next section.

2.3 Mapping and Defining Intersections and Segments Following Events

ArcGIS was used as a tool to map the identified car following events during braking and to classify each event into either an intersection or segment related event. The candidate events were imported to ArcGIS using the available position data of the host vehicle (i.e., latitude and longitude). Before starting the classification process, intersection and segment (non-intersection) events should be defined. The current literature has various definitions of the intersections influence area. For instance, both American Association of State Highway and Transportation Officials (AASHTO) and Highway Safety Manual (HSM) defines at-grade intersections by physical and functional areas (AASHTO 2010, 2011). The physical area is the area bounded by the stop lines of both intersecting roads. On the other hand, the functional area extends upstream the intersections on the intersecting roads. This area includes driver's decision and maneuver distances and queue storage distance. More specifically, intersection-related crashes are defined as the crashes which occur within 15 to 152 meters from the intersection center point (R.Stollof 2008, FHWA 2009). On the other hand, the 152 meters intersection influence area was considered to be suitable for intersections approaches according to the 40 mph speed limit only (Cottrell and Mu 2005). It is worth to mention that intersection crashes and intersection-related crashes have been defined as the crashes which occurred upstream or inside the physical area of the intersection (Wang et al. 2009). Therefore, similar to the definition of intersection crashes, the candidate events were identified as intersection events when they occurred either upstream or inside the physical area of the intersection.

Since speed is a main contributor to a driver's decision (i.e., perception-reaction) distance and maneuver (i.e., braking) distance, the speed limit should be considered when deciding the influence area (Fitzpatrick et al. 2000). Consequently, the candidate events were split into three categories based on the host vehicle speed, namely; low, medium, and high speeds. The boundaries of low, medium and high speeds were defined as less than 25 mph (40 km/h), between 25 mph and 55 mph (88 km/h), and more than 55 mph, respectively. These boundaries are chosen based on the speed limits in the City of Ann Arbor, Michigan (City of Ann Arbor 2013). Based on these categories, intersections influence area for each category was set and defined as shown in Figure 1.

Figure 1: Intersection influence area (shaded in yellow) and an example of the candidate intersection events (red dots) and segment events (blue dots).

The intersection influence distance (i.e., from the stop line and upstream the intersection) is calculated based on the driver's Stopping Sight Distance (SSD) which include driver's decision and maneuver distances as shown in the following equation.

[1] $SSD\ (meter) = 0.278 \times v \times PRT + \frac{v^2}{254(\frac{a}{g} \pm G)}$

where v is vehicle speed in km/h, PRT is driver's Perception-Reaction Time in seconds, a is vehicle deceleration in m^2/s, and G is roadway longitudinal slope.

The SSD was estimated for low, medium, and high-speed categories as 30, 80, and 150 meters respectively based on AASHTO recommendations for PRT of 2.5 seconds and deceleration of 3.41 m2/s (AASHTO 2011). It is worth to mention that this value of PRT is also recommended by Alberta's Highway Design Guide (Alberta Transportation 1999). Therefore, the intersection influence distance which starts from the intersection's stop line and upstream the intersection will be 30, 80, and 150 meters for low, medium, and high-speed categories. Based on this distance, the intersection influence area (shaded in yellow in Figure 1) was determined, and any candidate event within this area was classified as an intersection event or, otherwise, considered as a segment event as shown in Figure 1. All 44,383 candidate events were classified based on the position of the host vehicle as mentioned before.

3 RESULTS AND DISCUSSION

The identified events were mapped using ArcGIS to differentiate between intersection and segment events in three-speed groups as shown in Table 1. It is worth to mention that the high-speed group does not include any intersection events, as shown in Table 1.

Table 1: Events breakdown by event type and speed group

Speed Group	Intersection Events	Segment Events	Total Events
Low	11403	14962	26365
Medium	1995	12614	14609
High	0	3409	3409
Total	13398	30985	44383

The results were exported from ArcGIS to be processed in MATLAB. To characterize the driver behavior for each event type, the probability density functions (PDF) for minimum following distance and TTC were developed in each speed group for both intersection and segment events. The actual values of the measures were fitted to 17 continuous distribution types, namely; Beta, Birnbaum-Saunders, Exponential, Extreme value, Gamma, Generalized Extreme Value, Generalized Pareto, Inverse Gaussian, Logistic, Log-logistic, Log-normal, Nakagami, Normal, Rayleigh, Rician, *t* Location-Scale, and Weibull. Table 2 shows the distribution's name and parameters which best fits the actual values of the behavioral measures (i.e., minimum range and TTC) for each event type and speed group. The distribution was considered to best fit the actual values of the behavioral measures since the Bayesian Information Criterion (BIC), and Akaike Information Criterion (AIC) had the lowest values.

Table 2: The parameters of the fitted distributions to the actual PDFs

	Speed Group	Location (mean value)	Distribution Name	Parameters			*P-value*
				Location	Scale	Shape	
Minimum Range (m)	Low	Int.(9.01)	Generalized Extreme Value	7.127	3.153	0.020	<0.001
	Low	Seg.(11.05)	Generalized Extreme Value	8.496	4.071	0.047	<0.001
	Med.	Int.(19.5)	Gamma	-	4.430	4.403	<0.001
	Med.	Seg.(20.99)	Gamma	-	4.984	4.212	<0.001
	High	Seg.(29.07)	Birnbaum-Saunders	-	25.100	0.563	-
TTC (sec)	Low	Int.(10.48)	Generalized Extreme Value	3.556	3.081	0.739	<0.001
	Low	Seg.(17.23)	Birnbaum-Saunders	-	9.383	1.295	<0.001
	Med.	Int.(21.42)	Log-normal	2.549	1.087	-	<0.001
	Med.	Seg.(30.00)	Gamma	-	20.834	1.440	<0.001
	High	Seg.(31.81)	Birnbaum-Saunders	-	24.162	0.794	-

Since the distributions of all the measures were not normally distributed, a nonparametric statistical test (i.e., Kolmogorov-Smirnov (KS) test) was used to check the significance of the difference between the intersection/segment event distributions in each speed group. The null hypothesis is that the values for each measure of the intersection and segment events are from the same continues distribution. As shown in Table 2, the results of KS test showed that all the measures of intersection events were different from segment events at 95% confidence level. These results indicate that driver behavior at intersections is substantially different from driver behavior on segments. Such results are crucial when developing any driving assistance system because drivers will receive the warning differently in varied contexts. In other words, drivers could accept a warning relayed to them on midblock segments while considering the same warning at intersections as an early warning which will have an impact on the entire system acceptance.

Table 2 also shows the mean values of the selected behavioral measures in the third column between brackets. As shown in the table, measures' mean values for intersection events were less than segment events. For instance, the mean value of the minimum range of low-speed intersection events (i.e., 9.01 m) was less than the minimum range of low-speed segment events (i.e., 11.05 m). This indicated that the drivers tend to have larger following distance during braking when driving on segments than when driving on intersections. This observation was confirmed by the values of TTC where TTC of car-following during braking events at intersections were less than segment events. Moreover, the mean values of the behavioral measures in the medium-speed group were larger than the values of the low-speed group. In summary, drivers' following behavior during braking at intersections is statistically different from their behavior on midblock segments. In addition, drivers are more likely to follow closer to leading vehicles and to have lower TTC at intersections. Therefore, the results shown in Table 1 indicate that the drivers are more aggressive when approaching intersections. It is worth to mention that Morton et al., (2005) defined aggressive driving for teenage drivers based on speed and headway (i.e., high speed and smaller headway indicate an aggressive driving). In addition, drivers were identified according to their TTC and headway to prudent and aggressive drivers (Wang et al. 2010). Hence, in this paper, aggressive driver behavior is defined based on the minimum following distance and TTC. However, this aggressive behavior is defined here relative to driving behavior on midblock segments which means that the aggressive behavior in this paper is not necessary linked to risky driving.

4 SUMMARY AND CONCLUSIONS

This paper quantified four driver behavioral measures (i.e., minimum following distance and TTC) for car-following during braking situations. Using NDS, this quantification is carried out for events occurred at intersections and midblock segments. In addition, the selected behavioral measures for intersection and segment events were compared. To achieve the objectives of this paper, the car-following events during braking were extracted from the NDS. These events were mapped onto the road network using ArcGIS to identify intersection and segment events. The intersection influence area was defined for three speed groups (i.e., low, medium, and high) based on driver's SSD and speed. Then, the probability distributions for the behavioral measures for each speed group for intersections and segments were developed and fitted to continuous probability distributions. In addition, intersection/segment driver following behavior during braking was compared by checking the significance of the difference between the behavioral measures distributions in each speed group.

The results of this study revealed that there was a considerable difference between driver behavior during braking at intersections and on midblock segments. This difference was defined as the differences between the behavioral measures distributions of intersection and segment events. Using KS test, these differences were statistically significant at 95% confidence level. In addition, this study documented the distributions' names and parameters which best fitted the behavioral measures. By investigating the mean values of the behavioral measures, it was found that drivers approaching intersections had smaller minimum following distance and smaller TTC than when driving on segments. This finding implies that the drivers are more likely to exhibit relatively aggressive behavior when approaching intersections. This finding highlights the importance of considering the driver location when classifying the drivers based on their aggressiveness. According to the current literature, the classification procedures are usually based only on some selected behavioral measures (e.g., acceleration, TTC, jerk, etc.) regardless the location. Considering the findings of this study, the proposed procedures in the literature could misclassify a driver as an aggressive driver when most of the data were collected in high intersection density areas (e.g., urban areas or downtown areas). Therefore, as the main contribution of this study, the driving data should be split based on the location before classifying drivers' behavior. In addition, the driving assistance systems should consider the surrounding environment (i.e., proximity to intersections) by modifying these systems to context-aware driving assistance systems. Such context-aware systems are expected to be more robust and reliable since it will accommodate the changing drivers' surrounding environment.

One of the primary goals of understanding driver behavior is to develop innovative traffic safety solutions. Hence, the results of this study could be beneficial for the development of collision avoidance/warning systems such as FCW. Forasmuch as driver behavior at intersections is different from the behavior on segments, it is recommended to develop an FCW algorithm which accounts for the driver location in the network. To develop such FCW algorithm(s), further analysis of the results will be conducted to select events of interest based on the distribution of acceleration, jerk, and TTC.

In addition, the results could enhance the calibration process of microsimulation models by updating the distributions of drivers' behavioral measures in simulation packages. Also, the PDF distributions of the behavioral measures could be used in modeling driver following behavior mathematically as suggested by Gonzalez et al. (2014). The PDF distributions generated from this study is expected to lead to more robust simulation results because of the consideration of the driver location and the fact that these distributions were generated based on NDS. The NDS grants the recording of normal driving behavior without exposing the drivers to a test setting as in test track which will affect their behavior.

Moreover, this study shed light on the importance of integrating the NDS with the road network map to deeply investigate the driver behavior. The pairing of the NDS and the road network, using ArcGIS, LiDAR data, or other means, is an up-and-coming area of research which will allow insightful findings in driver behavior understanding. Therefore, as an extension of this study, an automated approach to differentiate between intersection and midblock segment events could be explored. This automated approach could enhance the definition of the intersection influence area which has a wide range of definitions in the literature (e.g., the suggested definition in this study which was based on the SSD and vehicle speed).

References

AASHTO. 2010. Highway safety manual, first edition, with 2014 supplement.

AASHTO. 2011. A Policy on Geometric Design of Highways and Streets. : 942.

Abe, G., and Richardson, J. 2005. The influence of alarm timing on braking response and driver trust in low speed driving. Safety Science, **43**(9): 639–654.

Alberta Transportation. 1999. Highway geometric design guide.

Ben-Bassat, T., and Shinar, D. 2011. Effect of shoulder width, guardrail and roadway geometry on driver perception and behavior. Accident Analysis and Prevention, **43**(6): 2142–2152.

Berndt, H., Wender, S., and Dietmayer, K. 2007. Driver braking behavior during intersection approaches and implications for warning strategies for driver assistant systems. Proc. IEEE Intelligent Vehicles Symposium,: 245–251.

Casucci, M., Marchitto, M., and Cacciabue, P.C. 2010. A numerical tool for reproducing driver behaviour: Experiments and predictive simulations. Applied Ergonomics, **41**(2): 198–210.

Chen, R., Sherony, R., and Gabler, H.C. 2016. Comparison of Time to Collision and Enhanced Time to Collision at Brake Application during Normal Driving. SAE Technical Paper, (2016-01–1448).

City of Ann Arbor. 2013. City of Ann Arbor Speed Limits Map. Available from http://www.a2gov.org/departments/engineering/Pages/default.aspx [accessed 12 April 2017].

Cottrell, W.D., and Mu, S. 2005. Utah Intersection Safety: Recurrent Crash Sites: Identification, Issues and Factors. University of Utah, Salt Lake City.

Cox, M., and Ellsworth, D. 1997. Managing Big Data for Scientific Visualization. ACM Siggraph, **97**: 1–17. Available from http://www.dcs.ed.ac.uk/teaching/cs4/www/visualisation/lectures/98-99/lect16ref.ps.gz.

van Driel, C.J.G., Hoedemaeker, M., and van Arem, B. 2007. Impacts of a Congestion Assistant on driving behaviour and acceptance using a driving simulator. Transportation Research Part F: Traffic Psychology and Behaviour, **10**(2): 139–152.

Elragal, A. 2014. ERP and Big Data: The Inept Couple. Procedia Technology, **16**: 242–249.

Farah, H., Koutsopoulos, H.N., Saifuzzaman, M., Kölbl, R., Fuchs, S., and Bankosegger, D. 2012. Evaluation of the effect of cooperative infrastructure-to-vehicle systems on driver behavior. Transportation Research Part C: Emerging Technologies, **21**(1): 42–56.

Feng, F., Bao, S., Sayer, J.R., Flannagan, C., Manser, M., and Wunderlich, R. 2017. Can vehicle longitudinal jerk be used to identify aggressive drivers? An examination using naturalistic driving data. Accident Analysis and Prevention, **104**: 125–136.

FHWA, U.S.D. of T. 2009. Manual on Uniform Traffic Control Devices.

Fitzpatrick, K., Carlson, P.J., Wooldrigge, M.D., and Brewer, M.A. 2000. Design factors that affect driver speed on suburban arterials. : 164.

Harding, J., Powell, G., Yoon, R., Fikentscher, J., Doyle, C., Sade, D., Lukuc, M., Simons, J., and Wang, J. 2014. Vehicle-to-Vehicle Communications : Readiness of V2V Technology for Application. Washington, DC.

Hashem, I.A.T., Yaqoob, I., Anuar, N.B., Mokhtar, S., Gani, A., and Ullah Khan, S. 2015. The rise of "big data" on cloud computing: {Review} and open research issues. Inf Syst, **47**(Journal Article): 98–115.

Henclewood, D., and Rajiwade, S. 2015. Safety Pilot Model Deployment – Sample Data Environment Data Handbook. **1.3**.

Jensen, M., Wagner, J., and Alexander, K. 2011. Analysis of in-vehicle driver behaviour data for improved safety. International Journal of Vehicle Safety, **5**(3): 197.

Kusano, K.D., Chen, R., Montgomery, J., and Gabler, H.C. 2015. Population distributions of time to collision at brake application during car following from naturalistic driving data. Journal of Safety Research, **54**: 95–104.

Kusano, K.D., Montgomery, J., and Gabler, H.C. 2014. Methodology for identifying car following events from naturalistic data. IEEE Intelligent Vehicles Symposium, Proceedings,: 281–285.

Luckow, A., Kennedy, K., Manhardt, F., Djerekarov, E., Vorster, B., and Apon, A. 2015. Automotive big data: Applications, workloads and infrastructures. Proceedings - 2015 IEEE International Conference

on Big Data, IEEE Big Data 2015,: 1201–1210.
Manyika, J., Chui, M., Brown, B., Bugin, J., Dobbs, R., Roxburgh, C., and Byers, A.H. 2011. Big data: The next frontier for innovation, competition, and productivity. McKinsey Global Institute, (June): 156.
Martinez, C.M., Heucke, M., Wang, F.Y., Gao, B., and Cao, D. 2017. Driving Style Recognition for Intelligent Vehicle Control and Advanced Driver Assistance: A Survey. IEEE Transactions on Intelligent Transportation Systems, **PP**(99): 1–11.
Messer, C.J. 1980. Methodology for Evaluating Geometric Design Consistency. Transportation Research Record: Journal of the Transportation Research Board, **757**: 7–14.
Mohammed (Mostafa H. Tawfeek), M., Ke, Y., Gao, J., Zhang, H., El-Basyouny, K., and Qiu, T.Z. 2016. Connected Vehicle V2I Communication Application To Enhance Driver Awareness At Signalized Intersections. *In* 11th International Transportation Specialty Conference, CSCE 2016.
Montgomery, J., Kusano, K.D., and Gabler, H.C. 2014. Age and gender differences in time to collision at braking from the 100-car naturalistic driving study. Traffic injury prevention, **15 Suppl 1**(November): S15-20.
Olia, A., Abdelgawad, H., Abdulhai, B., and Razavi, S.N. 2015. Assessing the Potential Impacts of Connected Vehicles: Mobility, Environmental, and Safety Perspectives. Journal of Intelligent Transportation Systems, **2450**(September): 1–15.
Osafune, T., Takahashi, T., Kiyama, N., Sobue, T., Yamaguchi, H., and Higashino, T. 2017. Analysis of Accident Risks from Driving Behaviors. International Journal of Intelligent Transportation Systems Research, **15**(3): 192–202.
R.Stollof, E. 2008. Intersection and Junction Fatalities in the Context of Access Management. 8th National Conference on Access Management, (July).
Rodriguez Gonzalez, A.B., Wilby, M.R., Vinagre Diaz, J.J., and Sanchez Avila, C. 2014. Modeling and detecting aggressiveness from driving signals. IEEE Transactions on Intelligent Transportation Systems, **15**(4): 1419–1428.
Simons-Morton, B., Lerner, N., and Singer, J. 2005. The observed effects of teenage passengers on the risky driving behavior of teenage drivers. Accident Analysis and Prevention, **37**(6): 973–982.
Staubach, M., Schebitz, N., Köster, F., and Kuck, D. 2014. Evaluation of an eco-driving support system. Transportation Research Part F: Traffic Psychology and Behaviour, **27**(PA): 11–21.
Sundbom, M., Falcone, P., and Sjoberg, J. 2013. Online driver behavior classification using probabilistic ARX models. IEEE Conference on Intelligent Transportation Systems, Proceedings, ITSC,: 1107–1112.
Tawfeek, M.H., and El-Basyouny, K. Forthcoming 2018. A Perceptual Forward Collision Warning Model using Naturalistic Driving Data. Canadian Journal of Civil Engineering,.
USDOT. 2018. ITS Public Data Hub. Available from https://www.its.dot.gov/data/ [accessed 3 January 2018].
Wang, J., Lu, M., and Li, K. 2010. Characterization of Longitudinal Driving Behavior by Measurable Parameters. Transportation Research Record: Journal of the Transportation Research Board, **2185**: 15–23.
Wang, W., Liu, C., and Zhao, D. 2017. How Much Data is Enough? A Statistical Approach with Case Study on Longitudinal Driving Behavior.
Wang, X., Abdel-Aty, M., Nevarez, A., and Santos, J.B. 2009. Investigation of Safety Influence Area for Four-Legged Signalized Intersections: Nationwide Survey and Empirical Inquiry. Transportation Research Record: Journal of the Transportation Research Board, **2083**(1): 86–95.
Zheng, Y., Chase, R.T., Elefteriadou, L., Sisiopiku, V., and Schroeder, B. 2017. Driver types and their behaviors within a high level of pedestrian activity environment. Transportation Letters, **9**(1): 1–11.

Building Tomorrow's Society
Bâtir la Société de Demain

Fredericton, Canada
June 13 to 16, 2018 / *13 au 16 juin, 2018*

VEHICLE ROUTING PROBLEMS: TRUE OPTIMAL SOLUTIONS

Christie, James[1,2]
[1] University of New Brunswick, Canada
[2] christie@unb.ca

Abstract: Optimizing vehicular use of the transportation infrastructure results in benefits for private transport companies, public departments of transportation infrastructure, and the general traveling public. Benefits include reduced operating costs, reduced energy use, decreased pollution, reduced demand for resources in terms of vehicles and infrastructure, improved safety in terms of saved lives and fewer injuries, improved level of service with extended economic life and sustainability for the existing infrastructure. Transportation engineers have been challenged for decades with finding a true optimal solution. Developing a methodology to find the optimal solution to the vehicle routing problem is the first step. This has been the primary goal of this research. The focus has been on an exact solution to the point-to-point vehicle routing problem. In the simplest situation this means that a single vehicle is used to deliver (or pick-up) goods to (or from) a number of customers that are located throughout a service area; the objective is to leave the depot and service each customer using the most efficient route from customer to customer and finally returning back to the depot – thus defining the traveling salesperson problem. If the capacity of one vehicle is exceeded then more vehicles are dispatched, and the general vehicle routing problem has to be solved such that the customer base is serviced most efficiently with the least number of vehicles. A novel way to merge integer linear programming and constraint programming has been developed that is efficient and effective in solving the vehicle routing problem.

1 INTRODUCTION

The benefits of the application of vehicle logistics optimization has been discussed by the current author and coauthors of supervised research projects in previous conference publications based on the related PhD Dissertation (Satir 2006) and MScE Thesis (Satir 2003). For example, Christie and Satir (2004 and 2006) identified a number of potential spin-off benefits to the transportation industry, both for those operating the vehicles and those operating the transportation infrastructure and for the general public. One paper focused on the benefits in terms of reducing energy use and reducing emissions with between 20 and 40 percent reduction in some cases with the application of computerized vehicle routing and scheduling optimization (Christie and Satir 2006). In another paper other traffic spin-off benefits were identified with those relating to pickup and delivery operations whereby up to 50 percent reduction in the related truck traffic could potentially be realized (Christie and Satir 2004). This leads to improved level of service of the infrastructure, extended economic life, reduction in truck related accidents and reduced load repetition on pavements leading to improved pavement life.

This identifies the need for continued improvement in the optimization methods available to transportation engineers and others involved in both the vehicle and transportation infrastructure systems. This is the primary motive for the current research with the goal to develop new and improved methods to solve the traveling salesperson problem (TSP) and the more general multiple vehicle routing problem (VRP). In terms of a truly optimal solution to the TSP, little further progress has been made since the formulations of the problem in 1954 (Dantzig, Fulkerson and Johnson 1954), denoted as DFJ, and the formulation presented in 1960 (Miller, Tucker and Zemilin 1960), denoted as MTZ. No exact solution methods to solve the VRP have been presented to date. This research presents a new formulation to provide an exact solution to both problems.

2 CURRENT EXACT ASIGNMENT FORMULATIONS TO SOLVE TSPs

For decades transportation researchers and practicing professionals in engineering and other transportation related fields have been grappling with finding exact solutions to the famous TSP and the more general VRP. Two assignment formulations were presented for the TSP. The conventional formulation for N cities identified by nodes 1,2,3 ..., n (Dantzig, Fulkerson and Johnson, 1954; Miller, Tucker and Zemlin, 1960) is:

Xij is a binary (0, not selected, or 1, selected) decision variable that represents a link connecting city i to city j with an associated cost of Cij, and a complete tour, starting arbitrarily from say node 1, continues sequentially to each of the other cities until the completed tour ends back at node 1 with all N cities included without any sub-tours. The objective is to -

[1] Minimize: $\sum_{i,j} CijXij$ for i = 1, 2 ... n; j = 1, 2 ..., n

[2] Subject to: $\sum_{i} Xij = 1$ for all j

[3] $\sum_{j} Xij = 1$ for all i

Both the DFJ and MTZ methods incorporate this assignment formulation initially, but sub-tours may result with this integer linear programming formulation.

The DFJ method eliminates sub-tours using an iterative procedure whereby the linear programming assignment procedure is run followed by a non-linear step where sub-tours are identified and sub-tour elimination constraints are added, and the assignment procedure is rerun with the additional constraints. This iterative process continues until no further sub tours can be identified. The sub-tour elimination constraint is presented in the following equation [4]:

[4] $\sum_{i \in S, j \in S} Xij \leq |S| - 1$ for all $S \subset \{2, \dots, n\}$

With the MTZ formulation, constraint variables are included to eliminate sub-tours. An integer variable, u_i represents the relative position of each node (city) in the complete tour, with node 1 arbitrarily set as the first position. The resulting constraints are:

[5] $u_1 = 1$

[6] $2 \leq u_i \leq n$

[7] $u_i - u_j + nx_{ij} \leq n - 1$ for i, j = 2, 3, ..., n

Both the DFJ and MTZ formulations have inefficiencies that grow with increasing problem size, and although the MTZ constraint formulation avoids an iterative application of integer linear programming the constraint formulation itself is weak (Pataki, 2000; Gutin and Punnen, 2007). Some improvements in the MTZ sub-tour elimination constraints have been made, for example, by M. Desroches and G. Laporte (Desroches and Laporte, 1991). The other shortcoming of the DFJ and MTZ formulations is that they are not designed to effectively solve the more general VRP. Heuristic methods cannot necessarily be proven to be exact optimal solutions to the TSP and are not generally considered in this paper. However, because the Concorde TSP Solver is fairly well known and has been used to solve some fairly large symmetric TSPs it is acknowledged here (Applegate, Bixby, Chvatal and Cook, 2006). Again Concorde is not designed to solve asymmetric TSPs nor the more general VRPs. Most recently Zamorano and Stolletz (2017) used branch-and-price methods to solve related multi-period technician routing and scheduling problems (MPTRSP). Mathlouthi, Gendreau and Potvin (2018) used mixed integer linear programming to solve a similar MPTRSP. Woods, Punnen and Stephen (2017) developed a linear time algorithm to solve the 3-neighbor TSP. A general linear programming formulation is still needed to solve both TSPs and VRPs.

3 PROPOSED METHOD TO SOLVE TSPs AND VRPs

The research presented here is designed to solve symmetric and asymmetric TSPs and VRPs using the mixed integer linear programming assignment formulation, as does the DFJ and MTZ methods. However, the assignment formulation has been revised to also accommodate VRPs. New sub-tour elimination constraints are considered that help alleviate the shortcomings of both the DFJ and MTZ sub-tour constraints by merging the standard integer linear programming methods with the approaches used in constraint programming. Although not noted in the literature, actually this is, in effect, the approach used with the MTZ method, but their sub-tour elimination constraints are weak. The current method presented here uses strong constraints. This requires the inclusion of constraint variables in addition to the usual decision variables found in standard linear programming solutions. Details are presented in the following.

3.1 Assignment-Transshipment Formulation to solve symmetric and asymmetric TSPs

As with the DFJ and MTZ methods, the integer programming assignment formulation is used with equations [1], [2] and [3]. The potential sub-tours are eliminated by using a transshipment formulation of the shortest route problem. The rationale is that, in the complete TSP tour, all N nodes or cities are connected from one to the other sequentially, in which case any individual node can be reached from say node 1. If there were sub-tours some of the nodes could not be reached from node 1, assuming some portion of the TSP tour is to be followed. Therefore a transshipment formulation of the shortest route problem is introduced for each node (> 1) to ensure a possible connection from node 1 to each of the other nodes. With this additional constraint introduced the TSP assignment formulation is solved to optimality resulting in a complete tour. Tables 1a and 1b depict this formulation of the TSP in an Excel spreadsheet format.

Table 1a: Assignment(Transportation)-Transshipment Formulation of the TSP(VRP): Part A

COST Matrix, C

From\To	j = 1	2	---	---	---	n		
i = 1	C_{11}	C_{12}	---	---	---	C_{1n}		
2	C_{21}	C_{22}	---	---	---	C_{2n}		
|	---	---	---	---	---	---		
|	---	---	---	---	---	---		
|	---	---	---	---	---	---		
n	---	---	---	---	---	C_{nn}		

SOLUTION Matrix, O **Assignment(Transportation) Formulation of TSP(VRP)**

From\To	j = 1	2	---	---	---	n	i-SUM =	Constraint
i = 1	X_{11}	X_{12}	---	---	---	X_{1n}	0	#Vehicles
2	X_{21}	X_{22}	---	---	---	X_{2n}	0	1
|	---	---	---	---	---	---	0	1
|	---	---	---	---	---	---	0	1
|	---	---	---	---	---	---	0	1
n	---	---	---	---	---	X_{nn}	0	1
j-SUM =	0	0	0	0	0	0	**TOTAL**	$\sum_{i,j} CijXij$
Constraint	#Vehicles	1	1	1	1	1	**COST =**	
							for all i, j = 1 to n	

CONSTRAINT MATRICES, Sk **Transshipment Formulation of Shortest Route Problem**

Shortest route from node 1 along TSP tour to node k = 2

Fr1\To2	j = 1	2	---	---	---	n	i-SUM =	Constraint
i = 1	X_{112}	X_{122}	---	---	---	X_{1n2}	0	2
2	X_{212}	X_{222}	---	---	---	X_{2n2}	0	1
|	---	---	---	---	---	---	0	1
|	---	---	---	---	---	---	0	1
|	---	---	---	---	---	---	0	1
n	---	---	---	---	---	X_{nn2}	0	1
j-SUM =	0	0	0	0	0	0		
Constraint	1	2	1	1	1	1		

Shortest route from node 1 along TSP tour to node k = 2, 3, ---, n

Fr1\To n	j = 1	2	---	---	---	n	i-SUM =	Constraint
i = 1	X_{11k}	X_{12k}	---	---	---	X_{1nk}	0	2
2	X_{21k}	X_{22k}	---	---	---	X_{2nk}	0	1
|	---	---	---	---	---	---	0	1
|	---	---	---	---	---	---	0	1
|	---	---	---	---	---	---	0	1
n	---	---	---	---	---	X_{nnk}	0	1
j-SUM =	0	0	0	0	0	0		
Constraint	1	1	1	2	1	1		

Table 1b: Assignment(Transportation)-Transshipment Formulation of the TSP(VRP): Part B

Multiplication Matrix, M, to eliminate diagonal values in Constraint Matrices

From\To	j = 1	2	---	---	---	n
i = 1	0	1	---	---	---	1
2	1	0	---	---	---	1
|	---	---	---	---	---	1
|	---	---	---	---	---	1
|	---	---	---	---	---	1
n	1	1	---	---	---	Mnn = 0

Product Matrix, P, to prevent diagonal assignments in Solution Matrix

From\To	j = 1	2	---	---	---	n
i = 1	1	0	---	---	---	0
2	0	1	---	---	---	0
|	---	---	---	---	---	0
|	---	---	---	---	---	0
|	---	---	---	---	---	0
n	0	0	---	---	---	Pnn = 1

Constraint = SUMPRODUCT (Product Matrix, SOLUTION Matrix) = $\sum_{i,j} PijXij = 0$

THIS CONSTRAINT ONLY NEEDED FOR VRP

for all i, j = 1 to n

CONSTRAINT MATRICES * Multiplication Matrix

Fr1\To2	j = 1	2	---	---	---	n
i = 1	$X_{112} * M_{11}$	$X_{122} * M_{12}$	---	---	---	$X_{1n2} * M_{1n}$
2	$X_{212} * M_{21}$	$X_{222} * M_{22}$	---	---	---	$X_{2n2} * M_{2n}$
|	---	---	---	---	---	---
|	---	---	---	---	---	---
|	---	---	---	---	---	---
n	---	---	---	---	---	$X_{nn2} * M_{nn}$

Constraints for k = 2 : $X_{ij} \geq (X_{ij2} * M_{ij})$

Fr1\To n	j = 1	2	---	---	---	n
i = 1	$X_{11k} * M_{11}$	$X_{12k} * M_{12}$	---	---	---	$X_{1nk} * M_{1n}$
2	$X_{21k} * M_{21}$	$X_{22k} * M_{22}$	---	---	---	$X_{2nk} * M_{2n}$
|	---	---	---	---	---	---
|	---	---	---	---	---	---
|	---	---	---	---	---	---
n	---	---	---	---	---	$X_{nnk} * M_{nn}$

Constraints for all k ≥ 2 : $X_{ij} \geq (X_{ijk} * M_{ij})$

In Table 1a, the Cost Matrix and the Assignment formulation of the Solution Matrix represent the standard formulation of the basic TSP used with both the DFJ and MTZ methods. The cell entries in the Solution Matrix noted as #Vehicles are replaced with the numeric value of 1 (for 1 person or vehicle), thus completing the Assignment formulation. The Constraint Matrices shown in Table 1a are introduced as sub-tour elimination constraints and are effective whether the cost matrix is symmetric or asymmetric. The inspiration for this approach arose from a review of the Handbook of Constraint Programming (Rossi, van Beek and Walsh 2006), particularly chapters 11, 12 and 15. The ideas of constraint formulations, constraint satisfaction problems, constraint logic programming and using operations research methods in constraint programming led to the idea of reversing the process and finding a way to use constraint programming methods and approaches in operations research methods to solve the TSPs and VRPs. This is the first time this approach has ever been presented.

The fact that a valid solution to the TSP allows for a viable routing from any given node to each and every other node in the complete tour then provided the fundamental system of constraints to eliminate any possible sub-tours. The analyst can choose any given node as the basic starting node, so node 1, representing the depot, has been arbitrarily chosen with no loss of generality. Then an integer linear programming formulation of the shortest route problem is required and is readily available in the transshipment problem formulation (see, for example, Ragsdale 2008; Hillier and Lieberman 2005). Now binary constraint variables, X_{ijk}, are introduced with a value 0 if not selected as a link (node i to node j) or 1 if selected as a link (i to j) forming a part of the shortest route open tour from node 1 to node k (see the Constraint Matrices in Table 1a where k represents the target node destination for the relevant shortest route problem). The diagonal values, where i = j, are eliminated to provide the final transshipment shortest route solution from node 1 to node k. To accomplish this, the individual cell values, X_{ijk}, of each of the Constraint Matrices, C_k, shown in Table 1a is multiplied by the corresponding cell values, M_{ij}, in the Multiplication Matrix, M, shown in Table 1b. The final sub-tour elimination constraints are then shown in Table 1b as the CONSTRAINT MATRICES*Multiplication Matrix for each k≥2 where each $X_{ij} \geq (X_{ijk} * M_{ij})$ represent the individual constraints. Unlike the typical integer linear program these constraint values are not constant; they are a function of the constraint variables that are solved for simultaneously in conjunction with the standard decision variables which determine the final TSP tour and the final objective value shown in equation [1] and in Table 1a. The constraint variables do not directly determine the objective value; however, the constraint variables, X_{ijk}, constrain the values in the Solution Matrix in Table 1a such that all $X_{ij} \geq (X_{ijk} * M_{ij})$ for all k ≥ 2 to n.

As one final, but important, note, no integer (binary) constraint is required for the constraint variables because the decision variables are already restricted to be integer (binary) values. By default then the constraint variables will have integer values, and there is no need to actually constrain them to be integer variables. Therefore they can be solved as continuous variables with standard linear programming procedures, which reduces the required computational time considerably.

3.2 Transportation-Transshipment Formulation to solve symmetric and asymmetric VRPs

There are only a few changes in the formulation when solving the VRP. The starting node has to be the depot node, which is node 1 in the current representation. Because that is the node from which all vehicles leave, each and every other node must be accessible in a valid solution. Again the # Vehicles denoted in the SOLUTION MATRIX now has to be entered in the two cells in Table 1a as a numeric value of 2 or greater (which represents the actual number of vehicles being used to service the nodes/customers/cities, which completes the Transportation formulation. Although not required for solving TSPs, the Product Matrix, P, in Table 1b is required for solving the VRP SOLUTION Matrix for the Transportation formulation in order to prevent diagonal assigned values of 1 in the SOLUTION Matrix. The additional constraint required is:

[8] $\sum_{i,j} PijXij = 0$, for all i, j = 1 to n

With this one additional constraint the Transportation-Transshipment formulation then provides an integer programming solution to the VRP for the number of vehicles provided to service all the nodes from the depot node. Again it is important to apply the binary constraint to the SOLUTION Matrix variables (decision variables) only and not to the CONSTRAINT Matrices variables (constraint variables). As with the TSP solution, the constraint variables are treated as continuous variables which will inherently solve to have binary integer values. This saves considerable computing time which is very important with larger problems.

4 APPLICATION OF THE PROCEDURE

The Transportation-Transshipment formulation of the VRP is demonstrated with a small example. Figure 1 shows a screenshot of the Excel spreadsheet with the Cost Matrix for an asymmetric problem with only 6 nodes. Node 1 represents the depot and the nodes are serviced by 2 vehicles. Table 1 can be referred to for descriptive details.

The SOLUTION MATRIX is the Transportation formulation of the solution. To prevent sub-tours in the SOLUTION MATRIX the CONSTRAINT MATRICES are shown in a Transshipment formulation. Note that the '2' shown in each of the CONSTRAINT MATRICES (for k= 2 to n) is to allow for transshipment movements, and it does not represent the number of vehicles dispatched from the depot node. For details describing the shortest route solution using the Transshipment formulation consult an operations research textbook (for example, Hillier and Lieberman 2005). The problem was solved using an Add-in to Excel called OpenSolver version 2.9.0 (Mason 2012) and the GUROBI mixed integer linear programming solver engine version 7.5 (GUROBI Optimization 2017). Figure 2 shows the OpenSolver user input screen used to identify the objective cell to be minimized, the decision and constraint variables, the various linear constraints and the solver engine (Gurobi) to be used. Figure 3 shows the OpenSolver model runtime statistics: number of continuous and integer variables, number of simplex iterations and processing time, and the final objective value along with the integer gap, if any.

COST Matrix

From\To	1	2	3	4	5	6
1	0	27	43	16	30	26
2	7	0	16	1	30	25
3	20	13	0	35	5	0
4	21	16	25	0	18	18
5	12	46	27	48	0	5
6	23	5	5	9	5	0

Multiplication Matrix, M

From\To	1	2	3	4	5	6
1	0	1	1	1	1	1
2	1	0	1	1	1	1
3	1	1	0	1	1	1
4	1	1	1	0	1	1
5	1	1	1	1	0	1
6	1	1	1	1	1	0

SOLUTION Matrix Depot = 1 TSP=1 Vehicle VRP=X Vehicles

From\To	1	2	3	4	5	6	SUM =		Constraint	
1	binary 0	0	0	1	0	1	2		2	=# of vehicles
2	1	0	0	0	0	0	1		1	
3	0	0	0	0	1	0	1	=	1	
4	0	1	0	0	0	0	1		1	
5	1	0	0	0	0	0	1		1	
6	0	0	1	0	0	0	1		1	
SUM =	2	1	1	1	1	1	TOTAL		min	
Constrain =	2	1	1	1	1	1	COST =		87	

=# of vehicles

Product Matrix, P

From\To	1	2	3	4	5	6
1	1	0	0	0	0	0
2	0	1	0	0	0	0
3	0	0	1	0	0	0
4	0	0	0	1	0	0
5	0	0	0	0	1	0
6	0	0	0	0	0	1

SumProduct = 0= 0

THIS CONSTRAINT ONLY NEEDED FOR VRP

CONSTRAINT MATRICES

Shortest route from node 1 along TSP/VRP tour to node k=2

Fr1\To2	1	2	3	4	5	6	SUM =		Constraint
1	1	0	0	1	0	0	2		2
2	0	1	0	0	0	0	1		1
3	0	0	1	0	0	0	1	=	1
4	0	1	0	0	0	0	1		1
5	0	0	0	0	1	0	1		1
6	0	0	0	0	0	1	1		1
SUM =	1	2	1	1	1	1			
Constrain =	1	2	1	1	1	1			

Shortest route from node 1 along TSP/VRP tour to node k=3

Fr1\To3	1	2	3	4	5	6	SUM =		Constraint
1	1	0	0	0	0	1	2		2
2	0	1	0	0	0	0	1		1
3	0	0	1	0	0	0	1	=	1
4	0	0	0	1	0	0	1		1
5	0	0	0	0	1	0	1		1
6	0	0	1	0	0	0	1		1
SUM =	1	1	2	1	1	1			
Constrain =	1	1	2	1	1	1			

Shortest route from node 1 along TSP/VRP tour to node k=4

Fr1\To4	1	2	3	4	5	6	SUM =		Constraint
1	1	0	0	1	0	0	2		2
2	0	1	0	0	0	0	1		1
3	0	0	1	0	0	0	1	=	1
4	0	0	0	1	0	0	1		1
5	0	0	0	0	1	0	1		1
6	0	0	0	0	0	1	1		1
SUM =	1	1	1	2	1	1			
Constrain =	1	1	1	2	1	1			

Shortest route from node 1 along TSP/VRP tour to node k=5

Fr1\To5	1	2	3	4	5	6	SUM =		Constraint
1	1	0	0	0	0	1	2		2
2	0	1	0	0	0	0	1		1
3	0	0	0	0	1	0	1	=	1
4	0	0	0	1	0	0	1		1
5	0	0	0	0	1	0	1		1
6	0	0	1	0	0	0	1		1
SUM =	1	1	1	1	2	1			
Constrain =	1	1	1	1	2	1			

Shortest route from node 1 along TSP/VRP tour to node k=6

Fr1\To6	1	2	3	4	5	6	SUM =		Constraint
1	1	0	0	0	0	1	2		2
2	0	1	0	0	0	0	1		1
3	0	0	1	0	0	0	1	=	1
4	0	0	0	1	0	0	1		1
5	0	0	0	0	1	0	1		1
6	0	0	0	0	0	1	1		1
SUM =	1	1	1	1	1	2			
Constrain =	1	1	1	1	1	2			

CONSTRAINT Matrices * Multipication Matrix

Fr1\To2		1	2	3	4	5	6
1		0	0	0	1	0	0
2		0	0	0	0	0	0
3	≥	0	0	0	0	0	0
4		0	1	0	0	0	0
5		0	0	0	0	0	0
6		0	0	0	0	0	0

Fr1\To3		1	2	3	4	5	6
1		0	0	0	0	0	1
2		0	0	0	0	0	0
3	≥	0	0	0	0	0	0
4		0	0	0	0	0	0
5		0	0	0	0	0	0
6		0	0	1	0	0	0

Fr1\To4		1	2	3	4	5	6
1		0	0	0	1	0	0
2		0	0	0	0	0	0
3	≥	0	0	0	0	0	0
4		0	0	0	0	0	0
5		0	0	0	0	0	0
6		0	0	0	0	0	0

Fr1\To5		1	2	3	4	5	6
1		0	0	0	0	0	1
2		0	0	0	0	0	0
3	≥	0	0	0	0	1	0
4		0	0	0	0	0	0
5		0	0	0	0	0	0
6		0	0	1	0	0	0

Fr1\To6		1	2	3	4	5	6
1		0	0	0	0	0	1
2		0	0	0	0	0	0
3	≥	0	0	0	0	0	0
4		0	0	0	0	0	0
5		0	0	0	0	0	0
6		0	0	0	0	0	0

Figure 1: Screenshot of Excel Spreadsheet Demonstrating VRP Solution for 2 Service Vehicles

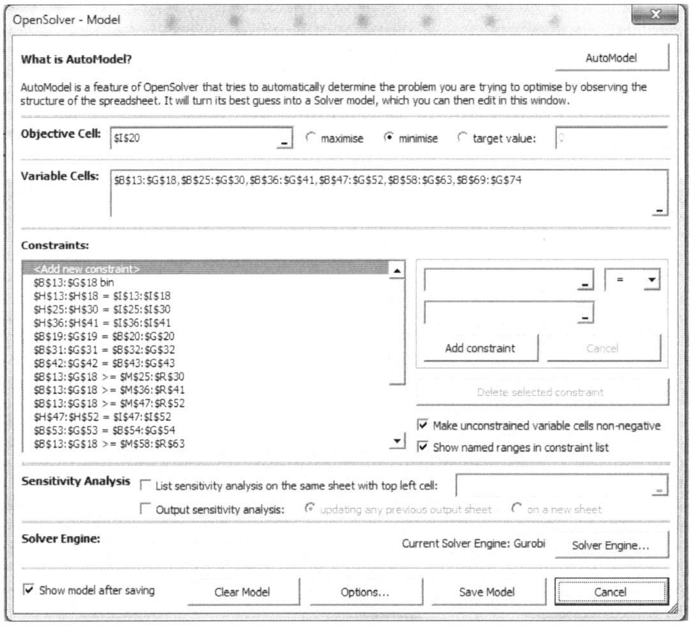

Figure 2: User input screen for the OpenSolver Linear Programming Model

Figure 3: OpenSolver model runtime statistics

Figure 1 depicts the objective value to be 87 (which represents the total cost, time or distance, depending on the user's chosen unit measure) when serviced by 2 vehicles. The same problem was run with 1 vehicle with a resulting objective of 63 and again with 3 vehicles with a resulting objective value of 119. Although this is a very small example it does demonstrate the relative efficiencies and importance of choice of number of service vehicles. The least cost choice is to provide the service with 1 vehicle, and the cost is almost doubled by using 3 vehicles.

For the same 6 node example, one may question which of the nodes should be the depot node to optimize the service cost. To determine this, two vehicles were again used to service the nodes. However, each node in turn was considered to be the depot node. The resulting total cost for each was: node 1, 87; node 2, 70; node 3, 70; node 4, 73; node 5, 73; and node 6, 56. For optimal service cost with two vehicles, the depot should be located at node 6 with a total service cost of 56.

Preliminary results have been generated for a larger problem consisting of 50 nodes with node 1 again assumed to represent the depot node. This is an ongoing research database where various other techniques for reducing computational time are being investigated, but to this point with a single vehicle servicing the nodes a solver engine time of about 2 minutes and 30 seconds has been achieved using a 2011 version of the GUROBI software on a 2006 DELL Precision 690 workstation. With the latest version of the software it is estimated the computational time would be under 10 seconds.

5 CONCLUSIONS

A novel method to successfully solve symmetric and asymmetric TSPs has been demonstrated in this paper by using an Assignment-Transshipment formulation of the TSP. The Assignment formulation contains the standard decision variables, while the Transshipment formulation part includes the addition of applicable constraint variables. A mixed integer linear programming solver can then be used to solve the minimization problem to an optimal solution providing the optimal sequence for servicing a given number of nodes, starting from the depot node.

With very few alterations the TSP formulation can be revised to solve the more general VRP. Basically a Transportation-Transshipment formulation is developed to solve symmetric or asymmetric VRPs. The Transportation Problem formulation contains the decision variables for the VRP, and an additional constraint is included to prevent diagonal assignments. Similar to the TSP situation, the Transshipment formulation part includes the addition of applicable constraint variables. Again a mixed integer linear programming solver can be used to solve the minimization problem, so that a set of nodes can be serviced by 2 or more service vehicles starting from the one depot node at the least total cost.

This research has developed a new way to combine constraint programming methods with linear programming to successfully solve historically difficult linear programming problems that have challenged transportation engineers in civil engineering for a number of decades. Although used to solve TSPs and VRPs here, there is potential to broaden the application of this approach within the civil engineering field and beyond.

Acknowledgements

The author wishes to acknowledge the financial support provided through Student Work Program Grants and the Retirees Research Fund Grant at the University of New Brunswick.

References

Applegate, D.L., Bixby, R.E., Chvatal, V. and Cook, W.J. 2006. The Traveling Salesman Problem: A Computational Study, Princeton University Press, Princeton, New Jersey.

Christie, J.S. and Satir, S. 2004. Transport Logistics: Potential Traffic Related Spin-off Benefits, *2004 Canadian ITE Annual Conference*, Moncton, NB, Institute of Transportation Engineers, District 7.

Christie, J.S.and Satir, S. 2006. Saving Our Energy Sources and Meeting Kyoto Reduction Targets While Minimizing Costs with Application of Vehicle Logistics Optimization, *2006 Annual Conference & Exhibition of the Transportation Association of Canada: Transportation without Boundaries*, Charlottetown, Prince Edward Island.

Dantzig, G., Fulkerson, D.R. and Johnson, S.M. 1954. Solution of a Large-Scale Traveling-Salesman Problem, *Operations Research* 2, Operations Society of America, pp. 393-410.

Desrochers, M. and Laporte, G. 1991. Improvements and Extensions to the Miller-Tucker-Zemlin Subtour Elimination Constraints, *Operations Research Letters* 10, 2-36.

GUROBI Optimization. 2017. GUROBI Optimizer Quick Start Guide, Version 7.5, Gurobi Optimization, Inc., http://www.gurobi.com/documentation/7.5/quickstart_windows/index.html

Gutin, G. and Punnen, A.P. (editors). 2007. The Traveling Salesman Problem and Its Variations, *Combinatorial Optimization, Volume 12*, Springer Science + Business Media, LLC, printed in the United States of America.

Hillier, F. S. and Lieberman, G. J. 2005. Introduction to Operations Research, Eighth edition, McGraw-Hill, New York City, New York, United States, p. 403.

Mason, A.J. 2012. OpenSolver – An Open Source Add-in to Solve Linear and Integer Programmes in Excel, *Operations Research Proceedings 2011,* eds. Klatte, Diethard, Luthi, Hans-Jacob, Schmeddars, Karl, Springer Berlin Heidelberg, pp 401-406, http://dx.doi.org/10.1007/978-3-642-29210-1_64, http://opensolver.org

Mathlouthi, I. Gendreau, M. and Potvin, J. 2018. Mixed Integer Linear Programming for a Multi-attributeTechnician Routing and Scheduling roblem, *INFOR: Information Systems and Operational Research*, 56:1, 33-49, DOI:10.1080/03155986.2017.1335047.

Miller, C.E., Tucker, A.W. and Zimmerman, R.A. 1960. Integer Programming Formulations of Traveling Salesman Problems, *Journal of the Associated Computing Machinery*, Volume 7 Issue 7, pp. 326-329.

Pataki, G. 2000. The bad and the good-and-ugly: formulations for the traveling salesman problem, *Technical Report CORC 2000-1*, Department of IE/OR, Columbia University.

Ragsdale, C.T. 2008. Spreadsheet Modeling & Decision Analysis, 5e, Thomson South Western, Louiseville, Quebec, Canada, pp. 184-189.

Rossi, F., van Beek, P. and Walsh, T. (editors). 2006. Handbook of Constraint Programming, 1st edition, Elsevier Science, Amsterdam, The Netherlands, pp. 378-406, 407-453 and 527-570.

Satir, S. 2003. *Applied Vehicle Routing and Scheduling Optimization for Pickup and Delivery Operations at a Trucking Terminal*, MScE Thesis, Graduate Academic Unit of Civil Engineering, University of New Brunswick.

Satir, S. 2006. *Improving the Transportation Operation Efficiency of Full Trailer Load Linehaul Movements for a Large Scale Company with Spatially Scattered Multi Terminals*, PhD Dissertation, Graduate Academic Unit of Civil Engineering, University of New Brunswick.

Woods, B., Punnen, A.P. and Stephen, T. 2017. A linear Time Algorithm for the 3-Neighbour Travelling Salesman Problem on Halin Graphs and Extensions, *Discrete Optimization*, DOI10.1016/j.disopt.2017.08.005

Zamorano, E. and Stolletz, R. 2017. Branch-and Price Approaches for the Multi-period Technician Routing and Scheduling Problem, *European Journal of Operational Research,* 257:55 – 68.

AUTHOR INDEX